Monitoring Dam Performance

Instrumentation and Measurements

Prepared by the
Task Committee to Revise Guidelines for
Dam Instrumentation of the
Committee on Water Power of the
Energy Division of the
American Society of Civil Engineers

Edited by
Kim de Rubertis, P.E., D.GE

ASCE AMERICAN SOCIETY
OF CIVIL ENGINEERS

Published by the American Society of Civil Engineers

Library of Congress Cataloging-in-Publication Data

Names: American Society of Civil Engineers. Task Committee to Revise Guidelines for
Dam Instrumentation. | de Rubertis, Kim, editor.
Title: Monitoring dam performance : instrumentation and measurements /
prepared by the Task Committee to Revise Guidelines for Dam Instrumentation of the
Committee on Water Power of the Energy Division of the American Society of
Civil Engineers; edited by Kim de Rubertis, P.E., D.GE, F.ASCE.
Description: Reston, Virginia : American Society of Civil Engineers, [2018] |
Includes bibliographical references and index.
Identifiers: LCCN 2017061624 | ISBN 9780784414828 (hardcover : alk. paper) |
ISBN 9780784480984 (pdf) | ISBN 9780784480991 (ePub)
Subjects: LCSH: Dams–Inspection. | Dam safety. | Dam failures–United States–
Prevention. | Hydraulic measurements.
Classification: LCC TC550 .A44 2018 | DDC 627/.80289–dc23
LC record available at https://lccn.loc.gov/2017061624

Published by American Society of Civil Engineers
1801 Alexander Bell Drive
Reston, Virginia 20191-4382
www.asce.org/bookstore | ascelibrary.org

MANUALS AND REPORTS ON ENGINEERING PRACTICE

(As developed by the ASCE Technical Procedures Committee, July 1930, and revised March 1935, February 1962, and April 1982)

A manual or report in this series consists of an orderly presentation of facts on a particular subject, supplemented by an analysis of limitations and applications of these facts. It contains information useful to the average engineer in his or her everyday work, rather than findings that may be useful only occasionally or rarely. It is not in any sense a "standard," however; nor is it so elementary or so conclusive as to provide a "rule of thumb" for nonengineers.

Furthermore, material in this series, in distinction from a paper (which expresses only one person's observations or opinions), is the work of a committee or group selected to assemble and express information on a specific topic. As often as practicable the committee is under the direction of one or more of the technical divisions and councils, and the product evolved has been subjected to review by the Executive Committee of the Division or Council. As a step in the process of this review, proposed manuscripts are often brought before the members of the technical divisions and councils for comment, which may serve as the basis for improvement. When published, each work shows the names of the committees by which it was compiled and indicates clearly the several processes through which it has passed in review so that its merit may be definitely understood.

In February 1962 (and revised in April 1982), the Board of Direction voted to establish a series titled "Manuals and Reports on Engineering Practice" to include the manuals published and authorized to date, future Manuals of Professional Practice, and Reports on Engineering Practice. All such manual or report material of ASCE would have been refereed in a manner approved by the Board Committee on Publications and would be bound, with applicable discussion, in books similar to past manuals. Numbering would be consecutive and would be a continuation of present manual numbers. In some cases of joint committee reports, bypassing of journal publications may be authorized.

A list of available Manuals of Practice can be found at http://www.asce.org /bookstore.

CONTENTS

DEDICATION

This manual is dedicated to John Dunnicliff, Dist.M.ASCE, upon whose shoulders stand all who work with instrumentation. From his 1988 classic *Geotechnical Instrumentation for Monitoring Field Performance* to his continuing efforts to advance instrumentation knowledge in his contributions and editing for "Geotechnical Instrumentation News," all have benefited. It is with this committee's recognition of how integral Dunnicliff's work has become woven into the state of practice that, with great appreciation, this manual is dedicated.

ACKNOWLEDGMENTS

ASCE Energy Division Executive Committee Contact Member
Jason Hedien, P.E., M.ASCE

ASCE Hydropower Committee Control Group Members
Kim de Rubertis, P.E., D.GE., F.ASCE
Steven Fry, P.E., M.ASCE
Christopher Hill, P.E., M.ASCE
Jeffrey Auser, P.E., M.ASCE

Committee Members
Kim de Rubertis, P.E., D.GE, F.ASCE, *Chair*
Steven Fry, P.E., M.ASCE, *Vice Chair*
Christopher Hill, P.E., M.ASCE, *Secretary*
Jeffrey Auser, P.E., M.ASCE
Bill Broderick, P.E., M.ASCE
Catrin Bryan, P.E., P.Eng.
Craig Findlay, Ph.D., P.E., G.E. M.ASCE
John France, P.E., D.WRE, M.ASCE

Ralph Grismala, P.E., M.ASCE
Justin Nettle, P.E.
Zakeyo Ngoma, P.E.
Minal Parekh, Ph D., P.E.
Elena Sossenkina, P.E., M.ASCE
Jay Statler, P.E., M.ASCE
Manoshree Sundaram, P.E., M.ASCE
Travis Tutka, P.E.
Jennifer Williams, P.E.
Karen Aguillard, P.E.

Organizations That Provided Support for Member's Committee Participation

AECOM	McMillen-Jacobs
ASCEA	Metropolitan Water District of
Avista Corporation	Southern California
Brookfield Renewable Resources	MWH
Colorado School of Mines	Paul C. Rizzo Associates, Inc.
FERC	Puget Sound Energy
Findlay Engineering	Steve Fry Consulting
HDR Engineering, Inc.	US Bureau of Reclamation
ICF	USACE

The committee expresses special appreciation to Dr. Pierre Choquet, without whose inspiration this committee might not have been formed and for his generous assistance in assembling and reviewing the manual.

PREFACE

This Manual of Practice for *Monitoring Dam Performance: Instrumentation and Measurements* is an addition to the series of publications by members of the Hydropower Committee of the Energy Division of the American Society of Civil Engineers (ASCE).

The Hydropower Committee of ASCE's Energy Division was formed to develop and distribute information on all aspects of hydroelectric power to the hydroelectric community. The Hydropower Committee prepares and publishes Manuals of Practice, Guidelines, and Technical Reports about engineering and scientific issues related to hydroelectric facilities. The Hydropower Committee seeks to serve an audience beyond civil engineers to include scientists, economists, and technologists with expertise in related areas. This focus is driven by the hydropower industry's desire to integrate science, environment, economics, operation, and maintenance into the scope of its activities.

In 1997, task committee volunteers studied and reported means and methods of monitoring dam performance with instruments and measurements. The outcome of that committee's work was *Guidelines for Instrumentation and Measurements for Monitoring Dam Performance* published in 2000. Since that publication, there have been many innovations in both means and methods of monitoring dam performance. This manual presents the current state of practice with the intent of providing a convenient reference for owners, engineers, regulators, and others with an interest in dam safety.

There are no simple rules or standards for determining the proper level of instrumentation and measurements to monitor dam performance. Each dam is unique. The consequences of failure, complexity of dam and foundation, known problems and concerns, and degree of design conservatism all require consideration in deciding both what to measure and how to measure. With that understanding, the purposes of this manual are to consider the following:

- Recognizing vulnerabilities affecting dam performance,
- Identifying performance indicators,
- Understanding means and methods of measuring performance indicators,
- Planning and implementing a monitoring program,
- Acquiring data,
- Managing, presenting, and evaluating data, and
- Making decisions and taking action.

Additional advances in monitoring technology will bring new and better means and methods for monitoring, but the principles outlined in this manual will have enduring value.

This manual is intended to be informative but not prescriptive. With its foundation and appurtenant works, each dam presents its own set of challenges. Deciding on how to monitor its behavior requires skill and judgment beyond the scope of this manual.

CHAPTER 1
INTRODUCTION

When you can measure what you are speaking about, and
express it in numbers, you know something about it. . . .
Lord Kelvin

Welcome to *Monitoring Dam Performance—Instrumentation and Measurements*. This manual reviews the current state of practice for monitoring dam performance as part of the American Society of Civil Engineers commitment to serve the worldwide community of dam engineering professionals.

Monitoring dam performance is a global enterprise. Almost every country in the world has dams that are monitored for performance by their owners and engineers. Many countries have agencies with monitoring guidelines and regulations for safe practice. These include agencies such as the CWC in India, U.S. Federal Energy Regulatory Commission (FERC), China's Ministry of Energy, Canadian Dam Association, The Environment Agency in England and Wales, France's Permanent Technical Committee, and ANCOLD in Australia. The intent of this manual is to capture the fundamentals and current state of practice of instrumented measurements for monitoring dam performance in the global dam engineering community.

Dam-performance monitoring is a blend of visual surveillance coupled with instruments to measure indicators that answer a basic question: "How is this dam performing?"

Visual surveillance is the backbone of all performance-monitoring programs.

Monitoring of every dam is mandatory because dams change with age and may develop defects. There is no substitute for

1

systematic intelligent surveillance. But monitoring and surveillance are not synonymous with instrumentation.

Ralph Peck

Instruments can supply measurements of performance indicators that
evade visual surveillance. It is difficult for the eyes to accurately estimate
the flow or amount of embankment settlement. There are instruments to
measure both. But are they required?

Certainly the fundamental rule today should be that no instrument should be installed that is not needed to answer a specific
technical question pertinent to the safe performance of the dam.

Ralph Peck

Specific technical questions are unique to every dam. Like people, no
two dams are exactly alike. The questions to ask rely upon an understanding of a dam's vulnerabilities. Designs are intended to provide adequate
defenses against perceived vulnerabilities. Monitored instrumentation is
intended to provide measurements to validate the design or point to performance that could lead to loss of reservoir control. This manual describes
the process of using instrumented monitoring to answer the basic question
of how a dam is performing.

- First: Understand what sequence of events could cause a dam to fail
  to retain its reservoir.
- Second: Select which performance indicators to measure and the
  appropriate instruments to measure them.
- Third: Plan and implement the program to make observations and
  measurements.
- Fourth: Gather, manage, and present measured data in a form amenable to decision making.
- Fifth: Evaluate the data.
- Last: Make decisions based on the monitoring results and take action
  as needed.

Chapter 2 reviews the importance of dam safety not only to protect the
public but also to sustain the benefits that dams provide. The case for developing a responsible performance-monitoring program is presented.

The vulnerabilities of different types of dams are the subject of Chapter
3. Modes of failure and their performance indicators vary for different types
of dams. Understanding a dam's potential modes of failure sets the stage
for asking performance questions for which answers may be provided by
instrumentation and measurements.

Chapter 4 describes how to plan and implement a monitoring program
tailored to provide answers to questions about a dam's performance.

Many new means of measuring have been added to performance-monitoring practice in recent years. Instruments to measure a variety of loads and responses are described in Chapter 5.

Chapter 6 introduces geodetic monitoring and its role in measuring a dam's response to loads.

Chapters 7 and 8 describe data acquisition, presentation, and management to provide a basis for evaluating the measurements from the performance-monitoring program.

Chapter 9 discusses the process of acquiring the measurements to answer the questions asked, evaluating those measurements, deciding what action to take, if any, and then acting on those decisions.

Typical instrumented monitoring of embankment, concrete, and other dams are the subjects of Chapter 10, 11, and 12, respectively.

Sample data forms and plots are illustrated in Chapter 13.

Monitoring dam performance has a long and colorful history. Chapter 14 captures that history.

Four appendices provide a

- List of references about dam-performance monitoring,
- Procedure for conducting failure mode analyses,
- Discussion of precision and accuracy, and
- Glossary.

This manual of practice is not intended to establish a standard for the design, installation, operation, or use of dam instrumentation systems. This manual presents the current state of practice. Each dam is unique in its geologic setting, physical loads, materials of construction, purpose, design life, and method of operation and presents its own distinct set of challenges in understanding its behavior.

Dam instrumentation and measurement provide a key part of the information that owners, engineers, and regulators consider to understand how a dam is performing compared to design estimates and historic performance, to identify any sequence of events that could cause a loss of reservoir control, and to inform decisions about dam safety actions.

CHAPTER 2
PERFORMANCE MONITORING

Engineers shall hold paramount the safety, health and welfare of the public and shall strive to comply with the principles of sustainable development in the performance of their professional duties.

Canon 1, ASCE Code of Ethics

Dam-performance monitoring is a key element in preserving public safety and sustaining the benefits that dams provide.

Every dam has an inherent potential for failure, no matter how well it is investigated, designed, constructed, operated, and maintained. Even the best engineering, construction, operation, and maintenance practices cannot completely eliminate the uncertainties associated with a man-made structure constructed on natural materials (the dam foundation), built with natural and manufactured materials, and subjected to the forces of nature. These uncertainties can manifest anywhere within the "system" of a dam (foundation, man-made structures, mechanical, electrical, operations, and maintenance) and can emerge at any point within the life cycle of a dam, as shown in Fig. 2-1.

Uncertainties can take the form of an undiscovered deficiency in the foundation, an incorrect estimate or mistake during design, an undocumented or inappropriate change during construction, poor maintenance, or improper operation. Each uncertainty, on its own or in conjunction with others, can produce a potential mechanism for loss of reservoir control. Although allocating more resources and care during each phase of a dam's life cycle can reduce unknowns, uncertainty is inherent, and no dam is completely without potential for failure.

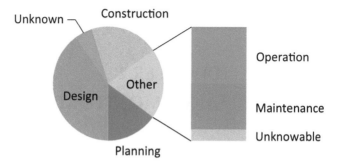

Fig. 2-1. Resolving uncertainty

In the planning phase for a new dam, even the most robust site investigation program cannot completely resolve the uncertainties of how the ground will respond to the dam and reservoir. Geologists and geotechnical engineers base judgments about the quality of a foundation on the evidence and tools available to them. Visually exploring and mapping a site provides surficial evidence of soil and rock quality. If only what is apparent from the surface informs the design, the level of uncertainty will be high, as subsurface conditions may differ greatly. Exploratory drilling and testing within the foundation can reduce the level of uncertainty. The level of uncertainty is reflective of the level of effort in site investigation; however, some uncertainty regarding the condition of the foundation is unavoidable because the resources allocated to an exploratory program are finite, and some uncertainties may defy resolution. At some point, the drilling and testing stop, and the unresolved uncertainties retain the potential to affect dam performance.

Additional uncertainties can arise at the design phase. Engineers not only must rely on the site characterization during the planning phase but also must apply judgment and make estimates regarding the anticipated loads that the structure must resist during its life. Designers must estimate the quality and strength of the materials available for construction. Future operation and maintenance requirements require attention. The accuracy of the engineer's estimates will depend on the quality of the data from the site investigation and the accuracy of the predictions of the loading conditions and the dam's response to imposed loads. Detailed flood studies and seismic hazard assessments may reduce the level of uncertainty regarding the loads a structure must withstand; however, flood and earthquake loadings cannot be estimated with certainty (unknowable), and their application in design must consider acceptable risk. Even with adequate resources and a high standard of care during design, some level of uncertainty remains, as the design relies on the quality of the estimates made to support the design. The goal at the end of design is to reduce the uncertainties (unknowns) to those manageable during construction.

Unforeseen conditions may be discovered during construction, and design changes may be required. An extensive quality control program that oversees and documents construction can greatly reduce the uncertainty associated with construction by reducing workmanship errors that have caused unsatisfactory dam performance.

As dams age, other factors such as inadequate maintenance, improper operation practices, or loading that is inconsistent with estimates made during design may result in unsatisfactory performance. Surveillance and monitoring can provide a basis for deciding whether or not a dam is performing as expected.

Monitoring the performance of dams is necessary because the uncertainties that could introduce a mechanism for failure cannot be fully resolved. The risk of failure, however remote, always exists. Failure has societal and financial consequences that include loss of the benefit a dam provides and the potential for loss of life and property damage.

Performance monitoring as part of a comprehensive dam safety program is a powerful tool for identifying and managing the risk of failure associated with a dam. Instrumented monitoring systems may be among the tools contained in a performance-monitoring program. As with any tool, proper understanding of the purpose and use of a monitoring tool maximizes its benefit.

The guidance contained in this chapter is intended to assist the reader in recognizing the uncertainties generated in the life cycle of a dam, identifying a dam's vulnerabilities associated with those uncertainties, understanding how unacceptable performance may develop, prioritizing those modes of failure by the level of risk associated with them, and developing a dam safety program that effectively incorporates performance monitoring for managing that risk.

2.1 DAM SAFETY PROGRAM

A dam safety program is a dam owner's road map to designing, constructing, monitoring, maintaining, and safely operating a dam. Without a structured dam safety program, the monitoring systems described in this manual lack context. A dam safety program is the framework for those responsible for the safety of a dam to identify and manage the risks associated with the structure. Instrumented monitoring is a valuable tool to reduce the probability that reservoir control will be lost (World Bank 2002).

There is no "one size fits all" dam safety program. The scope of a dam safety program depends on the size, type, complexity, and age involved and the potential consequences associated with the dam. A dam safety program seeks to establish the following:

- Performance criteria—a statement of the purposes and design requirements for the dam;
- Potential failure modes—identification of dam-specific failure modes;
- Performance-monitoring plan—a planned surveillance and monitoring plan tailored to the identified failure modes, including visual inspections and instrumented monitoring;
- Performance evaluation methodology—a comparison of the dam's actual performance to the performance expectations;
- Safety assessment methodology—analyses to determine whether the dam can safely respond to the imposed loads;
- Duties and responsibilities for those responsible for dam safety;
- Procedures for safe operation and maintenance;
- Procedures for identifying, planning, designing, and constructing dam safety improvements, modifications, and remedial measures;
- Access to sufficient technical resources and expertise as necessary;
- Training plan for staff involved in dam safety—the personnel who carry out the routine dam safety monitoring efforts at the dam site; and
- Emergency planning and contingency planning for intervention in the event of a developing failure.
- An effective dam safety program requires (1) a comprehensive understanding of potential sequences that could lead to failure of the dam or appurtenant structures to retain the reservoir and (2) an understanding of the likelihood that those sequences would develop.

2.2 DAM OWNER RESPONSIBILITY

Dam safety requires recognition on the part of the dam owner of the inherent risks associated with the impoundment of large quantities of water and the potential adverse consequences of an uncontrolled release. The owner of a dam has an ethical and legal responsibility to maintain and operate the dam in a safe manner. Failure to do so not only endangers lives, property, and the natural environment but also can expose the owner to substantial loss of revenue, legal costs, and remediation expenses.

Legal liability for a dam owner has typically been imposed under one of two different legal doctrines. Under the standard of strict liability, a dam owner is responsible for any damages caused by a dam incident or loss of reservoir control regardless of the cause. Under the standard of negligence, a dam owner would be liable if the owner did not exercise sufficient care in the design, construction, operation, or maintenance of the dam. In many cases, the appropriate standard of care will include instrumentation and measurements to monitor dam performance and the provision of sufficient resources for the performance monitoring.

2.3 FAILURE MODES

A potential failure mode is described by the U.S. Federal Emergency Management Agency as "a physically plausible process for dam failure resulting from an existing inadequacy or defect related to a natural foundation condition, the dam or appurtenant structures design, the construction, the materials incorporated, the operations and maintenance, or aging process, which can lead to an uncontrolled release of the reservoir" (FEMA 2004). The term "failure" means the loss of reservoir control and does not necessarily mean dam failure. An uncontrolled release of the reservoir such as the gate failure at Folsom Dam (Fig. 2-2) is an example of loss of reservoir control.

The Swift No. 2 embankment breach (Fig. 2-3) is an extreme example of loss of reservoir control—a dam failure.

The key to developing a dam safety program is identifying what could go wrong. In another sense, the term "failure" is a misnomer because loss of reservoir control often occurs as an incident associated with operation and maintenance. With an understanding of what could cause a loss of reservoir control, a surveillance and monitoring scheme tailored to

Fig. 2-2. Gate failure at Folsom Dam
Source: U.S. Bureau of Reclamation.

Fig. 2-3. Embankment breach at Swift No. 2

answering the relevant questions informs decisions about what to measure to judge a dam's behavior.

To be credible, a potential failure mode must define an initial condition and follow a progression of events to failure. The initial condition is any combination of material properties and loads that could progress to a failure. Design and construction flaws, operational errors, or unanticipated loads are potential contributors to progression. Changes may occur gradually or suddenly.

For example, the initial condition for a piping failure of an embankment dam could be the lack of a filtered zone downstream from the dam's core. Seepage flow could begin to erode and move soil particles, form a soil pipe into the dam, and erode a void in the dam's core such as the incident shown in Fig. 2-4.

Collapse of the void and subsequent erosion could cause loss of freeboard, breach of the embankment, and dam failure.

A reservoir release does not necessarily have to be a complete structural failure of the dam. Rather, an uncontrolled release may initiate at appurtenant structures such as a penstock rupture, shown in Fig. 2-5, canal break, gate failure, or failure of a diversion tunnel plug.

When evaluating potential failure modes for a dam, it is important to consider the dam as part of an overall system. This "system" includes the

Fig. 2-4. Piping failure at Fontenelle Dam
Source: National Park Service.

Fig. 2-5. Penstock failure
Source: Courtesy of Brookfield Renewable Resources, reproduced with permission.

dam, the foundation and abutments, appurtenant structures, and human factors:

- Pipelines, penstocks, and spillways that pass flow through or past a dam,
- Mechanical features such as gates, valves, and generators that control flow through a dam,
- Electrical systems that provide power to those mechanical features,
- Control and data-acquisition systems that monitor and control those mechanical and electrical features, and
- Human behavior during operation and maintenance, including opportunities to intervene when an emergency arises such as responding to uncontrolled turbid seepage at the toe of an embankment dam.

Every part of the system has the potential to introduce a failure to control the reservoir. Therefore, the entire system is worthy of scrutiny to identify potential failure sequences amenable to instrumented monitoring.

After identifying potential failure modes, the next step is to estimate their likelihood. Whether likelihood of occurrence is expressed as a numeric value or a relative category, the results provide the basis for understanding, communicating, prioritizing, and managing the risks to a dam by allowing the owner to allocate monitoring resources in the most efficient manner to protect public safety and sustain the benefits the dam provides.

The goals of failure mode analysis are to identify and judge the likelihood of any sequence of events that may lead to a loss of reservoir control and to suggest means and methods of surveillance and monitoring to ensure reservoir control. Guidance on how to perform failure mode analyses [potential failure mode analysis (PFMA) or failure mode effect analysis (FMEA)] is provided in Appendix B.

2.4 SURVEILLANCE AND MONITORING

A dam's performance depends on the decisions made to bring it from concept to physical reality. Those decisions were intended to provide answers to the following questions:

- Magnitude—How large are the loads to which the dam and its foundation will be subjected?
- Location—Where will the loads be applied? Where will water go (over, under, around, through)?
- Response—How does the dam support and distribute its load, and how does it manage water?

Assessing the performance of a dam involves evaluating responses to those loads with data collected by visual and instrumented monitoring and then comparing the data against expected performance.

Visual surveillance involves routine site inspections by personnel trained in dam safety and familiar with the design and operation of the dam and its appurtenant structures. That activity is the backbone of any surveillance and monitoring program. Each dam is unique, and visual monitoring must be tailored to capture all visible aspects of a dam's performance. For those aspects of performance that cannot be seen, instrumented monitoring can play an important role.

> Monitoring of every dam is mandatory, because dams change with age and may develop defects. There is no substitute for systematic intelligent surveillance. But monitoring and surveillance are not synonymous with instrumentation.
>
> Ralph Peck

Instrumented monitoring produces quantitative data regarding conditions surrounding the dam (foundation, reservoir, tailwater, precipitation) and within the dam—conditions that are not apparent from the ground surface and may not be detectible through visual inspection. Instrumented monitoring can provide long-term records of data, allowing for detection of time- or load-dependent trends in a dam's performance. The ability to detect a change over time is important because these changes may be indicative of a developing performance problem related to a potential failure mode. When used in conjunction with visual monitoring, instrumentation data can verify visual observations of change in a structure or vice versa, providing a level of redundancy and confidence to the monitoring program. The collection of data can be automated, allowing for monitoring on a more frequent (approaching real-time) basis. Automated monitoring systems can initiate alarms indicating unexpected performance.

Surveillance and monitoring seek to identify conditions suggestive of a developing potential failure mode by blending visual observations and instrumented measurements to form a coherent picture of a dam's performance.

2.4.1 Design

Field investigation is the primary information-gathering method used during the design phase to characterize the geology and materials at and around a dam site. Instrumentation installed as part of this investigation establishes baseline conditions for items such as groundwater levels and downstream spring flow rates. Monitoring during construction and first filling allows detection and evaluation of changes to baseline conditions.

2.4.2 Construction

Instrumentation during construction can confirm design predictions or recognize changed conditions. Instrumentation also provides information about conditions during construction that may affect construction quality. For example, (1) temperature is routinely monitored when placing and curing concrete, (2) measurements of material moisture content and compaction densities are important to confirm the design soil shear strength during embankment construction, and (3) pressure measurements are required during grouting operations. Instrumentation indicating unexpected performance during construction can identify unsafe working conditions. Additionally, instrumentation can be used during construction to help verify compliance with required environmental conditions such as water-quality monitoring.

2.4.3 Postconstruction and First Filling

Instrumentation during construction and first filling provides data regarding performance as the structure receives loads. Reservoir filling tests the seepage resistance of the dam, foundation, abutments, and reservoir rim for the first time. In addition, the reservoir load tests the structural stability of the dam. Instrumentation data during this phase provide early indications of unusual or unexpected performance and establish baseline measurements for future operating conditions.

2.4.4 Operation

Instrumented monitoring supports operations of the dam and provides feedback to operations personnel. For example, measurements of reservoir elevation, downstream flow, water temperature, and dissolved oxygen at a hydroelectric dam provide data to operate the generation equipment and to meet governmental regulatory or environmental requirements.

2.4.5 Maintenance

Some dams will experience unexpected performance, even after many years of uneventful operation. Instrumentation can help identify a problem and confirm the effectiveness of remedial actions. For example, if an embankment experiences an unexpected increase in the amount of seepage as indicated by a high phreatic surface and increasing weir flow measurements, the remedy might include construction of a reverse filter at the toe in the area of the seepage. Instrumentation and monitoring could then be used to evaluate the effectiveness of the remediation.

CHAPTER 3
FAILURE MODES

It's fine to celebrate success but it is more important to heed the
lessons of failure.

Bill Gates

Chapter 2 traced performance monitoring throughout a dam's service
life and introduced the importance of failure modes in the design of an
effective program for dam-performance monitoring. By first identifying the
most significant threats to a dam's performance, a performance-monitoring
program can be planned to address key indicators relative to the potential
causes of unsatisfactory performance.

This chapter describes the state of knowledge about how embankment, concrete gravity, and concrete arch dams perform. It outlines vulnerabilities that may lead to unsatisfactory performance, and it describes
indicators that may signal a developing failure mode.

Knowing how a failure may develop and asking appropriate questions
form the basis for planning and implementing an effective performance-
monitoring program, the subject of Chapter 4.

3.1 EMBANKMENT DAMS

Constructed using soil and rock, embankments are particularly vulnerable to (1) uncontrolled seepage, (2) overtopping, and (3) slope instability.

3.1.1 Design

Embankment dams rely on mass and foundation strength for stability.

Homogeneous embankments, shown in Fig. 3-1, are mixtures of soil and rock using the low permeability of those materials to keep internal pressure from rising to cause instability. Finer materials are placed in the interior and coarser materials in the shells.

Zoned (central or inclined core) embankments, shown in Figs. 3-2 and 3-3, utilize low-permeability (fine-grained) soil for the core, again to manage internal pressure. Because fine-grained soil in the core is susceptible to transport through the coarser downstream shell, an engineered filter zone is placed downstream from the core to prevent migration of the core. Design practice is to place a drainage layer adjacent to the filter zone to manage internal pressure in the downstream shell. Many embankments have a similar filter arrangement on the upstream slope to manage pressure in the event of a rapid reservoir drawdown.

Foundation seepage management often is achieved with a cutoff trench or grout curtain under the core and a toe drain contiguous with the drain zone behind the filter.

An embankment dam may be designed with an impervious upstream face. One popular design is the concrete-faced rockfill dam (CFRD), shown in Fig. 3-4. Other designs have employed different materials, mainly asphalt (Fig. 3-5), but clay has also been used. An upstream impervious face for a rockfill dam has the advantage of a free-draining downstream shell, thereby reducing or eliminating the potential for problems of uncontrolled seepage or slope instability.

Fig. 3-1. Homogeneous embankment dam

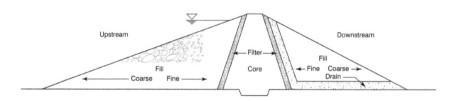

Fig. 3-2. Central core embankment dam

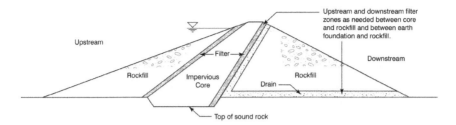

Fig. 3-3. Inclined core embankment dam

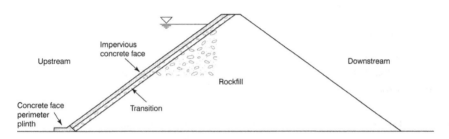

Fig. 3-4. Concrete-faced rockfill dam

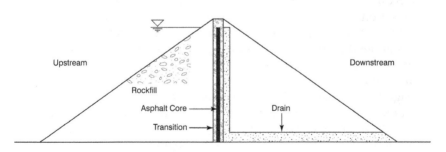

Fig. 3-5. Asphalt core embankment dam

3.1.2 Vulnerabilities

3.1.2.1 Uncontrolled Seepage. Internal erosion (piping) may initiate in many ways. Design may have failed to include appropriate provisions for filters and drains or to lengthen the seepage path. Construction practice may leave poorly compacted or pervious material, offering a privileged path of seepage. Abrupt changes in abutment slopes against which fill is placed may cause core cracking, allowing flow to overwhelm the drains.

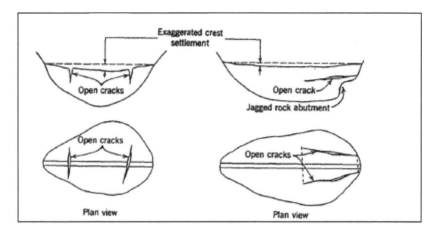

Fig. 3-6. Settlement-induced cracks that can initiate internal erosion
Source: ASCE Task Committee on Instrumentation and Monitoring Dam Performance (2000).

Foundation settlement can initiate core cracking as well. Poor compaction around conduit penetrations in the fill may allow a seepage path to develop. Or a conduit may leak, allowing seepage path development along the conduit exterior. Fill placed against rigid structures such as a spillway training wall requires care in design to include granular materials that can stop seepage from exploiting the soil/wall contact. The foundation may contain a defect such as an open joint that communicates with the reservoir and passes under the embankment, bringing reservoir pressure to the embankment toe, as shown on Fig. 3-6.

Internal erosion initiates when one of the causes of piping allows the seepage flow velocity to increase fast enough to transport soil. The first indicator is seepage on the downstream slope or at the toe. If the seepage force is strong enough, it begins to transport soil by eroding backward from the toe toward the core, creating a void that begins to enlarge provided that the soil is strong enough to support the upstream progression of the void. Fig. 3-7 illustrates a serious case of internal erosion. Rapid erosion follows that leads to embankment slumping and loss of freeboard and reservoir control, resulting in a breach and embankment failure.

Embankments with an impervious upstream face or with an asphalt central core resist piping through the embankment, but they may be vulnerable to uncontrolled seepage in the foundation, along the abutments, or by leakage along embedded conduits, as shown in Fig. 3-8.

Key indicators that signal development of a piping failure sequence are

- Increasing seepage along abutments, conduit penetrations, and/or at the toe,
- Turbid seepage,

- Change of embankment shape such as bulging, sinkholes, unexpected crest settlement, and
- Longitudinal or transverse cracking along the crest.

3.1.2.2 Overtopping. Flow over the crest of an embankment has caused many failures, as shown in Fig. 3-9. The design may lack sufficient freeboard to accommodate the inflow design flood (IDF).

All types of embankment dams are susceptible to failure by overtopping because downstream shells are likely to be eroded, ultimately leading to a loss of freeboard. Judgment is required to estimate whether the duration and depth of overtopping is likely to initiate failure.

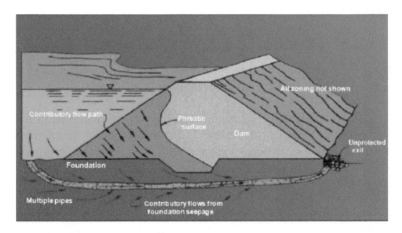

Fig. 3-7. Internal erosion through foundation
Source: FEMA (2015).

Fig. 3-8. Internal erosion along training wall
Source: Courtesy of ND State Water Commission.

Fig. 3-9. Overtopping failure sequence
Source: FEMA (2013).

The only indicator that signals development of an overtopping failure sequence is the progressive loss of freeboard.

3.1.2.3 Slope Stability. Embankment slopes are designed to remain stable under all loading conditions. Slope failure, either downstream or upstream, may reduce freeboard to the point where the reservoir is exposed, allowing discharge to erode the downstream slope until a breach forms and failure ensues.

Understanding the foundation conditions, properties of embankment materials, and quality of construction are key to identifying whether embankment slopes will remain stable under all loading conditions. Slope performance may be affected by

- Material quality (density, shear strength, moisture content, and plasticity),
- Reservoir operation,
- Slope aspect,
- Seepage gradient,
- Susceptibility to liquefaction (embankment and foundation),
- Time-dependent strain, and
- Strong shaking during an earthquake.

Slope failure may initiate because a slope is too steep to remain stable against a rising seepage gradient or rapid drawdown (upstream slope) without adequate time or provision for drainage or because of liquefaction in the foundation during an earthquake. Judging the adequacy of a slope to remain stable requires review not only of its design but also of its performance history. If vulnerability is discovered, then appropriate monitoring can be planned.

Figs. 3-10 and 3-11 show an upstream slope failure at Fort Peck Dam and a downstream slope failure at the Mount Polley tailings dam, respectively.

Figs. 3-12 and 3-13 show downstream slope erosion from seepage and upstream slope erosion from wave cutting, respectively.

Key indicators that signal development of a slope stability failure sequence are

- Increasing seepage gradient (elevated phreatic surface),
- Turbid seepage,
- Change of embankment shape—slumping, scarp development, unexpected crest settlement, and
- Longitudinal cracking along the crest.

Fig. 3-10. Upstream slope failure at Fort Peck Dam
Source: USACE Fort Peck, http://www.nwo.usace.army.mil/Missions/Dam-and -Lake-Projects/Missouri-River-Dams/Fort-Peck/.

Fig. 3-11. Downstream slope failure at the Mount Polley tailings dam
Source: Imperial Metals, reproduced with permission.

Fig. 3-12. Downstream slope erosion from seepage
Source: Courtesy of Texas Commission on Environmental Quality, reproduced with permission.

Fig. 3-13. Upstream slope erosion from wave cutting
Source: U.S. Army Corps of Engineers.

3.2 CONCRETE GRAVITY DAMS

Conventional and roller-compacted concrete gravity dams are vulnerable if they lack adequate resistance to sliding and/or overturning. Sliding and overturning or combinations of both motions are the dominant concrete dam failure modes. A failure sequence may initiate from a design error, poor workmanship, concrete deterioration, or loss of foundation integrity. A failure sequence may progress by exploiting the lack of appropriate operation or maintenance procedures.

3.2.1 Design

Concrete gravity dams rely on their mass and foundation strength for stability. The forces acting on a gravity dam are well understood. How those forces are resisted is also well understood. The weight of the dam must overcome all the applied forces. Typical designs employ features that promote stability, such as

- Proper foundation preparation and treatment to support the weight of the dam and ensure that no foundation rock blocks capable of displacement are stabilized or removed and replaced with concrete,
- Grout curtains or cutoff walls that lengthen the seepage path in both the foundation and the abutments,
- Drainage curtains downstream of the grout curtain or cutoff wall to reduce uplift on the base of the dam,
- Internal drainage within the dam body to relieve uplift that may form along lift lines,
- Provision of waterstops and/or keyed contraction joints, and
- Careful spillway layout to reduce scour potential.

Fig. 3-14 illustrates the importance of effective drainage to limit uplift on the base of the dam.

3.2.2 Vulnerabilities

3.2.2.1 Drain Design.   If drains are effective, headwater pressure will be reduced at the line of drains and provide a stabilizing effect. Similarly, effective drainage within the body of the dam serves to reduce the probability that a threat to stability will develop along a lift line.

If drains are ineffective, a crack may form at the heel, introducing headwater pressure across a larger area of the base and creating a destabilizing effect. Similarly, a failure sequence may initiate along a lift joint in the concrete if it cracks and uplift exceeds the tensile strength of concrete. Once a crack develops, seasonal temperature changes may "rock" the joint, allowing deeper uplift penetration along a lift joint. Drainage within the dam body can be effective in reducing the probability that tension will develop.

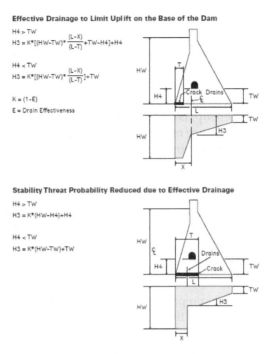

Fig. 3-14. *Uplift reduced by effective drainage*
Source: *ASCE Task Committee on Instrumentation and Monitoring Dam Performance (2000).*

Design errors may take many forms. The error most often encountered in aging dams is failure to provide for positive drainage at the foundation and within the dam body. An indicator of a design error is unexpected movement.

3.2.2.2 Workmanship.   Poor workmanship may be a contributing factor that allows a threat to progress. The failure of Gleno Dam, shown in Fig. 3-15, was attributed in part to the poor quality of cement used to produce the concrete.

Poor workmanship was also cited as a contributing factor in the failure of Teton Dam. A common indicator of poor workmanship is heavy weeping visible along lift joints on the downstream face.

3.2.2.3 Concrete Deterioration.   Concrete deterioration may contribute to failure if it adversely affects the concrete strength and reduce its resistance to sliding or overturning.

Freeze–thaw (F/T) damage, shown in Fig. 3-16, is common in older concrete placed prior to the advent of air-entrained concrete. F/T damage begins shallow and superficially. Left untreated, it will continue to reduce

Fig. 3-15. Gleno Dam post-failure
Source: Wikimedia Commons, en.wikipedia.org.

Fig. 3-16. Freeze–thaw damage

concrete thickness and strength, but it is not usually considered a contributor to a failure sequence.

Concrete produced with aggregates reactive with the alkalis and carbonates in cement is likely to develop through the entire concrete section, gradually causing expansion and cracking, as seen in Fig. 3-17. The expansive reaction is referred to as an alkali–silica reaction (ASR) or an alkali–aggregate reaction (AAR), and, depending on the degree of severity, may contribute to a failure sequence.

Fig. 3-17 shows an expanding aggregate particle and the cracks induced in the surrounding cement paste. An indicator of ASR is cracking and crazing pattern of cracks on the downstream face, as shown in Fig. 3-18.

3.2.2.4 Foundation and Abutment Integrity. A gravity dam failure sequence initiates when uplift reduces shear resistance at the foundation, allowing the dam to move or slide downstream, as shown in Figs. 3-19 and 3-20.

Leakage around and/or under the dam can initiate a sequence that allows uplift pressure to rise, causing displacement of the foundation or abutments, and then progress until the concrete loses support, cracks, and moves downstream, causing a breach that leads to failure.

Slope instability in an abutment from gravity alone (Fig. 3-21) may initiate without uplift and cause damage, threatening reservoir control.

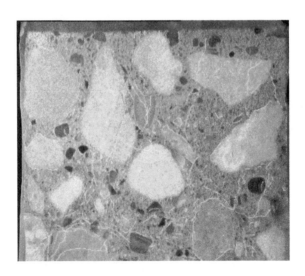

Fig. 3-17. Alkali–silica reaction (ASR) in aggregate
Source: U.S. Department of Transportation (2008).

The only known failure involving fault rupture and offset during an earthquake occurred at Shih-Kang Dam in Taiwan during the 1999 Chi-Chi earthquake, shown in Fig. 3-22. Judging the probability of such an event at another dam evades rational analysis. A fault in the foundation of a gravity dam is addressed during design and treated during construction. Instruments measuring movement may detect fault displacement for a dam in service.

Fig. 3-18. ASR in downstream face of a gravity dam

Fig. 3-19. Abutment failure, Camara Dam

Fig. 3-20. Foundation/abutment failure, St. Francis Dam

Source: Regan (2009, left); Wikimedia Commons (2018, right).

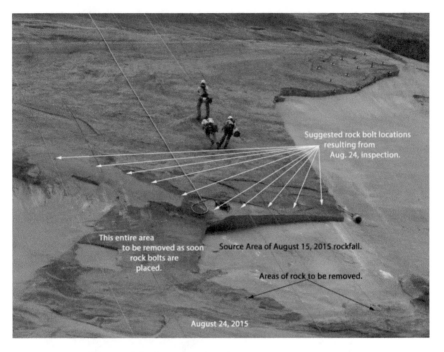

Fig. 3-21. Stabilizing left abutment at Glen Canyon Dam
Source: U.S. Bureau of Reclamation.

Indicators that may signal foundation or abutment instability are (1) displacement of abutment rock blocks, (2) leakage along the abutments, (3) loss of drainage capacity to reduce uplift under or within the dam, and (4) elevated uplift pressure at the heel.

Spillway arrangements vary widely. Spillways may be through the dam, over the crest of the dam on an abutment, or remote from the dam, as shown in Figs. 3-23 and 3-24.

Spillway discharges deliver enormous amounts of kinetic energy that must be dissipated at the foundation. In the process, erosion often scours the rock, creating a "scour hole."

Spillway discharge or overtopping of a concrete dam can cause scour that undermines the toe and reduces a dam's resistance to sliding and overturning, as shown in Fig. 3-25.

Spillways are concrete structures and can thus develop instability indicators similar to those of a dam. Cavitation of concrete surfaces exposed to high-velocity flow merit attention. Both gated (Fig. 3-23) and uncontrolled spillways are susceptible to scour. An indicator that scour threatens the toe of a dam or spillway requires interpreting profiles of scour depth. Scour undercutting the toe of a dam or spillway requires positive intervention.

Fig. 3-22. Fault rupture at Shih Kang Dam
Source: Courtesy of Robin Charlwood, reproduced with permission.

Fig. 3-23. Ogee-gated (left) and chute-gated (right) dams
Source: Wikimedia Commons (right); Courtesy of Seattle City Light, reproduced with permission (left).

Fig. 3-24. Uncontrolled overflow (left) and Morning Glory (right)
Source: U.S. Bureau of Reclamation (left); U.S. Army Corps of Engineers (right).

Fig. 3-25. Scour hole at base of arch dam

3.3 CONCRETE ARCH DAMS

3.3.1 Design

Concrete arch dams derive their strength by transferring the imposed loads to their abutments and foundations. Valley shape is a key consideration in design that seeks to drive the arch loads as deeply into the abutments as possible to accept the loads and limit the effect of those loads on

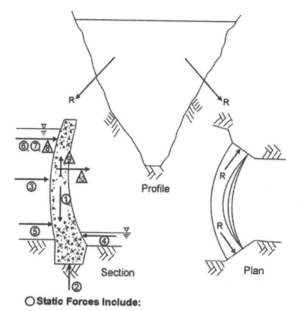

Profile

Section

Plan

O **Static Forces Include:**

1 - Weight of Concrete
2 - Uplift
3 - Hydrostatic Force - Upstream
4 - Hydrostatic Force - Downstream
5 - Thrust Created by Silting Material
6 - Ice Load
7 - Wave Setup

△ **Dynamic Forces Created by Seismic Activity Include:**

8 - Additional Hydrostatic Force
9 - Additional Vertical Component
10 - Additional Horizontal Component

R - **Force Resulting from Forces 1 to 10**

Fig. 3-26. Forces acting on an arch dam
Source: ASCE Task Committee on Instrumentation and Monitoring Dam Perfor-
mance (2000).

the abutment itself. A narrow valley makes that easier to accomplish. In addition to the reservoir, thermal stresses affect how the arch dam transfers load to the abutments seasonally. Arch dam stability relies upon the stability of the abutments and foundation, as indicated in the distribution of forces shown in Fig. 3-26.

Typical arch dam designs employ features similar to those of gravity dams to promote stability, such as

- Proper foundation preparation and treatment to provide adequate support for the dam,
- Grout curtain or cutoff wall that lengthens the seepage path,

- Drainage curtain downstream of the grout curtain or cutoff wall to reduce uplift on the base of the dam,
- Internal drainage within the dam body to relieve uplift that may form along lift lines,
- Provision of waterstops and/or keyed contraction joints,
- Grouted contraction joints, and
- Careful spillway layout to reduce scour potential.

Not every arch dam has either a drainage curtain or internal drains within the dam. An example of measuring drain discharge is shown in Fig. 3-27.

Because positive abutment support is required to resist arch loads, concrete thrust blocks may be required to accept the loads if an abutment cannot furnish adequate resistance.

Typical designs divide the structure into monoliths separated by contraction joints that are fitted with waterstops and often keyed between monoliths to provide shear resistance. The contraction joints are grouted to promote monolithic action from abutment to abutment. An example of this design is the Glen Canyon Dam, shown in Fig. 3-28.

Arch dams are resistant to earthquake forces, and there are no recorded failures of arch dams from earthquakes. The few recorded arch dam failures from any initiator have involved defects in the abutments or foundations. The most frequently cited is Malpassat Dam in France, shown

Fig. 3-27. Measuring toe seepage with V-notch weir.

Fig. 3-28. Glen Canyon Dam
Source: U.S. Bureau of Reclamation.

Fig. 3-29. Malpassat Dam after failure
Source: Michel Royon / Wikimedia Commons.

in Fig. 3-29, which failed during first filling because of a foundation defect that went undiscovered during design and construction.

3.3.2 Vulnerabilities

Unlike gravity dams, arch dam performance is not very sensitive to uplift; however, it may be a consideration if the base is thick. Abutment and foundation stability are keys to acceptable performance.

Indicators that may signal foundation or abutment instability are (1) displacement of abutment rock blocks and (2) leakage along the abutments.

A developing threat to stability may be indicated if loss of drainage capacity to reduce uplift under or within the dam is measured or if measured uplift pressure at the heel increases unexpectedly.

3.4 OTHER DAMS

This chapter described the most common dam types (embankment, concrete gravity, and concrete arch), their designs, vulnerabilities, and performance indicators that may signal a developing threat to safety and initiation of a failure mode. There are several other types of dams, each with its own design and vulnerabilities, and they are described in Chapter 11.

CHAPTER 4
PLANNING AND IMPLEMENTING
A MONITORING PROGRAM

Dam surveillance aims to detect, by visual observation and monitoring, any phenomenon that can compromise the structural and operating integrity of a structure or its related operating equipment.

ICOLD 1988

Chapter 3 explained how monitoring of performance indicators is linked to questions about a dam's performance relative to its potential failure modes. This chapter provides guidance for planning and implementing a monitoring program to provide data alerting dam owners of potential threats to safety by measuring key performance indicators. A well planned and implemented monitoring program recognizes the consequence of loss of reservoir control, not only for potential loss of life and property damage but also for the loss of the benefit the dam provides. Properly implemented, surveillance and monitoring reduce risk.

The two integral components of a dam-performance monitoring program are (1) visual surveillance and (2) instrumented measurements. Visual surveillance employs the eyes of informed and motivated inspectors. Instruments help fill the gaps of what cannot be seen by providing measurements of quantifiable responses such as pressure, flow, movement, stress, strain, and temperature.

Dam instrumentation by itself has no particular value. The best instrument does not enhance dam safety if it is not working, is recording irrelevant information, is in the wrong place, is not monitored with appropriate frequency, or if the acquired data are not evaluated. Monitoring is the collection, processing, and evaluation of the measurements recorded by the instruments, combined with information from visual observations.

The following are discussions of a program's overall objectives and key factors that need to be considered in each step to plan and implement an effective program that integrates the interests of the owner, regulator, engineer, and technician. The first step in planning is to ask the key questions concerning a dam's performance, i.e., what aspects of performance require answers? The next step is to decide what, where, and when to measure performance indicators to provide answers to those questions.

4.1 PLANNING

To the extent practical, the key to a successful program is simplicity in planning, installation, monitoring, data collection, and evaluation. Simplicity is served by limiting the number of interfaces between an instrument, data logger, transmitter, and database and through the use of simple, robust instruments and sensors that are easy to install, calibrate, operate, and maintain. There will be cases when delicate and sophisticated instrumentation is required; however, when given a choice: keep it simple!

Tailoring a monitoring program requires some soul-searching to answer another question:

"Just because I can measure a performance indicator, should I?"

Should dams exhibiting no indication of unexpected performance require monitoring other than visual surveillance? This manual cannot provide an answer to that question because (1) each dam is different, (2) each owner is different, and (3) each dam may be subject to regulations that have specific requirements. Many dams are monitored only by visual surveillance, a staff gauge, and crest monuments. Some dams have complex instrumented monitoring in addition to visual surveillance. Most dams fall in between those two extremes.

Agencies with jurisdiction and expertise throughout the world have different regulations for monitoring and reporting that require planning for compliance.

A systematic plan is required to design, procure, install, operate, and maintain the monitoring system; to acquire, evaluate, and interpret information; and to take appropriate actions in a timely manner to ensure safe performance of a dam.

Most dams have monitoring programs balanced to observe and measure only those performance indicators of concern. It is important to understand that the need for observations and measurements may change during the life of a dam. The monitoring plan anticipates those changes and provides sufficient flexibility to address unexpected changes.

During planning and design, measurements are helpful to establish baseline conditions such as existing groundwater levels and any evident ground movements. Site evaluation prior to the start of construction can provide a baseline that will be helpful in measuring responses during and after construction. For example, the potential for settlement may result from measurements and tests made during site investigation.

Monitoring during construction is required to understand ground response to changing loads, to provide construction safety, and to measure the quality of the work. The results can be used to confirm key design assumptions in comparison with actual site conditions, to comply with regulatory monitoring (where applicable), and to determine when specific conditions are met to allow construction to continue (e.g., hold points).

Filling of a reservoir following construction or following a prolonged drawdown is probably the most severe test of a dam and its appurtenant structures. Both visual and instrumented monitoring are vital because the structural stability of the project, as well as the seepage resistance of the dam, foundation, abutments, and reservoir rim, is tested. Observations made during filling can provide an early indication of unexpected performance and also confirm satisfactory performance. Frequent measurements are planned during filling and drawdown. Once steady-state conditions are reached and a stable set of baseline measurements is acquired, measurement frequencies are established for continued long-term monitoring.

Pumped storage projects present a special case of reservoir fluctuation because filling and drawdown occur frequently. Planning for measuring the performance and response of a pumped storage scheme may require continuous monitoring.

Planning for long-term monitoring requires flexibility to provide measurements of performance indicators that may change with time. Procedures are planned to detect trends in responses over the life of the project. Monitoring is tailored to answer the basic question, "Is this dam performing as expected?" Surprises are not uncommon. An effective monitoring plan describes the actions required in the event that measurements suggest monitoring error or unexpected behavior.

Often overlooked in the planning process is the need to provide adequate funds to acquire, install, maintain, and measure all the instruments for the service life of a dam.

Summarizing, the scope of a monitoring program depends on

- Type, purpose, and size of dam;
- Hazard classification;
- Consequences of failure;
- Age and condition of dam;

- Performance indicators to be measured;
- Regulatory requirements; and
- Commitment to funding program life-cycle costs.

How each of these factors affects overall program costs will differ from project to project.

The success of a monitoring program depends on developing objectives, verifying site conditions, complying with regulatory requirements, and understanding measuring requirements. A successful program will meet the objectives of the designers, owners, operators, and regulators. The program is reviewed and updated periodically. Data evaluation may reveal needed modifications to the program because of changes in estimated loading conditions (floods and earthquakes). Instruments may fail. Dedicated staff and regular training are required to maintain an effective program.

A monitoring plan can be a stand-alone document or incorporated into a set of standard operating procedures (SOPs) for a dam. In addition to defining the monitoring needs, effective planning requires specific and detailed answers to the following questions:

- What questions need to be answered?
- Which performance indicators are to be monitored?
- Why are they monitored?
- How are they monitored?
- Where will they be monitored?
- How often will they be monitored?
- How much will the measurement system cost?
- Who will do the work (install, maintain, operate, interpret, report, and take action and responsibility)?

Planning for the life cycle of a monitoring system, both visual and instrumented, begins with design, procurement and installation, and commissioning. Finally, it must address operation, maintenance, continual training, rehabilitation, replacement, placing on inactive status, and abandonment. These activities cost money. A successful monitoring program requires development of a realistic estimate of both initial capital costs and the operational life-cycle costs. An adequate budget is needed for design, engineering, and oversight during procurement and installation, as well as for calibration and commissioning. Annual budgets for operations, maintenance, and training are estimated during the initial planning phase to allow for a realistic assessment of total long-term facility operation and maintenance and capital costs for monitoring. Over the longer term, software updates may be needed. Eventually, instruments will reach the end of their operational life and require replacement.

4.1.1 Visual Surveillance

The performance of many small dams and dams with low consequences is often monitored visually with little or no instrumentation. Most dams are monitored by a combination of visual observations and instrumented measurements.

An instrument too often overlooked in our technical world is a human eye connected to the brain of an intelligent human being.

Ralph Peck

Visual surveillance is the most important element of a dam safety program. Thorough visual inspection by an individual knowledgeable about a dam's vulnerabilities can provide first warning of unexpected performance or response of a dam.

The frequency and extent of surveillance is a function of the complexity of the site, the consequences of failure, past performance, and staff requirements to perform other duties.

Results from routine and special visual inspections are recorded and captured on simple reporting forms, emails, or memoranda. Documentation of the observations made during inspections is important to develop a baseline understanding of the project performance and observable changes.

A trained operator on the crest of the dam might observe an area along the dam abutment contact where the color of the vegetative slope protection on the embankment has changed. This pattern difference may or may not indicate a potential initiator of a failure sequence. It may require additional evaluation as the cause might be simple intended differences in the vegetation type in that area, or it may reflect the development of a wet spot due to seepage through the dam or the abutment.

With high-quality cameras and the low cost of high-speed communications, it is possible to supplement on-site visual monitoring with monitoring via remote cameras. These tools are useful for monitoring areas of concern or providing surveillance of remote areas. However, cameras do not replace the need for regular observations by a qualified inspector.

4.1.2 Instrumented Monitoring

A simplified eight-step approach for planning and implementing a monitoring program, summarized in Table 4-1, follows:

Step 1: Review information about the project.

Review site investigations, siting studies, design and construction documents, construction photos, operations history, and monitoring

Table 4-1. Eight-Step Approach.

Steps	Description
1	Review information about the project.
2	Identify the vulnerabilities and questions that need to be answered.
3	Identify what measurements can and should be made.
4	Design appropriate monitoring system, including installation, calibration, maintenance, and data acquisition and management.
5	Procure, test, install, and commission program.
6	Operate and maintain instruments.
7	Collect, process, and evaluate data.
8	Take action when indicated.

data. Understand project conditions such as stratigraphy, foundation structure, material properties, groundwater conditions, ambient conditions, appurtenant structures, past construction issues, or proposed future construction.

Step 2: Identify sequences of events that could lead to a loss of reservoir control and the questions that require observations or measurements to answer. Develop an understanding of vulnerabilities and potential failure mechanisms.

Step 3: Identify performance indicators. Decide whether visual observations will be sufficient or whether instrumented measurements are required. Establish the frequency of measurements. Estimate instrument and dam behavior. Establish a range of expected values.

Step 4: Design an appropriate monitoring system including installation, calibration, maintenance, and data management. Provide sufficient financial resources for life-cycle monitoring costs.

Key factors requiring consideration during program design include

- Instrument purpose and location;
- Instrument range, resolution, accuracy, and repeatability;
- Environmental and climatic conditions;
- Instrument site accessibility and ease of installation;
- Availability of replacement parts;
- Reputation of vendor;
- Compatibility with other instruments in a dam-monitoring portfolio;
- Protection during and after installation;
- System power requirements;
- System reliability requirements;

- Need for redundancy and backup; and
- Training requirements.

Select instruments and optimal instrument locations based on the questions that need to be answered. Keep in mind accessibility of instruments for ease in reading, verification, repairs, or replacement. Choose locations that provide the best representation of the conditions of performance, whether at representative sections or specific potential problem areas. Consider the reliability of the instruments. Durable and simple instruments are sometimes the best choice. Establish data-collection procedures and frequencies. Consider the need for manual readings versus automated or semi-automated readings. Include the collection of complementary data to support response and performance interpretations. Establish threshold and action limits. Describe actions required when readings or observations reach these levels. Identify the type and amount of data to be acquired. Define who will be using the data and for what purpose.

Step 5: Procure, test, install, and commission the program. Chapter 5 contains a description of various instruments for monitoring. Once the goals of the program have been documented in a design and procurement package, the various components can be procured, tested, and installed per manufacturer guidance. Vendors also provide technical specifications, as well as operation and maintenance instructions.

Commissioning involves training of staff to implement the program from maintenance of the system to data collection to data evaluation.

Step 6: Operate and maintain instruments.
The instrumentation design report, an SOP, or other similar documentation contains the procedures to take readings, repair instruments, and reduce and evaluate data. Procedures are updated regularly as needed.

Plan to provide for labor and material costs. Acquiring data, scheduling maintenance, repair, or calibration, responding to instrument malfunction, instrument replacement, and technology and software updates, stocking backup supplies, preparing reports to meet reporting requirements, and evaluations and training all have life-cycle costs.

Step 7: Acquire and manage data.
Establish the means and methods for data acquisition and management. To the extent possible, avoid drowning in data.

Means for storage and retrieval of the data are dependent on the size and complexity of the program. Chapters 7 and 13 discuss data acquisition and presentation.

Step 8: Take action when indicated.
Configure the program to alert the owner when measurements suggest behavior of a developing failure sequence. Chapter 8 provides guidance for evaluating the data and taking action.

Overall simplicity is considered when finalizing the design for monitoring because an uncomplicated program is more likely to produce reliable measurements with fewer errors and less cost than a complicated system. This does not mean that having many instruments is a bad idea. The number of instruments is dependent on the number of performance indicators requiring measurement to answer questions about behavior. There may be many instruments, but they will produce best results if they possess as many of the following attributes as possible:

- Simple,
- Robust,
- Accurate,
- Reliable,
- Insensitive to environmental conditions, and
- Easy to read, access, and replace.

These attributes are applicable to small and large monitoring programs All instruments are manufactured to operate within a specified range, resolution, accuracy, precision, and repeatability. Instrument manufacturers and vendors supply their technical specifications. Selection of instruments is matched to program needs and weighed against cost, reliability, and ease of operation.

Instruments and appurtenant parts, including cables, cabinets, power sources, and connections, are designed to operate properly in the environmental conditions to which they will be exposed. In remote areas, batteries or solar collectors used to power instrument data loggers and transmitters may need to operate through dark, cold, hot, wet, dry, or corrosive conditions. For example, battery draw will be affected by sampling frequency data transmission. Incorporation of an automated system can provide continuous readings and the ability to continue to evaluate performance of the project under all conditions. Chapter 7 describes automated data-acquisition systems and their advantages and disadvantages.

Access to the instruments is an important consideration because installation, operation, and maintenance costs increase for sites that are difficult or dangerous to reach. The use of fully automated or semi-automated systems can reduce problems, but challenges will remain for calibration, maintenance, or repair if access is difficult. Other considerations include safety regulations when installing instruments in confined or enclosed spaces. Where instruments are needed in difficult locations, redundant instruments and special measures to protect appurtenances (cables and power) are con-

Fig. 4-1. Protecting cabling
Source: Courtesy of DG-Slope Indicator, reproduced with permission.

sidered to reduce the chances or consequences of unplanned maintenance and repair; however, power will still be required. An example of protecting cabling is shown in Fig. 4-1. Instrumentation in locations that are difficult to access requires careful evaluation to determine if there are alternative means to collect similarly useful data.

Instruments and their appurtenances require protection during and after installation. Damage during construction commonly occurs from traffic with heavy equipment. Locating instruments during construction requires consideration of the changes that will occur to the site over the course of construction. Care is needed to plan instrumentation to avoid heavy traffic and still serve areas that will change during construction. Durable installation materials will reduce the risk of damage.

Common protections include bollards, special coatings, exclusion fences, and burial in vaults.

Design of a system considers other instrumentation already installed. If there are similar instruments in an owner's portfolio, a sense of familiarity with types of instruments, instrument vendors, and possibly installation techniques are beneficial. Contact with other owners and operators or soliciting ideas through technical users' groups also may provide helpful information.

4.2 IMPLEMENTATION

Once the desired system has been designed, the next step is implementing the system to meet the established objectives. Procurement will be based on a set of drawings and specifications that reflect the design intent. Development of the estimate of the initial capital investment for procurement, installation, and initial calibration of the monitoring system is improved if consideration for costs includes the price of all system components and appurtenances such as cable, conduit, protective devices, hardware, software, and facility outages (lost revenue). Vendor mark-up and shipping costs are considered in the initial estimate, as well as labor, material, and equipment costs to properly install and calibrate the system.

A contingency is helpful in the planning-level cost estimate to provide flexibility for future refinements to the system or changes during installation to accommodate site conditions or other unanticipated events such as delays caused by bad weather. Examples of various approaches to procurement, as well as considerations in the development of the procurement package (or drawings and specifications) of the planned system, are discussed in the following sections.

4.2.1 Procurement Approaches

Items to consider as part of the procurement phase include identification of the parties responsible for

- Procurement of the instruments,
- Pre-installation review,
- Installation and calibration of the system, and
- Collection, management, and review of the data.

Procurement services typically involve instrumentation hardware and appurtenant equipment, software, and factory calibration provided by the instrumentation manufacturer or vendor. Construction oversight and review prior to installation are necessary to confirm that the intended and requested instruments will be appropriate for the site. Installation and calibration of the instrument system also includes pre-installation acceptance tests on both hardware and software, commissioning of the hardware and software, and calibration and development of a baseline set of data for the system to confirm successful operation. This, along with troubleshooting of the system to bring it in line with the services requested in the procurement package, is typically provided by the vendor and can also be supplemented or complemented by the owner's representative.

An important aspect of the planning process includes adequate budget and time to effectively schedule transition of the operation and maintenance aspects of the monitoring system after it is installed. A contractor or the owner may be responsible for instrument installation, maintaining installation records, and delivering operation and maintenance information.

If a contractor installs the system, an effective way to minimize challenges associated with the transition of a monitoring program to the owner is to involve the owner's personnel directly in the instrumentation installation and initial monitoring during construction. When direct owner involvement in the construction phase of the monitoring program is not possible, the contractor may provide on-site training for the owner's personnel. Either approach allows for the preparation of a detailed instrumentation operation and maintenance manual for future reference.

Once the system has been installed successfully, the responsibility for continued readings and maintenance is transferred to the owner. Depending on the extent and complexity of the system, some systems can be monitored on remotely accessed systems and others may be monitored locally. Some systems may be monitored by the owner or other entities as specified by the owner. These responsibilities need to be outlined in the plan for monitoring.

The same considerations apply for the installation of new instruments or for repairs to existing instruments at an existing dam. However, existing dams are being actively operated by their owners. When new instrumentation needs arise, direct contact between the owner and the instrumentation vendor is common. When instrumentation is to be added during a retrofit or other new construction phase, procurement through a general contractor can be used. For existing projects, the lessons learned from the prior plan should be taken into consideration.

Other procurement approaches include contracts through an engineering consultant or third-party design–install–operate firm. The advantage and disadvantages of any procurement method require evaluation based on the specifics of each particular project.

For complex monitoring requirements, it is recommended that a qualified instrumentation engineer be involved not only during the planning and design phase but also during procurement, installation, and initial calibration and testing of the system. Engineering expertise may be available from the owner's staff, engineering consultants, instrumentation vendors, or regulatory agencies. This collaboration is important for successful procurement and implementation. A contract package should be developed that includes specifications and drawings for the system. A part of the contract package may be a plan document to identify the specifics of the program, reading frequencies, reporting requirements, and requirements for backup equipment, replacements, and remedial actions. This type of documentation can be beneficial to the owner. Its inclusion in the contract package depends on the owner's intent to involve a third party in the overall monitoring. Also, bidding entities may provide comments and suggestions on the overall plan and program. It is suggested that at least two vendors be contacted during the acquisition process to confirm that the instruments specified will, in fact, be capable of measuring the required performance indicators.

4.2.2 Procurement Package—Drawings and Instrumentation Specifications

The preparation of drawings and specifications that clearly establish the system requirements and define the intent of the instrument monitoring system is important early in the process and should be developed in conjunction with the owner, designer, and instrumentation vendor(s). Early collaboration in the development of drawings and specifications allows for creative solutions and planning, as well as potential cost savings. As emphasized previously, site conditions at the dam such as remoteness, availability of power, dam size, access, and reporting needs must be considered. These criteria define the type and distribution of the instruments that will be used, the nature of the data-acquisition system required (manual, semiautomated, or fully automated), data transmission methods (cable, radio telemetry, fiber optics, distributed data loggers, etc.), data transmission regulations and security, power sources for the system (AC, batteries, solar, microhydro), and data collection, processing, and storage requirements.

In many cases, multiple vendors will be able to supply functionally equivalent instruments, and therefore the specification should not be so specific as to exclude qualified products on the basis of nontechnical issues or minor resolution or accuracy differences. Drawings and specifications should clearly describe the system requirements, as well as the installation, calibration, and testing to provide a working system that meets the objectives of the program. Many vendors include guidance specifications on the instruments themselves, as well as guidance documents on installation needs, procedures, and maintenance required for their systems. This information can be used in the development of the procurement package.

Most of the information that the dam owner requires to operate an instrument and apply its readings is contained in the specifications provided by the manufacturer. A complete set of specifications includes a description of the environmental and electrical conditions under which the instrument will operate and defines the characteristics of the signal that the instrument produces. The performance characteristics of each instrument's resolution, range, repeatability, and accuracy (outlined in Appendix C) belong in the specifications.

The performance characteristics provide the user with the conditions under which the instrument can be reliably used and the performance that can be expected. The manufacturer's specifications should be verified by performing a full-scale calibration check prior to installation; if this is not possible, the instruments' zero outputs should be checked.

Knowledge of instrument power requirements and signal output characteristics is also essential for ensuring compatibility with an automated data-acquisition system if one is needed. Each data-acquisition system (see Chapter 7) has its own signal input requirements, which must match the output of the monitoring instruments that will be connected to it. The hardware and software vendors will provide necessary advice and guidance in this area.

The following is an example of the key elements and considerations for an instrument specification; it follows Section 10 00 00 of the Construction Specifications Institute format:

- Establish general contract conditions.
- State known access restrictions and available utilities.
- Describe known special conditions related to dam safety, e.g., care with heavy equipment, access restrictions, etc.
- Describe the product.
- Provide either performance requirements (e.g., must report water level within 5-mm accuracy), the preferred instrument or equivalent, or the physical attributes of the instrument (amps, wiring, conversions).
- Require drawings to owner standards.
- Describe the execution.
- Require vendor presence during installation, calibration, and commissioning.
- Specify standards for acceptance of work including testing.
- Require instrument operation and maintenance manual that includes calibration and troubleshooting.
- Require instrument training in maintenance, readings, software use, and troubleshooting.

Installation and implementation of a new monitoring system at new or existing projects require adherence to the approved drawings and specifications, but they also include appropriate oversight and procedures to facilitate timely in-field decision making during installation as challenges arise.

4.2.3 Installation in New Dams

Installation of a new instrumentation system at a new dam or at an existing dam requires explicit procedures and should include interface with the instrumentation vendor(s). If possible, the instrument vendor should be present during instrumentation installation, testing, and commissioning and should also provide the owner or operator with training and documentation on the installed system that includes operation and maintenance procedures. If possible, the instrument or data management vendor should visit the site to acquire some familiarity with the site, its challenges, unique situations, limitations in access, or other issues to be able to tailor the system to the specific site. This is crucial not only for selection of proper instruments but also for proper installation techniques, accurate calibration, and testing, as well as the ability to troubleshoot anomalies or problems in the system with the vendor or manufacturer's representative present.

Continued contact with the vendor or manufacturer's representative once the system has been successfully installed will streamline troubleshooting and discussions on necessary repairs, upgrades, and maintenance issues over the life of the system. This will provide the necessary quality assurance and quality control for an effective system. Many vendors will provide training for correct installation, calibration, and testing of instruments. Vendors of data management systems often offer training so that the installed system is optimized for the specific project and its objectives. Successful operation of the instruments and data management system must be verified before components are buried or encased and inaccessible. Coordination with installation specialists, such as drillers for piezometers or inclinometers, also will need to be considered and managed.

Records of instrument installation should include the following information:

- Instrument type,
- Instrument identification,
- Readout type,
- Equipment and personnel responsible for installation,
- Initial measurements,
- Calibration results,
- Noted calibration constants,
- Site conditions,
- Changes from specifications,
- Changes from manufacturer's recommendations,
- As-built locations and dimensions,
- Complementary data and observations, and
- Anomalies in installation or construction, especially those that may affect data interpretation.

Responsibility for the collection and documentation of these data is often assigned to the vendor but it should be supplemented by the owner. This requirement should be included in the specifications as part of the procurement package and will become part of the instrumentation monitoring program, as it will be needed when recalibration or troubleshooting of the system is required.

4.2.4 Installation in Existing Dams

In some instances, instrument installation in an existing dam can be more straightforward than that at a new dam because identification of instrument location and type are dictated by the identified performance indicators. The need to install new instruments at an existing dam should be evaluated carefully and the actual installation and testing closely monitored, in addition to the overall performance and response of the dam and appurtenant features.

4.3 RESPONSIBILITY AND AUTHORITY

The following quotation captures the importance of owner engagement in dam safety monitoring:

> To provide the best basis for securing reliable and high-quality data, hence for securing best value, the people who have the greatest interest in the answers to the questions should have a major role in obtaining the data.
>
> John Dunnicliff

Clearly defined responsibilities are necessary for the development, implementation, and maintenance of a monitoring program. Responsibility should be linked to established decision-making authority for decisions related to dam safety.

The classification of personnel varies widely with different organizations. Small projects may assign several responsibilities (e.g., plan development, data collection, calibration and maintenance, and data analysis) to a single person. Larger organizations operating multiple projects with individuals for every facet of a monitoring program may including personnel responsible for

- Dam safety;
- Interfacing with regulatory agencies;
- Development, implementation, and maintenance of the program, including
  - Training and communication;
  - Review and analysis of information;
  - Operation and maintenance and data collection; and
  - Data entry, reporting, storage, and plotting.

The owner is ultimately responsible for the safe operation of the dam. This is an important concept for dam owners to understand when setting priorities, allocating funds, and assessing risk. The absence of specific regulatory oversight or instruction does not negate this responsibility. The owner establishes clear dam safety responsibilities and decision-making authority. Responsibilities can be delegated among individuals with diverse skills and backgrounds, but in all cases owner responsibilities are clearly understood.

Responsibilities include reporting and backup, which will be documented in a project's emergency action plan (EAP). Typical responsibilities are well defined by agencies with jurisdiction for dam safety and include decisions to notify the public, owners in key positions, dam safety engineers, and the media and on how to secure help in an emergency.

In general, regulatory agencies provide oversight and enforce rules, but they are not responsible for the safety of the structures they administer. For example, the responsibility of the U.S. Federal Energy Regulatory Commission (FERC) include

- Issuance of licenses for the construction of a new project;
- Issuance of licenses for the continuance of an existing project (relicensing);
- Oversight of all ongoing project operations, including dam safety inspections and environmental monitoring; and
- Enforcement of dam safety regulations.

Other agencies may operate differently. Whatever the degree of regulatory oversight and requirements, it is important to have the dam safety responsibilities outlined, assigned, and understood to minimize confusion or redundancy.

Regardless of the size of the project, it is best if a chief dam safety engineer is responsible for the program. Responsibilities include plan development, selection, procurement, and installation and commissioning of instrumentation and automation equipment design, as well as training and delegation of responsibilities when required. Large organizations will typically have internal staff to manage these activities, whereas smaller dam owners may rely on engineering consulting teams or another third party to provide guidance, recommendations, evaluation, and, in some cases, oversight.

The chief dam safety engineer responsible for the success of the monitoring plan oversees monitoring tasks such as data collection, instrument calibration, training, and reporting to confirm the program is being completed as planned. Communication of the content and intent of the program to other staff also is part of this responsibility. Delegation of responsibilities to a third party, whether to an engineering consultant or otherwise, does not relieve the owner of his or her dam safety responsibility to the public.

The dam safety engineer may also assist in data interpretation and analysis and provide technical guidance to other staff responsible for final data analysis. Depending on the complexity of the monitoring program, the owner may either have staff on hand for detailed technical evaluation or consider a partnership with consulting engineers to provide this analysis. Regardless, the data collected must be reviewed in a timely manner to identify possible problem areas and allow enough time to address these issues. Regular data review, report preparation, and distribution to appropriate personnel are required.

Data collection, field inspections, and reporting are assigned to staff familiar with the entire project. The benefits of using staff familiar with the project and areas of concern are familiarity with site conditions and the ability to notice changes over time. For these reasons, visual inspections are often assigned to current operating personnel, dam tenders, or dam operators. For projects with multiple staff levels, senior-level technical personnel train project personnel responsible for collecting and reporting of data, maintaining and calibrating instruments, and identifying action(s) to take based on field observations. Emphasis is placed on the importance of the task and the link between monitoring and dam safety.

Some owners with multiple projects choose to use a group of knowledgeable staff that rotate between projects and who then develop a familiarity and understanding of the performance of each project. This is an acceptable method of project staffing and, as long as staff training is kept current, it has the advantage of redundancy and multiple experienced observers.

In all cases, it is important to have continuity in personnel so that knowledge of historic trends is not lost. Often those with detailed knowledge of project conditions can quickly identify data-collection or observable anomalies in the field. Personnel responsible for data collection in the field must be reliable, dedicated, and motivated. Attention to detail is required because mistakes or omissions can result in lost information or even delays in warning of developing conditions. The opposite is also true where lack of experience or knowledge leads to the idea that an irregular reading automatically suggests a dam safety issue. Field personnel need to understand how the dam functions, its expected behavior, and the purpose of instrumentation and monitoring. The maintenance and calibration checks of an instrument and other automated equipment during their service life are best done by data-collection personnel who are able to notice potential problems, malfunctions, deterioration, or damage.

Data-acquisition personnel also may perform data entry. This can prevent data entry errors or misinterpretations. The need for data entry by hand is decreasing as more and more instruments are automated. However, automation of data collection is no substitute for thoughtful review and analysis.

Large organizations with a portfolio of projects may be able to assign a single person to the job of storing data, developing data plots, and preparing reports. Such a person must work closely with other individuals responsible for surveillance and monitoring. Tasks can include coordination between software programs, data transfer from remote sites, simple programming, creating reports, and making presentations. It is recommended that such a person also be trained in data collection and field maintenance to better understand not only the instruments and automation equipment in use but also to understand the data being evaluated.

Avoid drowning in data! Critically evaluate sampling frequency, storage, and processing. Swollen data-collection files can become filled with large quantities of partially processed data or useless and redundant data. Implementing the surveillance and monitoring plan requires adequate time and funds for data storage, maintenance, and evaluation following installation of the system. This requires staff or external sources to collect, store, and evaluate data and to assess polling frequency.

4.4 OPERATION AND MAINTENANCE

An effective monitoring program requires an adequately funded operation and maintenance program with well-defined responsibilities and procedures. It is often incorporated into the day-to-day project operation.

The basis for developing an operation and maintenance program takes into account several important factors that affect reliability and sustainability, such as

- Frequency of inspections (visual monitoring);
- Frequency of inspections of equipment (especially automated systems that may not regularly be inspected visually);
- Maintenance and service tasks especially calibration, testing, repairs and correction for drift or other malfunction that can affect performance or accuracy such as settlement, vandalism, or impact;
- Performance tests;
- Spare parts or repair/replacement of aging systems/components;
- Upgrades to software technologies and equipment;
- Labor and time required to record, review, and analyze the collected data; and
- Training needs.

Redundancy in staff knowledgeable in the systems is important to account for absences or additional staff needed in time of unusual conditions such as large floods.

The operation and maintenance program is needed for the life of the project. It is regularly reviewed for changing conditions. Reductions require justification. A summary of typical maintenance activities for common instruments is provided in Table 4-2.

The decision to repair, replace, remove, deactivate, or abandon an instrument or component of a monitoring system is based on an evaluation of

Table 4-2. Common Maintenance Activities.

Common maintenance activities for instrumentation

Instrument type	Maintenance activity	Frequency of maintenance
Borehole extensometer	Inspection for corrosion or damage	Annually then less frequently
Inclinometer (manual)	Recalibration	Per manufacturer direction
	Maintain clean tube	As required
Inclinometer (in-place)	Recalibration	Per manufacturer direction
Pressure transducers	Recalibration	Per manufacturer direction
Joint or crack meter	Inspection, cleaning	Annually
Load cells	None	N/A
Weirs or flumes	Remove debris, vegetation, sediment, and algae	Monthly to annually (or more frequently as required)
	Remove chemical precipitation or iron bacteria sludge	Annually (or as required dependent on conditions)
	Clean gauge orifice	Annually (or more frequently as required)
Piezometer (standpipe)	Flush standpipe to remove accumulated sediment (exercise care to prevent hydraulic fracturing)	As required
	Repair any damage to aboveground protective casing or cap	As required
Piezometer (pneumatic)	Check connection valve fitting and replace as required	Annually
	Check air vent from pressure test for clogging	Annually

Table 4-2. (*Continued*) Common Maintenance Activities.

Common maintenance activities for instrumentation

Instrument type	Maintenance activity	Frequency of maintenance
Piezometer (vibrating wire)	Check lightning protection device in terminal box or cable leads holder	Annually (or following a lightning storm)
Plumbline, inverted pendulum	Maintain full damping oil reservoir	As required
	Remove calcium deposits from line	As required
Settlement gauge	Check for impacts, settlement	As required
Strain gauge	None	N/A
Thermometer	Calibrate	N/A
Tiltmeter, beam sensor	None	N/A
Total pressure cell	None	N/A
Readout instruments	Recalibration	As required by manufacturer
	Replace batteries	As required by manufacturer
	Check fluid tank of pneumatic readout	As required
General	Check strip heaters in readout boxes for proper functioning	Annually and prior to onset of cold weather
	Check for vandalism	Regularly
	Maintain access to monitoring locations	As required
	Power source maintenance	As required

the overall program, the purpose of the specific instrument, the added value of maintaining the instrument, the possibility of the instrument being useful at some future date, and the effect of possible removal or abandonment on the overall monitoring program. These decisions will likely arise in the life cycle of any monitoring program. Consult the instrument manufacturer for instructions prior to disabling an instrument if the possibility of reactivation in the future is anticipated. Instruments that may be reactivated are not considered abandoned and can be referred to as inactive.

Instruments will need to be replaced when they reach their design life. Instruments that consistently malfunction because of wear or site conditions such as settlement, corrosion or other environmental damage, vandalism, vehicular impact, or even irreversible instrument drift should be replaced. Careful review of and familiarity with the historic performance of instruments will help identify when an instrument can no longer be relied upon to provide accurate measurements. Before replacing an instrument, consider how technology evolved since the original installation. Better instruments may be available.

Replacement of instruments or appurtenant components including cables, readout devices, and power sources can be economical. Updates or improvements to the instrument systems require regular review. Instrument manufacturers and vendors can help with a review.

Dam safety concerns may require the abandonment of an instrument that might create a source of a potential failure mechanism (e.g., an abandoned piezometer casing that provides a piping path in an embankment).

Instruments may be retired by simple removal of the device or by abandoning it in place. For example, resistance thermometers, thermistors, joint meters, and other similar instruments are often abandoned in place in concrete dams because their measurements are no longer required.

Next, Chapter 5 presents an array of instruments capable of measuring performance indicators to complete the implementation of a successful performance-monitoring plan.

CHAPTER 5
INSTRUMENTATION AND
MEASUREMENT TOOLS

The expectations of life depend upon diligence; the mechanic
that would perfect his work must first sharpen his tools.

Confucius

Chapter 4 described planning and implementation of a performance-
monitoring program. This chapter provides information to select appropriate instrumentation for a monitoring program based on the measurements
required to answer performance questions.

Instruments used for measuring different dam-performance parameters
are described by physical description, measured quantity, sensor options,
advantages and disadvantages, output options, operation and maintenance
requirements, and typical application in dam monitoring. References are
made to manufacturer websites that provide additional information, including photographs and figures.

Explicit direction or equations on how to read or calculate the measured
parameters are left to other references, including manufacturer materials, which are easily obtained. The following instrument descriptions are
intended to provide the reader with enough information to select instruments but not necessarily how to install, read, or evaluate the data.

Qualities such as precision, accuracy, robustness, cost, reliability, and
longevity are discussed; however, the ever-changing nature of instrument specifications and costs suggests that the reader refer to current
manufacturer-published data on those qualities that are also discussed in
greater detail in Appendix C.

The emphasis here is on the most commonly installed contemporary
instruments. Less attention is given to "legacy" instruments that may still
be in use or that may serve to provide redundancy to their contemporary

counterparts. A legacy instrument is one that was used in previous eras and may still be active on sites today; however, it is not commonly installed in new-use applications in modern practice. An example of a legacy instrument is a pneumatic piezometer.

The next section begins with the present state of practice and future trends in instrumentation and measurement tools. Subsequent sections are organized by measured performance indicators.

5.1 INSTRUMENTATION

Instruments may measure a dam's geotechnical, structural, hydraulic, or geohydrologic performance indicators. A generalized schematic of the flow of information from the physical world to the instrument user is shown in Fig. 5-1. Instrumentation is a general term used to refer to mechanical, electromechanical, or electronic instruments. A mechanical instrument measurement is converted to a visual reading by purely mechanical means. Some mechanical measuring methods are extremely simple, such as the use of a bucket and stopwatch to measure volumetric flow rate or seepage. Electromechanical devices convert mechanical movement into a measurable electronic signal. One example is an extensometer, in which movement of a rod anchored in a borehole operates an electronic transducer. In purely electronic instruments, the measured parameter generates an electrical signal without an intervening mechanical step. A thermocouple is an example of an electronic instrument.

Most new instruments today are electronic or electromechanical. The electronic part of these instruments combines a sensing element (sensor and transducer) and an electronic circuit (signal-conditioning electronics), as illustrated in Fig. 5-1. The transducer converts the sensed parameter into an electrical signal. The electronics excite the transducer and transform its signal into an amplified output that can be recorded at the end of a cable. The signal-conditioning electronics are typically packaged as close as possible to the transducer in the same housing.

Some transducers, such as vibrating-wire or fiber optic types, produce a signal that is relatively immune to electrical noise even when transmitted over cable lengths of many hundreds of feet. Where such transducers are used, the signal-conditioning electronics may be located near or within the recording or display unit.

Instrument readings may be recorded manually (by writing down or photographing the value displayed on a mechanical scale or digital readout), mechanically (typically by pen-and-ink lines on a circular or strip chart recorder), or electronically by a data logger or computer. The trend today is away from manual measurements and mechanical recorders and increasingly

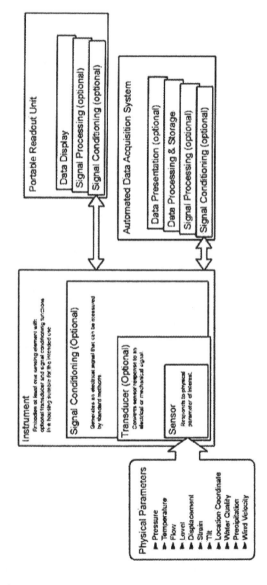

Fig. 5-1. Block diagram of generic instrumentation systems
Source: ASCE Task Committee on Instrumentation and Monitoring Dam Performance (2000).

toward electronic measurements and Automated Data-Acquisition Systems (ADASs). Potential benefits of automated recording and reporting include

- A current and continuous data record;
- A variety of data-processing options to improve accuracy;
- Lower costs to collect data and reallocation of labor resources to the more valuable functions of analysis and decision making;
- Ability to assess real-time data remotely in an office that may be located hundreds of miles from the site through web-based applications; and
- Ability to automatically initiate alarms and other actions if critical thresholds are exceeded (refer to Chapters 7 and 8 for more information).

The features and application of each instrument as a whole are the focus of this chapter. The instruments described are as inclusive as possible, but some devices may have been omitted in the array of instrumentation for dams. This oversight is unintentional. Furthermore, as technology advances, valuable new instruments will enter the market. For these reasons, the choice of instrumentation for a particular project will continue to evolve and is not necessarily limited to the devices described in this chapter.

The names used for the instruments described in this chapter are the common names used in the trade. Many of the names have been assigned by the instrument manufacturers or by long tradition. In some cases, more than one name may be used for the same type of instrument. Different names may refer to the specifics of the application and not to differences in the instruments or the fundamental measured parameter. For example, the terms "tiltmeter," "clinometer," and "inclinometer" all describe an instrument that measures angular rotation with respect to the vertical gravity vector. "Piezometers," "pressure transducers," and "pressure gauges" all measure pressure.

5.2 FUTURE TRENDS

Beyond the long-standing manual methods that are still in use, dam monitoring has embraced the technologies of sensors, electronics, computers, and information technology, all of which continue to rapidly advance. These current and future advances will more than likely affect the ways that dam monitoring is performed in the future. A few "crystal ball" predictions of future trends follow; however, it is certain that new technologies will emerge, changing the methods employed by our successors.

5.2.1 Sensors and Electronics

Most electronic instruments used on dams today output an analog signal, a voltage, current, or frequency that is proportional to the measured pressure, tilt, displacement, or other parameter. This output is transmitted in analog form and is then digitized and recorded at a data logger or computer.

However, in an emerging trend over the last decade, more instruments on the market have the required electronics to process and digitize the signal at the transducer level and produce an output in a standard digital protocol such as RS-485, allowing the measurements to be transmitted without being affected by electrical noise and electromagnetic interferences over distances that can reach several hundred feet. An additional advantage of digital transmission is that several sensors can be mounted along a single 4-conductor cable in what is often called a sensor array, as discussed in the next section. This development is further accelerated by the rapid development of micromachined sensors, known by the acronym MEMSs (microelectromechanical systems). In these devices, the mechanical components of the sensor are photo-etched into a silicon chip, along with much of the signal-conditioning circuitry. Accelerometers, strain gauges, and pressure sensors have already been produced using this method. Interestingly, MEMS accelerometers can now be produced with a low full-scale range of $+/- 1g$ or less, which makes them suitable to be used as high-accuracy inclination and tilt sensors in inclinometer probes, in-place inclinometers, and tiltmeters when read in static mode. The main advantages of MEMS sensors are their low temperature influence, high resolution, linearity and repeatability, and low zero drift, in addition to high-impact resistance (Sellers et al. 2008). A MEMS sensor itself can sustain impacts of several hundred g's without damage.

5.2.2 Fiber Optic Sensors

Fiber optical data transmission is now widespread in the telecommunications industry, and it is being introduced and used in other industries such as transportation and water supply; however, cost-effective fiber optical sensors have been slower to develop. As a result, fiber optical sensors are not commonly used in dam monitoring because other technology can currently monitor similar parameters more cost-effectively. Exceptions are found in the use of fiber optics to monitor water-temperature changes and to locate movement in long structures such as dikes. As explained in more detail further in this chapter, optical fibers can be used for distributed temperature and strain sensing. This opens the possibility for long fiber optic cables to be embedded in earth and concrete dams and to be used for distributed sensing at numerous locations in the dam. Practical considerations arise because the installation of long fiber optical cables can obstruct construction, and there is a risk of losing all measurements if the cable is accidentally cut or sheared in the structure at an inaccessible location where it cannot be repaired. Nevertheless, fiber optic sensors will probably be used more widely for dam safety purposes in the future. As an example of an application that could benefit from these techniques, piping detection in embankment dams lends itself to distributed-temperature fiber optic instrumentation, using water-temperature changes sensed by the optical fiber as an indicator. Accumulation of information and data from this approach could help to develop a better understanding of piping mechanisms and improve its prediction. Another advantage of fiber optic sensors is their immunity from electrical noise and transients. With continuing advances and price reductions, fiber optic sensors are likely to be used more frequently for dam-monitoring work in the future.

5.2.3 Sensor Arrays

Sensor arrays can be defined as strings of multiple sensors along the same signal cable, which can be electric or fiber optic. With the numerous recent technological advances in sensors with digital output and fiber optic sensors discussed previously, along with price reductions, the logical evolution is to integrate more sensing points along the signal cable. Examples of sensor arrays are strings of tilt inclination sensors based on MEMS accelerometers that can be used as in-place inclinometers, strings of digital thermistors points, strings of pressure sensors or piezometers, and optical fibers read using distributed strain and temperature-sensing methods. Strings of mixed types of sensors could certainly be feasible as well.

5.2.4 Computers and Data Acquisition

Although data transmission from instrument to acquisition system has generally been performed by copper wires, distribution through radio telemetry has become much more commonplace. Radio telemetry reduces the infrastructure traditionally used for networking the instruments to data-acquisition systems, especially signal cables in trenches. Lower-power and lower-cost radios, which make it possible to transmit on license-free spread-spectrum frequencies, together with lower-cost data loggers, have become more available and make it now very cost effective to transmit signals wirelessly between loggers as small as single-channel loggers, enabling the placement of data loggers at each instrument or small cluster of instruments. Additionally, within the past five years, the trend has moved toward wireless internet protocol communication between instruments and acquisition systems. Wireless internet protocol (IP) routers allow for more seamless communication of instrument data to the IP database system.

Automated collection and dissemination of monitoring data are now common practice. The continued increase in power and affordability of personal computers, the availability of digital instruments, the proliferation of low-cost data loggers with networking capabilities, and the availability of wireless telecommunication have all contributed to the trend toward real-time digital data acquisition and PC interfacing of monitoring data.

Use of digital cellular and satellite technology is starting to become a popular means of data transport also. Chapter 6 expands on the various current practices and trends in data-acquisition technologies.

5.2.5 Information Technology

The term "information technology" refers to how information is processed, distributed, and used. This is an area where some of the most important advances in dam monitoring have been made, and they are expected to continue to develop rapidly in the future. Advancements in computing power, internet-based databases, graphics programs, and higher internet speeds have greatly reduced the inefficiencies in the analysis and decision-making processes of previous monitoring systems. Databases now allow monitoring data to be organized, filtered, and graphed automatically—putting the data into forms that can be quickly and meaningfully evaluated by all concerned parties. Readily available internet communication schemes make secure dissemination of these results nearly instantaneous. Recent developments have allowed more efficient and effective automated warning systems and decision-making processes.

Over the past decade, continual upgrades in mobile communications have led to the development of powerful mobile peripherals. Smartphones and tablets have changed the way people send and receive information, and they allow for more remote usage. At present, mobile broadband is accessible at many dams, and the broadcasting infrastructure is growing rapidly on a global scale. It is expected that the trend toward handheld peripherals will continue, with a corresponding trend in databases and graphics programs that run on new operating systems. The portability of the tablet computer allows for convenient input of data directly into a digital database from the field. Data can then be wirelessly transmitted to a digital database. Although this trend is just emerging, it will continue to grow in the industry.

5.2.6 IP-Based Video Cameras

IP-based video cameras are being used more frequently for dam monitoring and remediation work. The cost of this technology has decreased in recent years, and improvement in the bandwidth of communication systems from microwaves to wireless local area networks (WLANs) has made this a much more common instrument for dam monitoring. IP-based video cameras are digital, frequently with tilt–pan–zoom (TPZ) capability, and they can be used to make manual backup readings (e.g., read a staff gauge on a weir).

5.2.7 Residual Load and Integrity of Tendons and Anchors

Concrete dams and their appurtenant structures often incorporate anchors to enhance the stability of the structure. The anchors are typically made of solid rods or 7-wire strands, also called tendons, which are tensioned and grouted in boreholes. Anchors have been used in dams for more than 50 years, and the technology regarding installation and long-term corrosion protection has improved over the years. Because of the inability to directly inspect the anchors, questions are arising regarding the long-term integrity of aging anchors.

The traditional approach to evaluating residual load (i.e., the current load capacity of the anchor) is to conduct a lift-off test of the anchor using a hydraulic jack. However, this method is costly and also risky when taking into account the fact that the anchor can fail during the test. In addition, the test may not always be feasible, as the anchor head may not be accessible or very little grip space may be available to mount the jack.

For this reason, there have been a number of attempts at using nondestructive testing (NDT) methods to evaluate existing anchors and tendons. Among others, Holt et al. (2013) described the results of a research project funded by the U.S. Army Corps of Engineers. The resulting method, which was based on dispersive wave propagation, was successfully used at two dams to determine the load in anchor rods used to secure trunnion gates.

In another approach, in 2002, a research effort was funded by the Transportation Research Board under its National Cooperative Highway Research Program (NCHRP) led to a comprehensive report titled, "Recommended Practice for Evaluation of Metal-Tensioned Systems in Geotechnical Applications" (Withiam et al. 2002). This report presented the findings of a research project to evaluate procedures for estimating the design life of metal-tensioned systems in new geotechnical installations and determining the conditions and remaining service lives of systems already in place. It presented a recommended practice for assessing the present conditions and remaining service lives of metal-tensioned systems with NDT techniques and an appropriate prediction model. The report identified several electrochemical tests, including measurement of half-cell potential and polarization current, which can detect the presence of corrosion and gauge the integrity of any corrosion protection systems. However, it found that mechanical NDTs, principally wave propagation methods such as impact and ultrasound techniques, must also be used to determine whether corrosion has caused loss of element cross section in the metal-tensioned system.

The report included a critical literature review on NDT methods and prediction models, as well as a detailed description of the NDT methods selected for use with the associated recommended practices. It should be noted that the Transportation Research Board sponsored additional research afterward on the same theme of evaluation and improved design of metal-tensioned systems.

The other approach is the ASTM standard D 5882–07 titled "Standard Test Method for Low Strain Impact Integrity Testing of Deep Foundations" (ASTM 2016), on which pile test procedures are based. Specifically, this standard covers the procedure for determining the integrity of individual vertical or inclined piles by measuring and analyzing the velocity (required) and force (optional) response of the pile induced by an impact device such as a handheld hammer. The test methods covered are the pulse echo method (PEM), where the pile head motion is measured as a function of time and the time-domain record is then evaluated for pile integrity, and the transient response method (TRM), where the pile head motion and force (measured with an instrumented hammer) are measured as a function of time and the data are evaluated usually in the frequency domain. Although not originally intended for rods and tendons, the methods described in the standard or variations thereof certainly hold promise for successful use in this type of application.

Methods for evaluating the residual load and integrity of anchors are not yet mainstream, but there is certainly a need and a demand that will keep increasing with time, especially if rather simple and very efficient methods can be developed and brought to the market.

5.3 CRITICAL PERFORMANCE INDICATORS

5.3.1 Internal Hydraulic Pressure

5.3.1.1 Piezometers - General. Piezometers are used to measure pore water pressures and are by far the most common of all instruments installed in dams. Piezometers installed in the abutments, foundations, and embankments of a dam are used to monitor phreatic levels and uplift pressures and to interpret seepage regimes. Piezometers installed in an earth embankment under construction can also be used to monitor excess pore pressures caused by changes in stress from the weight of the fill. For dam-performance monitoring, a reading accuracy to one-tenth of a foot is generally acceptable for piezometers.

The most common types of piezometers in use today are the open standpipe (open hydraulic) and the vibrating-wire piezometer. Other types include liquid-level, hydraulic, and pneumatic piezometers; however, these are not often installed in current practice. They are covered in this section because working instruments of these types may still be in use. Newer technologies such as resistance strain gauge piezometers, fiber optic piezometers, and quartz pressure sensors are also available but not yet as widely used. The following section describes these various types of piezometers and their advantages and disadvantages.

In situations when knowing the temperature of pore or joint water pressure is required, thermistors may be installed with along with piezometers in most applications.

5.3.1.2 Installation Considerations.

Pressure Sensors.　The user needs to keep in mind several considerations regarding pressure sensor installation. The instrument will need to be installed such that the total pressure sensor will always be submerged and that the barometric pressure sensor is never submerged. Should the water level drop below the total pressure sensor, data loss will occur. Should the barometric sensor become submerged, the barometric compensation will be incorrect and the data will be unreliable. Because of drift in pressure measurements as the sensor ages, it is recommended to periodically manually measure the height of water for sensor calibration.

Filters.　Piezometers buried in the ground or grouted into boreholes are often surrounded by a sand envelope. In these cases, it is necessary to incorporate a filter into the piezometer housing to permit the entry of water while excluding solid particles. Filters can also prevent the entry of air into the piezometer. Air inside a piezometer causes a lag in the response to a change in pore water pressure. This is especially problematic with hydraulic piezometers, which have long, liquid-filled lines.

Two types of filters are used: High air-entry (HAE) value filters have a pore size of 0.5–2 µm and an air-entry value of approximately 14.5 psi (1 bar). Low air-entry (LAE) value filters have a pore size of approximately 40–60 µm and an air-entry value approaching zero. Air-entry value is defined as the differential pressure that must be applied across a saturated filter before the filter allows air to pass through. The smaller the filter's pore size, the greater the effect of surface tension, which causes a higher air-entry value. HAE filters are generally used in unsaturated soils of low permeability to prevent soil particles from entering the piezometer. In contrast, LAE filters are generally used in coarse-grained saturated soils, where pore gas pressure is not present.

Piezometers installed in wells have the filter replaced by a mesh screen, which is less susceptible to blockage by scaling or crystallization of dissolved salts, especially where the standpipe may periodically become dry.

Venting.　Piezometers with pressure-sensitive diaphragms react to changes of barometric pressure in the same way that they react to changing groundwater pressures. Barometric pressure acts directly on the water surface in wells and standpipes, and small fluctuations in barometric pressure affect readings. Barometric effects are eliminated by venting the piezometer. Vented piezometers naturally compensate for barometric pressure changes by allowing the ambient barometric pressure to act on the backside of the diaphragm. Venting is achieved by inserting a small plastic tube within the cable all the way to the inside face of the pressure-sensitive diaphragm. The outer end of the tube is open to the atmosphere.

Vented piezometers are used in standpipes and observation wells, in which the piezometer measures the height of a water column. Nonvented piezometers are normally buried and sealed in fills or boreholes, in which the soil acts as a baffle and prevents pore water pressure from reacting to minor changes in barometric pressure.

A disadvantage is that the vent tube provides a path for moisture to migrate into the inside of the piezometer, which usually shortens the instrument's life. To prevent moisture from entering the vent tube, the open end has a chamber containing desiccant capsules. The desiccant capsules need to be replaced regularly to prevent corrosion and damage to the piezometer. For these reasons, venting is not always advisable, and it is sometimes preferable to correct for barometric pressure based on readings from a barometer. Barometric recordings can be automatically accounted for in digital data collection.

Casing Materials.   To retard corrosion, piezometer housings are made of noncorrosive materials. Plastic, stainless steel, and titanium are the most common materials in use. Soil chemistry (e.g., pH and corrosivity) are considerations when choosing the housing materials. Casing material choice requires care if sensors are installed in high-temperature environments (e.g., tailings dams).

Comparative Installations.   The four most common installation configurations for piezometers are open standpipe, observation well, zoned, and fully grouted; a comparison of these configurations is given in Table 5-1. All four installation methods can be instrumented to allow automated (digital) readings using pore-pressure sensors, such as vibrating-wire (most common) piezometers, but only two of the installation methods allow manual/physical confirmation of the reading by using a water probe.

The open standpipe piezometer and observation well installation offer the added benefit of supplementing the pore-pressure measuring device with the ability to directly measure or observe the water level in the standpipe. Open standpipe piezometers and observation wells are often installed with a zoned backfill to prevent vertical flow in the area surrounding the pipe or to target a specific subsurface layer.

Table 5-1. Comparison of Piezometer Installation Techniques.

Installation method	Advantages	Disadvantages
Open standpipe and observation well	• Open well allows access to physically confirm water level • Simple to install • Common practice • Longevity • Reliability	• Time lag in pressure equalization
Zoned backfill	• More robust than open standpipe	• No physical confirmation of data
Fully grouted	• Generally lower cost • Simple installation • More immediate response (less lag time to stabilize) in saturated soils • Easier to install multiple piezometers within one well	• Sensitive to grout mix and installation procedures • May not give reliable readings in certain formations or at interfaces with low and high permeable strata • No physical confirmation of data • Separate instrument (i.e., open well) required to confirm reading • May be more susceptible to damage due to settlement of surrounding strata • Requires barometric correction • Likely unsuitable for partially saturated soils • Cannot be replaced

A zoned backfill configuration employs the backfill characteristics of open standpipe piezometers and observation wells without using a pipe. The pore-pressure measuring device is installed in a borehole at a desired depth, with the electric leads, cables, and hoses guided up the borehole to the ground surface. The immediate area above and below the device is then backfilled with a porous material (sand pack). The area immediately above the sand pack is then backfilled with bentonite pellets or chips and hydrated to form a seal above the sand pack. The remaining borehole is then grouted to the ground surface.

Fully grouted installation is a simplified method where the pore-pressure device is lowered into a borehole at a desired depth, with the electric leads, cables, and hoses guided up the borehole to the ground surface. The entire borehole is backfilled with a nonshrinking cement–bentonite grout to prevent vertical flow. Methods for installation and grout mix design are detailed by Mikkelsen and Green (2003). Some of the key considerations are summarized here.

Piezometers have been installed using all four methods since the advent of diaphragm pore-pressure devices. In general, the open standpipe piezometer configuration has been most popular; however, there has been increased use of the fully grouted configuration by some members of the dam-monitoring community. Recent research and reviews by Contreras et al. (2012) and Mikkelsen and Green (2003) have advocated the increased use of fully grouted piezometer systems because they simplify construction and reduce costs, and their research indicates that such systems provide comparable results to the other installation methods. In contrast to this research, there have been practical applications of fully grouted piezometer systems that have reported erroneous or questionable data without the ability to confirm or calibrate the readings. Although no formal case histories or studies have been published regarding these issues, the dam-monitoring community remains divided over the suitability of fully grouted piezometer systems. Perhaps until more confidence is developed in fully grouted piezometers, standard practice might consider installing open-well piezometers adjacent to some of the fully grouted piezometers in such a way that these piezometers record the same data, so that comparisons can be made. This was done during remedial construction at the Saluda Dam in South Carolina (SCE&G Company), and good comparisons were obtained. It is unknown if the comparisons have continued over the years since installation.

If an owner or engineer wishes to capitalize on the cost savings of a fully grouted installation, the following guidance requires careful consideration:

- Proper grout mix is critical to the successful performance of grouted piezometers:
  o Grout needs to be of a creamy or "milkshake" consistency.
  o Cement is added first, then bentonite. It is important to use the proper water/cement ratio.
  o Grout needs to be designed for proper permeability. This ensures the pressures read in the grout reflect a response to hydrostatic pressure similar to that of the surrounding strata. Laboratory and analytical evaluations have indicated that the permeability of the grout can be up to three orders of magnitude greater than the surrounding strata (Contreras et al. 2012).

- Grouted piezometers may have questionable readings when placed near interfaces with strata of significantly different permeability characteristics (i.e., near the interface of a highly pervious filter or foundation zone and a low-permeability core or a concrete structure).
- The settlement or deformation potential of the surrounding soil needs to be considered.
- Grouted piezometers can generally withstand vertical strains of up to approximately 15%.
- Grouted piezometers in partially saturated soil may not provide the intended response.
- Industry experts should be consulted if you plan to install grouted piezometers within partially saturated soils or rock.
- If the system will include numerous grouted piezometer locations, consider installation of one or more open standpipe wells for comparison and data confirmation.

5.3.1.3 Open Standpipe and Observation Well. An open standpipe (also known as an open well or Casagrande piezometer) is the simplest type of piezometer. An example is shown in Fig. 5-2. It consists of a steel or plastic pipe, with a filter at its lower end, installed inside a borehole. The annular space adjacent to the filter is backfilled with sand; backfill above the filter is generally an impermeable material, such as bentonite grout or bentonite chips. Water entering the pipe through the filter rises to a height above the filter, which is equal to the pore water pressure at the filter elevation divided by the average density of the water in the standpipe.

Fig. 5-2 also shows a typical observation well. This differs from the open standpipe piezometer in that the full height of backfill material is permeable, usually sand or gravel. Observation wells provide a vertical connection between all the strata through which they are installed. If artesian

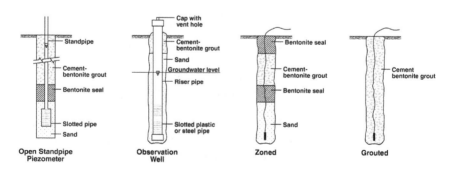

Fig. 5-2. Common piezometer installation techniques

Fig. 5-3. Water-level indicator (left) and pressure gauge (right)
Source: AECOM (left only), reproduced with permission.

conditions occur in one or more strata, interpretation of the water level in the well can be ambiguous. In addition, rain and/or irrigation water can seep in and accumulate in the bottom of the observation well in soils that decrease in permeability with depth. This gives a false indication of the piezometric surface. If infiltration is entirely from one unconfined aquifer, observation wells usually give a reliable indication of the groundwater level and are well suited to measure uplift at the base of a concrete dam.

Open standpipe piezometers and observation wells are not used in clay and other low-permeability soils because the time lag is too long between pore water pressure changes and the corresponding change of water level inside the well.

The water level inside an open standpipe piezometer is easily measured with a water-level indicator (dipmeter) or a pressure gauge, as shown in Fig. 5-3.

Reading an open standpipe piezometer can be automated by installing a pressure transducer or vibrating-wire pressure sensor inside the pipe. It is wise to choose a standpipe with an internal diameter greater than 0.8 in. (20 mm) because many pressure sensors have a diameter of 0.75 in. (19 mm). However, the larger the standpipe, the greater the amount of water that must flow into or out of it for the water level to reach equilibrium with the pore water pressure in the ground. If the ground around the filter has low permeability, the time lag to reach equilibrium could be excessive.

Another method to measure the water level in an open standpipe piezometer or observation well is with a bubbler system, as shown in Fig. 5-4.

The bubbler system consists of a pipe connection to the water level in the standpipe, a pressure transducer, and a small compressor. The compressor

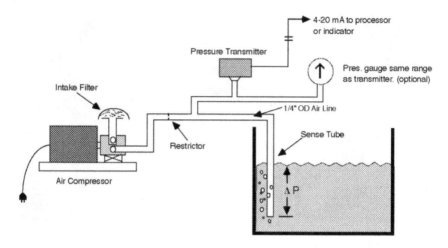

Fig. 5-4. Bubbler system
Source: Kele Associates, reproduced with permission.

pumps air into the well through the pipe, and the "backpressure" is measured. The pressure measured in the air pipe can be correlated to the water level. The advantage of bubblers is that the sensor is located out of the water where it can more easily be protected and maintained. They can also be used to retrofit small-diameter standpipe piezometers (less than 5/8 in. inside diameter), which were very commonly used in earth dams from the 1940s to the 1970s. Bubblers are also suitable for dirty water.

Similar to other manual instruments, the advantages of open standpipe piezometers include their low cost, longevity, simplicity, and reliability. They are typically installed in drilled holes. Major disadvantages are lack of real-time, automated readings. Although the use of pressure sensors can allow remote sensing, the lack of real-time response of the instrument in a low-permeability soil is a disadvantage for embankments in which changes in water levels and pore pressures are important factors in evaluating dam performance.

5.3.1.4 Pneumatic and Hydraulic Piezometers.

Pneumatic Piezometers. A typical pneumatic piezometer and its operating principle are shown in Figs. 5-5A, 5-5B, and 5-5C.

Water entering the body of the piezometer through a filter exerts pressure on a flexible membrane, closing two valve ports. Nitrogen from a pressurized tank at the readout location is metered at a controlled flow rate into one of two plastic tubes. When the nitrogen pressure at the piezometer tip equals the pore water pressure, the flexible membrane lifts off the valve ports, allowing nitrogen to flow into the second tube and vent to the atmo-

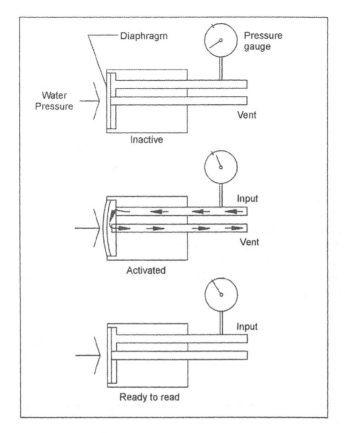

Fig. 5-5A. Operating principle of a pneumatic piezometer
Source: ASCE Task Committee on Instrumentation and Monitoring Dam Performance (2000).

Fig. 5-5B. Pneumatic piezometer sensor
Source: RST Instruments, reproduced with permission.

Fig. 5-5C. Pneumatic piezometer sensor with protective slotted casing
Source: RST Instruments, reproduced with permission.

sphere. The pressure at the inflow tube when the valve opens can be read on a pressure gauge. Alternatively, the nitrogen supply can be shut off, allowing the valve at the tip to close and the static pressure at no flow to be read on the pressure gauge.

The advantages of the pneumatic piezometer include its low cost and its immunity to lightning damage. Disadvantages include the prolonged times required to read piezometers with long tubes, the bulky readout box, the need for a nitrogen supply, and the difficulty of connecting to a data logger. Also, inaccuracies can be caused by leaks in the tubing. Water or dirt entering the tubing may foul the valve seats and cause reading errors.

Pneumatic piezometers are fairly uncommon for new dams, except in some cases where an owner may ask for a redundant technology (e.g., in addition to a vibrating-wire piezometer) that has no risk of being affected by lightning. Most applications for pneumatic piezometers are for short-term embankments such as pre-loads of compressible layers.

Twin-Tube Hydraulic Piezometers.   In the past, this type of piezometer was quite common but in recent years it has generally been replaced by pneumatic or, more commonly, vibrating-wire piezometers. The essential components are shown in Figs. 5-6A and 5-6B. Groundwater enters the tubing through a filter element designed to prevent air from entering the tubing. The tubing is usually made from two individual nylon tubes bundled inside a PVC outer jacket. The tubing leads to a readout terminal at a lower elevation, where water pressure in the tubing is measured by a Bourdon tube pressure gauge or a pressure transducer. Because the pressure reading is sensitive to the elevation of the readout terminal, this must be accurately surveyed, and any subsequent settlement or heave of the reference elevation accounted for.

The apparent simplicity and reliability of the system is compromised by operational difficulties. The main problem is the accumulation of air bubbles in the tubing and the need for periodic flushing with de-aired water. A pressure gauge with high accuracy and resolution and made of noncorrosive material is the best choice.

An additional concern about these piezometers is transmission of pore pressure along horizontal reaches of tubing, and, as the tubing ages and breaks, it may also cause piping of fine material.

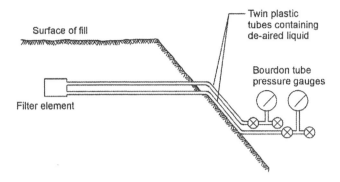

Fig. 5-6A. Twin-tube hydraulic piezometer schematic
Source: ASCE Task Committee on Instrumentation and Monitoring Dam Performance (2000).

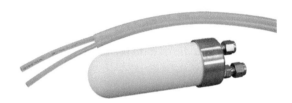

Fig. 5-6B. Twin-tube hydraulic piezometer
Source: RST Instruments, reproduced with permission.

5.3.1.5 Resistance Strain Gauge Piezometers. Resistance strain gauge piezometers and pressure sensors employ stainless steel, titanium, or ceramic diaphragms to sense pressure. Strain of the diaphragm resulting from the application of pressure produces a change in the resistance of an attached strain gauge, as shown in Fig. 5-7. Today most strain gauges are sputtered or plated onto the diaphragm, which results in more economical and reliable instruments. The resistive properties of silicon semiconductors are also used for pressure sensing by bonding a silicon chip to the diaphragm. Such sensors are then called piezoresistive pressure sensors.

Strain gauge and piezoresistive-type pressure sensors are available with internal signal-conditioning electronics that produce 0–5 V output or 4–20 mA output. They are also available without electronics, in which case the output is a millivolt-level signal. Millivolt-level signals are more easily affected by electrical noise and require external bridge completion circuitry. Outputs of 0–5 V or 4–20 mA are readily recorded by almost all automated data-acquisition systems.

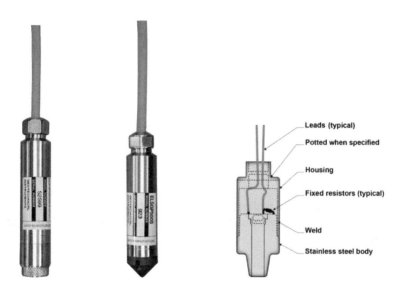

Fig. 5-7. *Diaphragm strain gauge piezometers with filter nose and ported nose*
Source: RST Instruments, reproduced with permission.

Many commercial pressure sensors of this type do not include a filter in the housing and are therefore best suited for use in open standpipes. These are generally referred to as borehole pressure gauges or piezometers with ported noses. Most have a vent tube in the cable to equalize barometric pressure changes. Models with filters can be buried in the ground and are true piezometers in that they directly measure pore water pressures. Most of these instruments have stainless-steel housings. Titanium housings are also available and have superior corrosion resistance in salty or brackish water.

An advantage of this type of pressure sensor is that it is widely used in the industrial market and is therefore available from numerous suppliers. These sensors are also suitable for dynamic measurements. The main disadvantages are susceptibility of the internal electronics to lightning damage and to moisture that enters the instrument through the vent tube. These problems are most critical in piezometers that are buried and therefore nonretrievable. However, some manufacturers include a lightning protection component (gas tube) in the piezometer to improve its reliability.

5.3.1.6 Vibrating-Wire Piezometers. Vibrating-wire piezometers, as shown in Figs. 5-8A and 5-8B, have been used extensively on dams throughout the world and have achieved an excellent reputation for long-term stability and reliability (Bordes et al. 1985; McRae et al. 1991; Choquet et al.

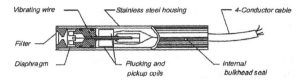

Fig. 5-8A. Vibrating-wire piezometer schematic
Source: ASCE Task Committee on Instrumentation and Monitoring Dam Performance (2000).

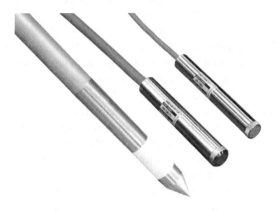

Fig. 5-8B. Vibrating-wire piezometers (standard, heavy-duty, and drive-point models)
Source: RST Instruments, reproduced with permission.

1999). The pressure-sensitive diaphragm is coupled to the vibrating wire, which is plucked and read by an electromagnetic coil. Various styles are available, including small-diameter versions for insertion into small-diameter standpipes, heavy-duty models for use with armored cable in earth dams, and standard models for burial in fill or installation in boreholes.

Vibrating-wire piezometers have the advantage of long-term stability, and the frequency signal can be transmitted accurately over very long cables. Additionally, they can be used in open standpipes, zoned piezometers, and fully grouted piezometers. There is some limitation to the dynamic capabilities for the measurement of rapidly changing pressures (e.g., earthquake effects), although their capability is improving with new data loggers, and they are susceptible to lightning damage but not to the same extent as resistance strain gauge types. Most manufacturers include a lightning protection component (gas tube) in the piezometer to improve its reliability.

5.3.1.7 Fiber Optic Piezometers. Fiber optic piezometers consist of MEMSs (microelectromechnical sensors) or MOMSs (micro-optical mechanical sensors) that send signals to an optical cable. The instrument is generally housed in a stainless-steel tube and is separated from the environment by a porous filter material composed of ceramic or stainless steel. Housings for fiber optic piezometers are generally tubular, with typical diameters around 0.3 to 0.8 in. (8 to 20 mm) and typical lengths of 2.2 to 4 in. (55 to 100 mm). These instruments are relative newcomers to the geotechnical field. As yet, they do not have an extensive history of use and are somewhat expensive. An example using Fabry–Perot Interferometry technology is shown in Fig. 5-9 (Choquet et al. 2000).

Major advantages of fiber optic sensors include their immunity to vibration, lightning damage, and radio and electromagnetic noise interference. Fiber optic piezometers can be used in harsh environments, including gaseous environments, and their optical signals can be sent over several kilometers with little signal loss. As such they can be considered for use in ADASs. Fiber optic piezometers are expected to have good long-term stability, but there are still few case histories to support that claim. Their simplicity of design and small number of parts hold the promise of reduced sensor unit costs in the future. Readout units can also be expected to follow a similar cost reduction trend with further advances in research and number of applications. Fiber optical cables are already less expensive than their copper counterparts with a similar degree of mechanical reinforcement and waterproofing.

5.3.1.8 Quartz Pressure Sensors. The resonant frequency of a quartz crystal changes as a function of the pressure applied to it. This property is used in submersible quartz pressure sensors available from several manufacturers. They are mainly used in dam monitoring for pool-level measurements. These instruments have very high accuracy and precision, but they are typically more expensive than the other types of piezometers discussed previously.

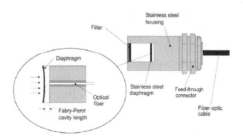

*Fig. 5-9. Fiber optic piezometer based on Fabry–Perot interferometry
Source: Courtesy of Roctest, reproduced with permission.*

5.3.1.9 Multiple Piezometers in a Single Borehole. Occasionally it is desirable to install more than one piezometer in a borehole. This can be difficult in standpipe and zoned backfill-type piezometers. A sand envelope is required around each piezometer tip. Furthermore, the tips need to be isolated from one another by sealing the borehole space between them with a bentonite grout or other impervious material. Various methods of installation have been used and specialized equipment exists for placing the various materials in controlled amounts in their proper locations (Dunnicliff 1993). The impervious plugs must be longer than imperfections in the embankment or joints in rock that would allow water to bypass the plugs. Fig. 5-10 shows an example installation detail for nested piezometers with multiple piezometers in the same borehole. Alternatively, a number of piezometers can be mounted on a multiconductor cable to ease installation, as shown in Fig. 5-11.

Fully Grouted Piezometers. An advantage of fully grouted piezometers is that they can easily accommodate several pore-pressure measuring devices in a single borehole. The grout column surrounding the instruments allows for the isolation of subsurface layers. The borehole can also accommodate other types of instruments alongside pore-pressure measuring devices, such as inclinometers. Fully grouted piezometers have become the most common form of nested multipoint piezometers. Details of fully grouted

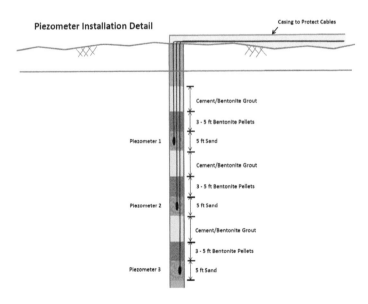

Fig. 5-10. Nested piezometers installation detail
Source: AECOM, reproduced with permission.

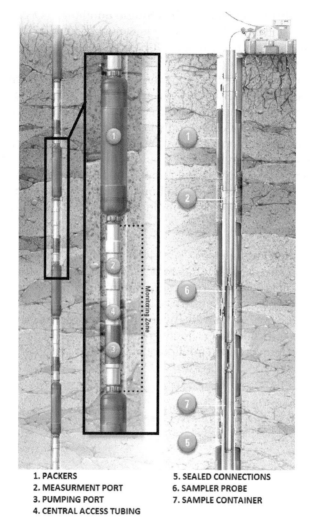

1. PACKERS   5. SEALED CONNECTIONS
2. MEASURMENT PORT   6. SAMPLER PROBE
3. PUMPING PORT   7. SAMPLE CONTAINER
4. CENTRAL ACCESS TUBING

Fig. 5-11. Multiple piezometers in one hole—the Westbay System
Source: Westbay Instruments, reproduced with permission.

piezometers would be similar to those shown in Fig. 5-10; however, the entire column would be grouted.

The Westbay System. The Westbay System, shown in Fig. 5-11, consists of casing, couplers, and independently-inflated packers permanently installed in a borehole. Packers isolate the couplings from their neighbors. Measurement port couplings have a valve, which can be located and operated by a wireline-deployed probe with a pressure transducer lowered into the casing. This port is used for measurement of formation fluid pressure, performance of hydraulic tests, or collection of formation fluid samples.

The Waterloo System. The Waterloo System, shown in Fig. 5-12, uses a series of special double packers connected by a watertight casing. The packers inflate chemically upon exposure to water and remain inflated permanently. The sample entry ports can be connected to pressure transducers located inside the casing (Solinst 2015).

5.3.2 Flow and Water Level

Measurement of seepage quantity is important to estimate the amount of water seeping and leaking through, beneath, or around a dam. Water clarity and quality are also important because high turbidity is a performance indicator that material is being removed from the dam or its foundations, raising the possibility that piping may lead to failure of the dam. Flow quantity is often measured indirectly using water-level devices. Seepage measurement devices described in this section focus on flow quantity and include calibrated containers, Parshall flumes, and weirs. Applications in dam engineering primarily include measurement of flow through seepage collection channels, toe drain outlets, and stream discharges. A reading accuracy of these flows to the nearest gallon is generally considered adequate for dam-performance monitoring. Also included in this section is a description of various liquid-level measuring devices. These are used to measure water levels in reservoirs and also to indirectly

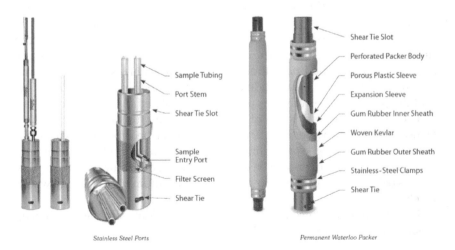

*Fig. 5-12. Multiple piezometers in one hole—the Waterloo System
Source: Solinst Waterloo Systems, reproduced with permission.*

calculate seepage flows by measuring the water level in a flume bay or weir box.

5.3.2.1 Flow Measurement.

Calibrated Containers. The simplest method of measuring seepage flows is to use a stopwatch to measure the time required to fill a container of known volume. The container volume is then divided by the time to fill the container to calculate a volumetric flow rate.

Parshall Flumes. Parshall flumes are frequently used to measure flow in open channels. The Parshall flume is a specially shaped section of a channel that restricts the flow and creates a gradient at the water surface, as illustrated in Figs. 5-13A and 5-13B. The water surface elevation is measured at two locations using stilling wells. The flow rate can then be calculated based on the difference in water surface elevations in the stilling wells.

Water levels are commonly measured with staff gauges fixed to the walls of a stilling well or the flume itself. Water levels are also measured using ultrasonic-level sensors, bubbler tubes, and submerged pressure sensors. These sensors are described in other sections of this chapter. Several sensor technologies provide the potential for connection to ADAS. The advantages of the Parshall flume are its simplicity, low maintenance, long service life, self-scouring design, and the potential for integration into ADAS for continuous, real-time remote sensing.

Weirs. Weirs, shown in Figs. 5-14A and 5-14B, are often chosen to measure seepage rates. They work better at low flow rates than Parshall flumes and have better precision. The shape of the weir may be square, trapezoidal, or V-notch. The shape and size of the weir depend mainly on the volume of flow to be measured. The depth of water is typically measured with a staff gauge installed near the weir. For best accuracy, vent pipes are attached to the downstream side of the weir plate (not shown) to ensure that the pocket beneath the nappe is fully ventilated. Also, flushing ports can be provided at the bottom of the weir plate and upstream baffles may be used to calm the water flow. They can, however, become filled with sediment or fouled by algae or floating debris, requiring periodic maintenance and flushing. Weir covers can be effective in controlling algae growth. The figures below also show an arrangement for the automatic reading of water levels using a vibrating-wire force transducer.

For large flow volumes, a rectangular or trapezoidal weir may be needed, as shown in Fig. 5-15.

River Flow Gauges. River flow gauge stations are often used for monitoring inflow and outflow from dams and their structures and to calibrate hydrologic predictions for the site. River flow gauges downstream of a dam

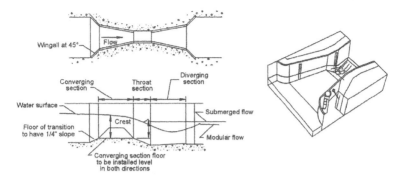

Fig. 5-13A. Parshall flume schematic
Source: ASCE Task Committee on Instrumentation and Monitoring Dam Performance (2000).

Fig. 5-13B. Parshall flume
Source: AECOM, reproduced with permission.

structure can provide an indication of unexpected releases and may be set to automatically trigger an alarm if an unusual spike occurs in river flows. River flow gauges upstream of a structure can be used as an early indicator of changes in inflow, which may warrant proactive operation of water conveyance features. The U.S. Geological Survey (USGS) maintains a network of publicly available river gauge data that can be accessed on their website. Owners may also decide to install their own gauges in proximity to their site. The USGS provides guidance on stream flow measurement in their manual *Techniques of Water-Resources Investigations of the USGS*, Chapter 8, "Discharge Measurement at Gaging Stations." This chapter can be downloaded for free at http://pubs.usgs.gov/twri/twri3a8/pdf/TWRI_3-A8.pdf.

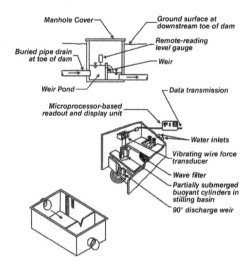

Manhole Cover
Ground surface at downstream toe of dam
Remote-reading level gauge
Buried pipe drain at toe of dam
Weir
Weir Pond
Data transmission
Microprocessor-based readout and display unit
Water inlets
Vibrating wire force transducer
Wave filter
Partially submerged buoyant cylinders in stilling basin
90° discharge weir

Fig. 5-14A. V-notch weir schematic
Source: ASCE Task Committee on Instrumentation and Monitoring Dam Performance (2000).

Fig. 5-14B. V-notch weir photo
Source: AECOM, reproduced with permission.

River flow gauge measurements require a staff gauge to measure the river stage, and then the cross section at the gauging site must be calibrated for stage-discharge.

Velocity Measurement.   Several instruments that directly measure the volumetric flow rate in pipes are available commercially. Flowmeters use

Fig. 5-15. Rectangular weir

a variety of sensing principles and techniques, including pitot tubes, propeller vanes, paddle wheels, electromagnetic, ultrasonic velocity through the fluid, Doppler, Coriolis mass, and vortex shedding. They typically are built into a short section of a pipe of specified size with attached signal conditioning and display module and sold as a unit. All flowmeter types have inherent advantages and disadvantages. Two well-proven general-purpose types are ultrasonic and vortex shedding.

The vortex shedding-type flowmeter, shown in Fig. 5-16, is suitable for use over a wide temperature range. It is simple, rugged, and maintenance-free. There are no moving parts; no sensor parts are exposed to the water.

When using any type of flowmeter fitted to a pipe, it is essential that the pipe be completely full at the time of measurement if the velocity is to be converted accurately into a volumetric flow (gal./min or L/min). This is sometimes challenging in applications such as toe drain outfalls, which do not always flow full. This stipulation applies to all types of flowmeters and is not only limited to vortex shedding.

Flowmeters may also be used in open channels to measure linear velocity (ft/s, etc.); however, the dimensions of the channel and elevation of the water must be known to obtain a volumetric flow rate. This can be readily accomplished in a hardened channel of known cross section with water-level sensing technology to determine water depth. Portable flowmeters have been used for this purpose, but the accuracy is not great because velocities vary across the stream profile. Paddle wheel- or propeller-type flowmeters require more maintenance than noncontacting types or vortex-shedding types.

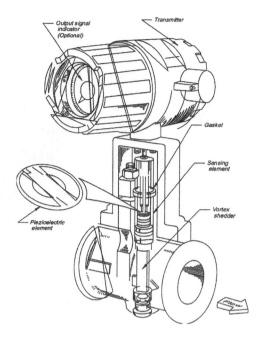

Fig. 5-16. Flowmeter, vortex shedding type
Source: ASCE Task Committee on Instrumentation and Monitoring Dam Performance (2000).

5.3.2.2 Water-Level Measurement.   Reservoir and tailwater levels are usually recorded even on the smallest dams. Unanticipated rapid increases in tailwater or decreases in reservoir level or both are often linked to activation triggers for operation of gates or valves or the emergency action plan (EAP). There are several ways to measure these water levels. Typically, a stilling well is used to house a measuring device to attenuate wave action. Level-monitoring instruments are located away from the nappe effects of spillways, powerhouses, and other intake and outlet structures. Common devices placed upstream and downstream of the dam to measure reservoir and tailwater, respectively, include staff gauges, float systems, bubblers, or submerged pressure transducers. As described previously, water-level measurement devices are also used to measure water levels in flumes and weirs in order to calculate flow.

Low-Range Pressure Sensors.   Low-range pressure sensors are available for measuring shallow fluid depths. The major advantage in using these instruments is greater accuracy when measuring the shallow depths often encountered in low-flow flumes and weirs. One sensor measures the total pressure at the base of the flume or weir and another sensor measures barometric pressure. The net fluid pressure is calculated and then converted to a height of water. The barometric pressure transducer may either be integrated into the top of the flume or weir or housed externally. In some cases, the barometric pressure is conveyed to the instrument through a data cable. In such instruments, extra care must be taken to protect the cable from damage. Pressure sensors are available for a wide range of pressures (i.e., flow depths) and accuracies.

Vibrating-Wire Weir Monitors. The vibrating-wire weir monitor, shown in Fig. 5-17, uses a vibrating-wire force transducer to monitor water levels upstream of a weir. The main component is a cylindrical weight hung from a vibrating-wire force transducer. The cylinder is partially submerged in the basin upstream of the weir. The water level behind the weir is related to the buoyant force on the cylinder, which is reflected in the tension in the wire and its resonant frequency. The depth of water in front of the weir can be used as an input to open-channel flow equations to calculate the estimated volumetric flow of water through the weir.

Staff Gauges. The simplest and most common device for monitoring reservoir elevation is the graduated staff gauge, which can be mounted on any vertical surface and bolted directly to a concrete, timber, or steel structure. Staff gauges are also commonly used to measure water levels within Parshall flumes and weirs. The gauge requires indelible graduations and markings so that it is resistant to sun bleaching, rusting, or other forms of deterioration. The positioning of the gauge is carefully surveyed so that it is easily visible and permits accurate water surface elevation readings. The

Fig. 5-17. Automated vibrating-wire V-notch weir

best location is one that is not influenced by wave action. In cold climates, protection of exposed staff gauges against ice loads may be necessary.

Many reservoirs, weirs, and flumes that are fitted with automatic water-level sensors are also equipped with a staff gauge that provides a convenient visual check on automatically read data and serves as a backup system in the event of electronic system failure. Also, a staff gauge for visual observation is often required by many regulatory agencies because of its reliability and accuracy. Frequent documented checks of automatic water-level sensors using manual staff gauges are required, including documented cross checks with readings at remote control centers or powerhouses.

A standard staff gauge used by the U. S. Bureau of Reclamation is shown in Figs. 5-18 and 5-19. The gauge is staff graduated in increments of 0.01 ft (5 mm). Typical materials include fiberglass, laminates, type 316 stainless steel, or steel with a coating of porcelain enamel. Staff gauges are simple, but they cannot be automated and must be protected from ice and debris.

Manual Tape Measurement. Steel tapes used for reservoir gauges are typically custom-made and graduated precisely to unique reservoir elevations. A common configuration is a steel tape coiled onto a drum and

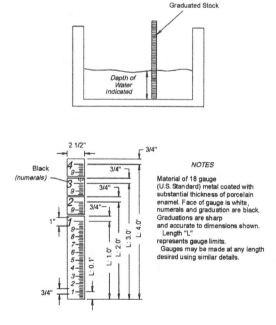

Fig. 5-18. Standard staff gauge
Source: ASCE Task Committee on Instrumentation and Monitoring Dam Performance (2000).

Fig. 5-19. Staff gauge in measuring weir
Source: AECOM, reproduced with permission.

mounted on the top of a standpipe. The standpipe is open to the reservoir and serves to attenuate wave action. Some steel tape reservoir-level monitoring devices have a tape that is electrically charged with a low current and low DC voltage and an in-line voltmeter to indicate contact with the water. To measure the reservoir elevation, the tape is lowered into the standpipe until contact is made (indicated by the voltmeter) and the tape is read.

Float Systems.   A typical float system, which uses a float attached to a chain leading over a sprocket to a counterweight, is shown in Fig. 5-20. As the float rises, the sprocket turns, moving a pen that draws a line on a cylindrical paper chart or activating a rotary encoder for electronic measurements. In the former case, the paper chart is driven by a small battery-powered electric motor or by mechanical clockwork.

Bubbler Systems.   Bubbler systems are commonly used to measure reservoir level. This type of system was described earlier in this chapter under the open standpipe section.

Submerged Pressure Transducer.   Water levels can be measured by using a submerged pressure transducer to measure the head of water above the transducer elevation. Pressure transducer gauges are usually mounted in a stilling well or a pipe with an open top and connected to the reservoir

Fig. 5-20. Water-level float recorder
Source: ASCE Task Committee on Instrumentation and Monitoring Dam Performance (2000).

at low elevation. The stilling well damps most wave action. Electrical or vibrating-wire pressure transducers require protection from lightning and may become fouled in dirty water. Corrosion of the transducer can be a problem. Transducers made entirely from titanium with very high corrosion resistance are available commercially if needed. Pressure transducers can be vented to eliminate barometric influences, but moisture must be prevented from entering the vent line. This requires the use of desiccant capsules, which must be replenished at regular intervals. Alternatively, nonvented transducers can be used and the barometric pressure can be recorded separately. Pressure transducers are relatively delicate and susceptible to damage. The locations for these gauges are selected strategically to limit the risk of damage due to impact, debris, and environmental exposure. Two different types of transducers are shown in Figs. 5-21 and 5-22, respectively.

Ultrasonic and Radar Sensing. Ultrasonic and radar technologies provide two methods of electronic-level sensing. The methods provide means for continuous level measurement of a fluid surface. The sensors for these technologies may be contacting (sensor is in contact with the measured fluid) or noncontacting (sensor is not in contact with the measured fluid). Noncontacting methods may be complicated by the presence of foams or liquid/liquid interfaces (such as an oil–water interfaces) within the measured fluid. The sensors can be connected to data loggers or ADASs.

The most common contact method is guided wave radar (GWR), which may also be known as time-domain reflectometry (TDR) or micro-impulse radar (MIR). In GWR, a probe extends into the measured fluid. A pulse of

Fig. 5-21. Transducer in secure location

Fig. 5-22. Submerged pressure transducer
Source: Teledyne Isco, reproduced with permission.

microwaves traveling at the speed of light is sent down the probe. When the pulse reaches the air/water interface, a large portion of the microwave energy is reflected back to the transmitter. This technique works well in applications where the liquid surface may not be perfectly flat, such as cases where waves are present. Changes in pressure and temperature do not affect the accuracy of this technique. The equipment typically has no moving parts and involves minimal or no maintenance.

Two techniques are available in noncontacting radar-level sensing (Fig. 5-23): pulse radar and frequency-modulated continuous wave (FMCW). Typical ranges of measured distances are up to approximately 100 ft. The principle in noncontacting pulse radar sensing is to send out a microwave signal that bounces off the surface of the liquid and returns to the gauge. Using the measured time for the return signal, onboard electronics calculate the distance to the liquid surface.

The FMCW technique also sends microwaves toward the liquid surface and receives signals that are bounced back. However, the transmitted

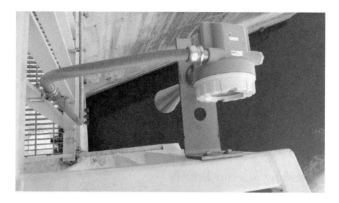

Fig. 5-23. Radar-level transmitter at Roosevelt Dam
Source: U.S. Bureau of Reclamation.

signals are of continuously varying frequency. Electronics within the sensor calculate differences between the transmitted and received frequencies, which are proportional to the distance to the liquid.

Ultrasonic-level sensing is similar to the noncontacting radar-level sensing except that the energy pulse is ultrasonic, which means that the energy travels at the speed of sound. The sonic transducer works by reflecting sound waves off the surface of the fluid back to the instrument. The time traveled for the sound wave is proportional to the distance the wave travels. This distance is subtracted from the known elevation of the transducer yielding the fluid elevation. The advantage of this system is that it can be located above the fluid surface, therefore providing better access for maintenance. Temperature affects the pulse speed, but this is compensated for by onboard temperature sensors and calculations.

In dam engineering, level sensors may be used to continuously measure and record reservoir levels in real time. For contacting sensors, a shield or stilling well may be required to prevent damage to probes by floating debris, waves, and so on. A shield or stilling well could also be provided below a noncontacting sensor to minimize the effects of waves that create a nonlevel water surface. These technologies may also be used as a component part to automate other instruments that require the elevation or depth of water to be known, such as a weir.

5.3.3 Water Quality

5.3.3.1 Turbidity. Turbidity is a measure of the degree to which light traveling through a water column is scattered by suspended particles. A variety of turbidity meters exists to measure this property. Instruments of this type are useful in detecting any changes, sudden or otherwise, in the

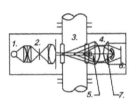

1. Light source
2. Optics
3. Process stream
4. Condensing lenses
5. Direct beam signal
   detector
6. Scatter light signal
   detector
7. Light trap

Spatial Filtering System : Condensing lenses (4) send light scattered from the area closest to the left hand window to the light trap (7), and scattered light from near the right hand window to infinity. The scattered light from the real area of interest falls on the detector (6). This is ratioed to the direct beam signal obtained by the direct beam detector (5).

Fig. 5-24. Turbidity meter
Source: ASCE Task Committee on Instrumentation and Monitoring Dam Performance (2000).

amount of material being eroded from a dam embankment or a dam's foundation.

Conventional turbidity meters, shown in Fig. 5-24, must be recalibrated for light transmission versus sediment concentration when the sediment-size distribution changes. For heavy sediment loads, more advanced turbidity meters obtain the size distribution of suspended particles by infrared laser multi-angle scattering (laser diffraction). These can measure the total suspended particle concentration and a mean sediment size.

5.3.3.2 Quality. Water-quality measurements at dams are typically made to indirectly assess conditions within the dam, its abutments, and its foundation. For example, increases in the turbidity of seepage water can indicate internal erosion and piping. On rare occasions, changes in water chemistry can indicate changes in the flow paths of seepage water. Changes in seepage water quality generally indicate a condition that requires closer investigation using other, more direct methods.

The most common water-quality measurements made at dams are turbidity, conductivity (resistivity), pH, total dissolved solids (TDS), and salinity. Instruments used today are commonly "multiprobe" instruments, an example of which is shown in Fig. 5-25, that make more than one type of measurement simultaneously. Water-quality instruments currently on the market are capable of measuring temperature, dissolved oxygen, percent saturation, specific conductance, salinity, resistivity, TDS, pH, redox, and depth.

These parameters can be measured by a combination of multiprobe instruments or by a single instrument. Water-quality instruments are typically equipped with their own digital displays. They also are available with

Fig. 5-25. EXO1 Multiparameter Sonde
Source: Photo courtesy of YSI Incorporated, reproduced with permission.

analog or digital outputs that can be recorded by ADASs. Chemical analysis is typically done by collecting water samples and performing laboratory tests.

5.3.4 Strain

Strain is defined as the change in the length of a solid material over a base length divided by the base length. Strain is a nondimensional value and is commonly expressed in units of micro-inches per inch (μ-in./in.) or microns per meter (μm/m). These units are commonly termed "microstrain units" or "microstrain."

Strain normally is the result of applied stress or thermoelastic deformation. The time-dependent strain (creep) can also be generated in soil, rock, and concrete, where strain increases with time under a constant stress. Because it is generally difficult to measure stress directly, strain is usually measured and stress is then calculated using Hooke's Law. Hooke's Law relates strain (ε) to stress (σ) using the Young's modulus (E) of the material ($\sigma = E \times \varepsilon$). There are, however, some exceptions, where stress can be measured directly using pressure cells, stress cells, or stiff inclusions in the solid material.

Instruments for strain measurement are usually referred to as strain gauges or strain meters. They are based on a variety of principles: some are purely mechanical and others are electrical.

5.3.4.1 Strain Gauges for Concrete and Soil. Mechanical strain gauges can only be used at the surface of materials, usually concrete or rock surfaces. They employ two reference studs, fixed a certain distance apart, and a removable micrometer of sufficient accuracy, usually 0.004 to 0.0004 in. (0.01–0.001 mm), to measure the change in distance between the two studs.

Electrical strain gauges can be used at the surface or internally in concrete, rock, and embankment materials. The most common types are strain, optical, and vibrating-wire gauges for use in concrete, where the strain values to be measured generally do not exceed 1,000 microstrains, and potentiometric or vibrating-wire gauges for use in soils where the expected strain values can be higher (i.e., several thousand microstrains). In this latter case, the instrument is generally referred to as a soil extensometer.

Vibrating-wire strain gauges, an example of which is shown in Fig. 5-26, are also used for embedment in concrete. The length of the measuring base of a vibrating-wire strain gauge is defined as the distance between the two end flanges. It is normally recommended that this length be at least three to four times the maximum size of the aggregate. For this reason, these instruments are most often manufactured in lengths of 4–10 in. (10–25 cm). Their measuring range is usually ±1500 microstrains.

Another instrument commonly used as an embedment strain gauge in concrete is the sister bar, shown in Figs. 5-27 and 5-28. A sister bar is a short, usually 3.28-ft (1-m)-long steel reinforcement bar (rebar) with a small diameter in the range of 0.5 to 0.6 in. (12–15 mm) containing a coaxially mounted vibrating-wire strain gauge. The sister bar is installed parallel to a larger-diameter rebar that is part of the normal reinforcement of the concrete. Strain measured in the sister bar, which is cast into the concrete, is then

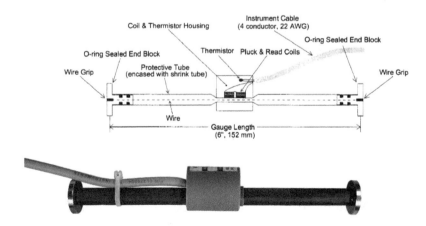

Fig. 5-26. Vibrating-wire strain gauge for embedment in concrete
Source: RST Instruments, reproduced with permission.

Fig. 5-27. Sister bar
Source: RST Instruments, reproduced with permission.

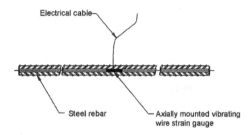

Fig. 5-28. Small-diameter sister bar to be installed parallel to large-diameter rebar
Source: ASCE Task Committee on Instrumentation and Monitoring Dam Performance (2000).

assumed to be the same as that in the larger-diameter rebar and in the concrete itself. It is assumed that the small-diameter sister bar has a negligible effect on locally modifying the strain or stress state in the vicinity of the larger rebar.

Figs. 5-29A and 5-29B illustrate a soil extensometer used to measure strain in soils. These are also described later in this chapter under the settlement section. It consists of a central tube with a telescopic coupling in the middle and two end flanges that define the base length. A displacement transducer, either potentiometric, linear variable differential transformer (LVDT) or vibrating-wire, is mounted longitudinally in the tubular piece and attached to the two end flanges to measure the change of distance between the flanges. Soil extensometers have a long base length when used in embankment materials, usually in the range 40 in. (1 m) and can be easily modified for greater base lengths. Measuring ranges of the displacement transducers are generally on the order of 1 to 12 in. (25 to 300 mm) and can be modified for greater ranges. A number of soil extensometers can be installed end-to-end to get a complete strain profile along a measuring section.

5.3.4.2 Multiple Strain Gauges (Rosettes). Concrete-embedded strain gauges are installed either singly, to measure strain in one direction, or in rosettes of two or more gauges to determine the principal strain directions and to obtain the complete state of strain in two or three directions.

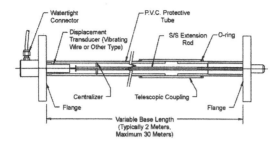

Fig. 5-29A. Soil extensometer schematic
Source: ASCE Task Committee on Instrumentation and Monitoring Dam Performance (2000).

Fig. 5-29B. Soil extensometer
Source: RST Instruments, reproduced with permission.

In Fig. 5-30, configurations A and B are for simple situations where it is already known that the principal stresses are vertical and horizontal. Configurations C and D are used to determine the state of strain in one plane. The three strain measurements are used in formulae derived from Mohr's circle to calculate the two principal strains in the plane and their orientations.

Configuration E is similar to those of C and D, except that one strain gauge is added in the same plane as the other three to add some redundancy and also in case of malfunction of one of the gauges, to still be able to determine the two principal strains and their orientations. In addition, a fifth strain gauge is installed perpendicular to the main plane of the measurement, where this perpendicular direction is known to be a principal strain direction. This happens when the geometry of the dam or the estimated loading directions make it apparent to the design engineer that a certain direction is a principal strain direction, such as a strain value of zero close to an outer boundary of the dam. Finally, configuration F involves six strain gauges installed in different directions in a pyramid geometry. This configuration is used in the more general case where no principal direction is known. The six strain measurements can be used to derive the complete state of strain at a given location, namely, the three principal strains and their orientations.

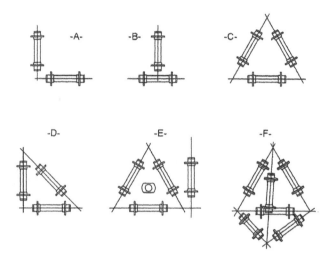

Fig. 5-30. Rosette configurations for installation of strain gauges in concrete
Source: ASCE Task Committee on Instrumentation and Monitoring Dam Perfor-
mance (2000).

5.3.4.3 Concrete Dummy Gauge.   It is generally a good practice to install one or more dummy gauges near concrete-embedded strain gauges. The purpose of a dummy gauge is to differentiate between measured strains due to hydration of the concrete and strains resulting from true loading of the concrete. This effect is greatest during the first 28 days of curing of the concrete, but it can be significant over longer time periods, especially in dams where large quantities of concrete are used and the heat of hydration persists longer than 28 days.

An alternative to dummy gauges is simply to accept that the initial state of zero strain occurs 28 days after the date of the concrete pour. However, this may be risky because construction is likely to have advanced enough in 28 days to have applied significant loads at the gauge location.

A dummy gauge is a simple device. It consists of a conventional strain gauge installed in a container with padded sides to prevent external loads from being applied to the dummy gauge. A no-stress enclosure for a dummy gauge is shown in Fig. 5-31.

The dummy gauge is typically installed just far enough from the embedded primary strain gauges to avoid modifying the stress field surrounding the primary gauges. The box is first solidly attached to a support so that it will not move during concrete placement. It is then filled with concrete from the same pour that embeds the primary strain gauges. The open end of the box allows the concrete in the box to experience the same hydration effects as the concrete of the pour, but the dummy strain gauge in the padded box

Fig. 5-31. No-stress enclosure for dummy gauge
Source: RST Instruments, reproduced with permission.

is isolated from the mechanical effects to which the primary strain gauges are subjected.

The usual way to apply dummy gauge readings is to subtract them from the readings of the closest embedded primary strain gauges to obtain the true mechanical strain in the concrete. Usually, strain in the dummy gauge increases significantly in the first few days after concrete placement because of the heat of hydration. The rate of strain then decreases and stabilizes after a few weeks, although hydration effects can continue for months and even years. Therefore, it remains a good practice to always subtract readings of the dummy gauge from the readings of the conventional gauges. A final advantage of this procedure is that it cancels the error induced by the temperature coefficient of the vibrating-wire sensor (the coefficient of thermal expansion of vibrating-wire strain gauges is about 5–10 °C, close to that of concrete).

5.3.4.4 Distributed Strain Measurement Using Optical Fibers. Distributed strain measurement using optical fibers allows the measurement of strain at many points along a single strand of a very long (up to many miles) fiber optic cable. The technique takes advantage of the sensitivity of signals within fiber optic cables to changes in temperature and strain. In short, the technique involves attaching the fiber to the medium to be monitored and sending light pulses of known wavelength into the fiber. Small fractions of the light are scattered at each point along the fiber. The nature of the scattered light varies with local changes in the fiber's temperature and strain. Time- and frequency-domain analyses of the scattered light are performed on the return signal, allowing the determination of strains and temperatures. The result is a profile of strain and temperature along the fiber.

The technology is based on three types of scattering that occurs within the cables: Raman, Rayleigh, and Brillouin. Raman scattering is inelastic and occurs as light is scattered by thermally activated molecular vibrations, allowing a measure of temperature within the fibers. Rayleigh scattering occurs when light is elastically scattered in all directions due to variations in a fiber's refractive index, which is affected by local strain and temperature changes. Brillouin optical time-domain analysis is performed by generating sound waves or introducing two counter-propagating light waves with a frequency difference equal to the Brillouin shift. Using the time of flight of the short light pulse, local variations of strain can be measured along the fiber.

Similar to a strain gauge, strain-sensing fibers must be firmly attached to or embedded into the monitored medium to ensure that fiber strains are similar to strains in the monitored media. Specialized fibers are available with increased resistance to crushing and more effective rodent protection for use in harsh environments, including subterranean, submarine, and tunnel environments.

The technology provides the potential for precise measurements of strain and temperature at many points along great lengths of fibers using a single transducer. A single cable may replace thousands of discrete point sensors, resulting in lower installation and maintenance time and costs. Full sets of measurements can be made in minutes, resulting in decreased costs when compared to manual sensors.

Terms related to the performance of an optical fiber measurement system include distance range, spatial resolution, sampling interval, distance precision, measurement uncertainty, and measurement resolution or repeatability.

Distance range describes the maximum distance over which the system can perform given a set of performance criteria. For optical fiber measurement technologies, this extends for kilometers.

Spatial resolution describes the ability of the system to accurately measure two adjacent locations with different temperatures or strains. The system will be able to measure temperatures or strains that spread over distances greater than the spatial resolution of the system with 100% accuracy. Temperatures or strains that do not span the minimum spatial resolution will not be measured with full accuracy. Optical fiber measurement technologies are capable of providing full accuracy for spatial resolutions on the order of 20 in. (0.5 m).

Sampling interval describes the distance between two measurement points along the fiber and is determined by the number of points along the sensor length. For optical fiber technology, sample intervals on the order of about 4 in. (0.1 m) are possible.

Distance precision describes the precision of the location of the measured point. For optical fiber measurements, this can be on the order of about 39 in. (1 m) or less.

Resolution is the smallest value or change in value that the measurement system to able to detect. Resolution has no relation to the accuracy of measurement. Current technology allows resolutions of about 0.1°C for temperature measurements and 1 to 2 microstrains for strain measurements. Techniques used to measure temperatures via optical fiber are similar.

Potential applications in dam engineering include measuring strains in tensile faces of concrete dams. To achieve this, a single length of fiber could be mounted across the dam face at different elevations. Other uses include measuring temperatures in concrete dams, particularly during early phases of concrete curing during peak thermogenesis; measuring distributed strains along tunnels and pipes; and real-time strain measurements of deflections during critical operations, such as first filling of a dam or for reservoir levels above design levels.

Inaudi and Glisic (2006) describe cases in which temperatures were monitored using fiber optic sensing for a 56-ft (17-m) high concrete dam raise on the Luzzone Dam (Switzerland) and bitumen joint monitoring at the Plavinu Dam (Latvia).

5.3.5 Stress and Load

The most common types of instruments for measuring bearing or load stresses directly without first measuring strain and then multiplying by elastic constants are stress inclusions, total pressure cells, and flat jacks. Load cells are used to measure long-term loadings.

5.3.5.1 Stress Measurement.

Stress Inclusions. Stress inclusions are instruments designed to be used for measurement of stress changes in rock or concrete. They enable measurement of stress impacted on elastic rocks and concrete. They consist of a stiff steel annulus incorporating a vibrating-wire across its diameter, as shown in Fig. 5-32. The instrument is wedged into a small-diameter borehole, usually 1.5 in. (38 mm) in diameter, by means of a special installation tool that can be used at depths reaching 65 ft (20 m) or more. The wedge is located on one side of the steel annulus and a load-spreading shoe is diametrically opposed on the other side. The wedged instrument monitors changes of borehole diameter occurring parallel to this diameter.

Laboratory tests have shown that the wall thickness of the steel annulus is great enough for the wedged inclusion to behave as a stiff inclusion. In other words, its stiffness is high enough for its change of diameter to be

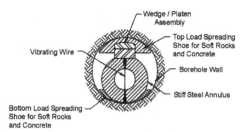

Fig. 5-32. Schematic of a vibrating-wire stress inclusion
Source: ASCE Task Committee on Instrumentation and Monitoring Dam Performance (2000).

directly related to the change of stress in the enclosing medium assuming that the Young's modulus of the medium is in the range of the usual values for rock and concrete. Commercial stress inclusions are available with different load-spreading shoes for either low-modulus or high-modulus rock and concrete. Correlation curves to assess change of stress are also available for different types of materials.

An alternative to the stiff stress inclusion is a strain-measuring cell of similar design but with thinner wall thickness. These also incorporate a vibrating wire along one diameter, but they behave as soft inclusions, meaning that they are used to measure the change of diameter of the hole in which they are installed caused by a change of stress in the host rock. Stress change is calculated from the measured strain and the formula for deformation of a circular hole in an elastic medium.

Total Pressure Cells and Concrete Stress Cells. Total pressure cells are commonly used to measure the earth pressures within fill and against concrete structures. Total pressure cells, also known as earth pressure cells, can be of two types: diaphragm or hydraulic. In a diaphragm cell, the load resulting from external pressure applied by a soil or fill deflects a circular steel plate that is supported at its periphery by a steel ring. The deflection of the plate is sensed on its inner side by a strain gauge (vibrating-wire or Carlson-type resistive). Diaphragm cells are best suited to situations where stresses on the diaphragm are uniform, without point loads or localized arching effects. They are less appropriate for earth pressure monitoring in embankment dams, where nonuniform loading can be expected as a result of granularity of the fill material and variations in its degree of compaction.

Hydraulic total pressure cells, a schematic of which is shown in Fig. 5-33A, are generally preferred for use in embankment dams because their design holds the potential of being less influenced by nonuniform loading. Hydraulic total pressure cells consist of two thin circular steel plates in the range of 8 in. (20 cm) in diameter or more welded together at their periphery with a thin gap between them. The gap, not exceeding about 0.008 in. (0.2 mm) in thickness, is filled with a hydraulic fluid such as oil or glycol to form a "pressure pad." This pad is then connected by a steel tube, also filled with hydraulic fluid, to a pressure transducer using vibrating-wire or another sensing principle (Fig. 5-33B). Stress acting in the fill or concrete in a direction perpendicular to the pressure pad is transmitted to the hydraulic fluid through the steel plates and the resulting pressure is measured by the pressure transducer.

Hydraulic total pressure cells can be embedded in the fill of an embankment dam or in rock or concrete. In the latter case, they are often referred to as concrete stress cells and may have a rectangular shape instead of a circular one. The generally accepted interpretation is that the measured pressure is equal to the stress acting on the pressure pad. Although true in principle, experience has shown that a number of precautions must be taken for this interpretation to be valid, both in the design of the total pressure cell and in its installation.

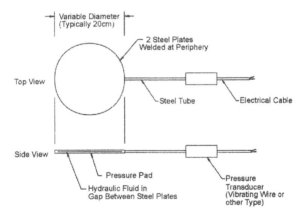

Fig. 5-33A. Hydraulic total pressure cell schematic
Source: ASCE Task Committee on Instrumentation and Monitoring Dam Performance (2000).

Fig. 5-33B. Hydraulic total pressure cell with vibrating-wire pressure transducer
Source: RST Instruments, reproduced with permission.

The stiffness of the pressure pad in the direction of the measured stress should be about the same as the stiffness of the medium in which it is embedded. In particular, the pad stiffness should not be significantly lower than that of the medium, as the soil would then create an arching effect in which the stress is transferred around the cell instead of acting on it. For this purpose, manufacturers can modify the width of the gap between plates to increase or decrease the thickness of the fluid layer and thereby adjust the overall stiffness of the cell. Total pressure cell stiffness is not a great concern for cells embedded in fill but is more important for rock or concrete because of their higher modulus.

Precautions pertaining to installation relate to the quality of the compaction around cells installed in fill. A common cause of total pressure cell readings that are difficult to interpret is poor compaction of the fill around the cell or a different degree of compaction than in fill compacted by mechanized equipment elsewhere in the dam. Portable compaction equipment must be used near the total pressure cells. Even so, the challenge is to achieve the same degree of compaction as with mechanized equipment and to verify it by appropriate compaction quality control tests. In spite of good construction practice, pressure cells have often provided strange and erroneous data.

As shown in Fig. 5-34, it is common to install total pressure cells in clusters of up to five at the same location. Best practice is to start with a flat, compacted fill surface and then excavate a trench with faces inclined at the angle to which the cells will be installed. The cells are then placed and the trenches backfilled and compacted with the original fill material (ISRM 1981).

For concrete stress cells, the main precaution is to restore contact between the plates of the pressure pad and the concrete after the concrete has cured and shrunk. For this purpose, the stress cell can be equipped during manufacture with a fluid-filled repressurization tube. The end of this tube is left protruding from the concrete after installation. After the concrete has cured, the repressurization tube is gradually compressed, forcing fluid into the cell, to inflate the pressure pad until full contact is restored with the concrete. The pressure transducer will show a pressure increase at the moment when the contact is reestablished.

Flat-Jack Testing.   A flat jack is a thin bladder that can be pressurized with hydraulic oil. The instrument is made to be inserted into a mortar joint where a slot is formed to determine the in situ stress, the deformability, and the compressive strength. Flat-jack testing has been used for decades for in situ experiments in rocks and existing structures such as masonry buildings and concrete dams. This method initiates with cutting a slot into concrete. The displacement within a slot by cutting is then measured. Next, a flat jack is inserted into the slot, as shown in Fig. 5-35. The flat jack esti-

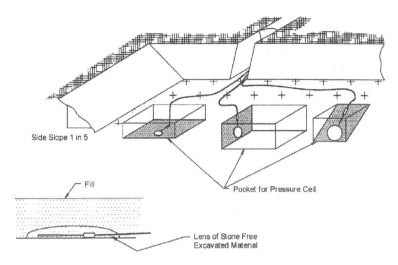

Fig. 5-34. Installation of total pressure cell cluster in fill
Source: ASCE Task Committee on Instrumentation and Monitoring Dam Performance (2000).

Fig. 5-35. Flat-jack testing device

mates stress using an oil pressure jack when the slot returns to its original width.

5.3.5.2 Load Measurement. Load cells measure long-term loads in tie-downs, tiebacks, posttensioned anchors, and rockbolts commonly used for improving the stability of a dam, its foundations, and abutments.

Increasing loads may be taken as an indication of rock movement or movement between the dam and its foundation, (i.e., of decreasing stability). Diminishing loads, on the other hand, may be an indication of anchor slippage and a need to retension the load member. A typical load cell installation is shown in Fig. 5-36. Load cells are also sometimes used to measure loads at the ends of piles and between structural members. Two predominant types are those using electrical resistance strain gauge sensors and those using vibrating-wire sensors. Less commonly used are hydraulic load cells.

Load Cells.
Electrical Resistance Load Cells. The most common style of strain gauge load cell is an annular ring-style electrical resistance cell, shown in Fig. 5-37. Strain gauges are attached to the outer periphery of the load-bearing annulus or cylinder and surrounded by a sealed protective cover. Greater accuracy (less influence from end effects) can be achieved by placing the strain gauges on either side of a shear beam. This latter design has the added advantage of being quite short in height.

Electrical resistance strain gauges may be connected together in a full Wheatstone bridge network. This has the advantage of combining the output from all the strain gauges, thus providing automatic averaging of eccentric and/or uneven loading.

Greater accuracy can also be obtained by using remote-sensing techniques to minimize cable effects. Input voltages at the strain gauges may fluctuate because of varying cable resistances caused by temperature changes, contact resistance changes, cable splicing, and other factors, altering the output voltages. Remote sensing eliminates the problem by using another pair of con-

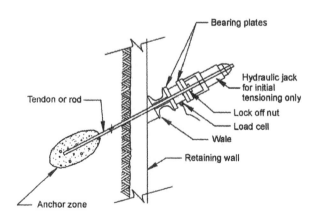

Fig. 5-36. Typical load cell installation on a tieback
Source: ASCE Task Committee on Instrumentation and Monitoring Dam Performance (2000).

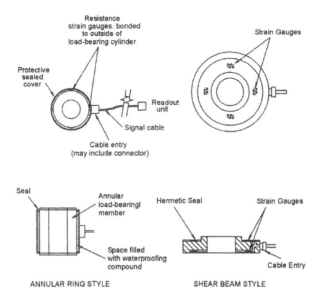

Fig. 5-37. Electrical resistance-type load cells (annular ring type)
Source: ASCE Task Committee on Instrumentation and Monitoring Dam Perfor-
mance (2000).

ductors to measure the input voltage right at the load cell. The readout box
is then set up to measure the ratio of input to output voltages.

Vibrating-Wire Load Cells. A typical vibrating-wire load cell is shown in
Figs. 5-38A and 5-38B. Three or more vibrating-wire strain gauges are located
inside holes drilled along the center of the annulus wall. To measure the load,
each strain gauge must be read individually and the readings summed. In
practice, it is common to see wide disparities in the output from individual
sensors, which are caused by eccentric and uneven loading. In extreme cases,
one or more sensors may go out of range and cease reading. The multiplic-
ity of sensors increases the reading difficulty, although multiplexers are
available. Multiplexers automatically cycle through the sensors and display
an average or total reading on the readout box. Data loggers use additional
channels but make the reading of multiple sensors much less difficult.
Vibrating-wire load cells have demonstrated long-term stability and are useful when load monitoring is required over a period of several years.

Hydraulic Load Cells. Hydraulic load cells, shown in Fig. 5-39, are made
by welding two steel plates together at their edges and leaving a gap
between them to be filled with hydraulic fluid. Applied loads squeeze the
plates together and build up a pressure in the fluid, which is measured by
a Bourdon tube pressure gauge with a dial-type display and/or by an electrical pressure transducer. Bourdon tube gauges are much less accurate and

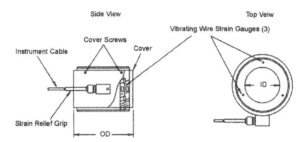

Fig. 5-38A. *Vibrating-wire load cell schematic*
Source: *ASCE Task Committee on Instrumentation and Monitoring Dam Performance (2000).*

Fig. 5-38B. *Vibrating-wire load cell*
Source: *RST Instruments, reproduced with permission.*

precise than pressure transducers. Also, it is necessary for the dial gauge to be clearly visible. In some situations, the simplicity of reading a load measurement directly from a dial gauge is a benefit.

Design Considerations.   Load cells are typically annular or cylindrical in shape. Annular load cells have an internal diameter sufficient to fit around the tieback bolt or tendon bundle and an outer diameter large enough to provide the necessary bearing capacity. Less commonly used in dams is the shear beam type.

Over-Range Capacity.   Because of the high probability of eccentric loading, load cells for tiebacks are designed so that the stress at the maximum design load is no more than 25% of the material yield stress (15% in the case of vibrating-wire load cells, where excessive loads could cause one or more of the vibrating-wire sensors to go completely slack and cease reading).

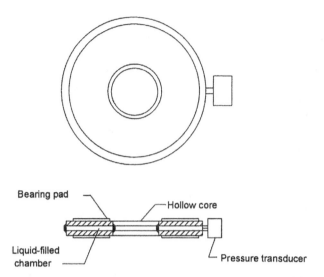

Fig. 5-39. Hydraulic load cell
Source: ASCE Task Committee on Instrumentation and Monitoring Dam Performance (2000).

Material. Materials used for load cells have high strength and high elastic modulus. Typical materials are high-strength alloy steel and aluminum. Titanium load cells have been tested successfully but are not commonly available.

Height. Shorter load cells of annular or cylindrical design suffer more from end effects. Load cells therefore need to be as long as possible consistent with space limitations and buckling propensities. An overall height-to-width ratio in the range of 1:1 to 2:1 is normal. For annular cells, the height-to-width ratio of the wall of the annulus normally falls in the range of 4:1 to 6:1.

Special designs, such as the shear beam type, allow shorter heights without sacrificing accuracy. For a given load capacity, these types of load cells have a larger diameter.

Temperature Effects. Vibrating-wire load cells temperature effects are minimized by matching the material of the sensor to the material of the cell itself so that the thermal coefficients of expansion match. Load cells with resistance strain gauge sensors automatically correct for temperature changes by using a temperature-compensated Wheatstone bridge. With hydraulic load cells, temperature changes cause expansion and contraction of the hydraulic fluid and may have a small effect on the load reading.

Waterproofing. Load cells need to be capable of operating while completely submerged in water. They are hermetically sealed or sealed by O-rings. Particular attention must be paid to any cable entry gland, which must prevent water from entering the load cell, even if the cable jacket is cut or damaged. Additional waterproofing of the load cell is sometimes obtained by enclosing it inside a grease-filled box.

Electrical resistance-type load cells require the most rigorous waterproofing techniques, because the slightest amount of water penetrating the electric circuitry will render the load cell useless. Vibrating-wire types are more forgiving. The vibrating-wire sensors are themselves fully sealed, and, because their output is a frequency rather than a change of millivolts, they are relatively unaffected by moisture penetration into the wiring or connections.

Hydraulic load cells with non-electrical outputs are the least susceptible to moisture effects. Bourdon tube-type pressure gauges, if used, must be resistant to corrosion. Glycerin-filled types, made from stainless steel, are recommended.

Installation.

Bearing Plates. Loads are usually applied through steel bearing plates placed on both sides of a load cell. Bearing plates need to be big enough to cover the entire load-bearing surfaces of the load cell. Surface finish is not as important as flatness: a surface finish of 160 μm is sufficient as long as the plates are flat.

Load cells of electrical resistance or vibrating-wire types are very rigid. Consequently, any distortion of the bearing plate will cause uneven loading of the load cell and gross calibration errors. Plates cut from cold rolled steel are frequently warped and may become even more warped if any welding is performed on them. Therefore, load cells with bearing surfaces that are machined flat after cutting or welding are preferred.

Bearing plates can become warped because of the uneven surfaces of the structure being loaded. When this occurs, it is beneficial to smooth the surface using quick-setting cement or epoxy or by using layers of deformable material such as hard rubber or plastic sheets. If deformable material is used, it needs to be included in a laboratory calibration. If the bearing surfaces are not parallel, spherical seats or wedge-shaped shims are used.

Calibration. Calibration depends heavily on the way the load is applied. In the laboratory, loads are applied uniformly through flat, parallel platens. In the field, loads are applied through surfaces that are neither flat nor parallel and by hydraulic jacks which, because of a size mismatch with the load cell, may cause the bearing plates to either wrap around the load cell or push through the middle of it. Spherical seats and layers of lubricant or sheets of deformable material such as copper or plastic may be used to make the loading more uniform. Nevertheless, these factors can affect the calibration, such that load values in the field can differ from the laboratory calibrations by as much as 20%. To avoid these discrepancies, it is advisable to duplicate the field conditions in the testing machine by using the same combination of bearing plates and jack as will be used in the field. Alternatively, a steel ring with the same dimensions as the jack may be used.

5.3.6 Temperature

Temperature measurements perform a variety of functions in monitoring the performance of dams. Many of the instruments discussed in this chapter have temperature sensors within them, allowing for correction of their temperature coefficient of output change. Concrete dams exhibit deformation from the temperature coefficient of expansion of the concrete itself. Rotations and displacements of concrete dams and appurtenant structures result from temperature changes caused by conductive heat transfer from the reservoir water, conductive heat transfer from the ambient air in contact with the dam, and radiative heat transfer from sunlight on the exposed concrete.

Temperature measurement devices, summarized in Table 5-2, are installed within concrete dams during construction for several reasons. Monitoring concrete temperature during placement indicates the rate of curing and the effectiveness of cooling, especially if cooling coils for cold water or ice are part of the design. Contraction joints in arch dams require grouting when the concrete cools sufficiently to allow the joints to open and accept grout. Grouting contraction joints in gravity dams also may require knowledge of the temperature within the dam body. Linking temperature to structural behavior promotes understanding of seasonal deflections.

Combined strain gauges and temperature sensors allow for correcting the temperature coefficient of strain in addition to measuring the temperature of the concrete. Temperature sensors are also used to locate seepage through embankments, foundations, and existing concrete dams. Temperature gauges (thermistors) integrated within piezometers can be used to detect reaction time to seasonal variations in reservoir temperature, which can be an indication of continuity of seepage paths within the embankment and the reservoir.

The most common temperature-measuring instruments are thermistors, thermocouples, resistance temperature detectors (RTDs), and Carlson instruments. Vibrating-wire and integrated circuit temperature sensors are also used although less frequently. Table 5-2 shows typical applications, the relative accuracy of the device, whether or not the device type is interchangeable with another device of the same part designation without a unique calibration, and the type(s) of signal conditioning required for measurement.

The accuracies of the devices presented in the table are stated in relative terms because families of devices having different accuracy specifications can be selected within each general type. However, each general type of device has an inherent accuracy limitation.

5.3.6.1 Thermistors.   The salient characteristic of a thermistor, shown in Figs. 5-40 and 5-41, is that it has a very high temperature coefficient of

Table 5-2. Temperature Measurement Devices.

Device type	Where or how used	Relative accuracy	Device interchangeable?	Signal conditioning
Thermistor	(1) Within instruments for temperature compensation (2) seepage mapping on embankment dams	Medium	Yes	2-terminal resistance measurement
Resistance temperature detector (RTD)	(1) Concrete dams for construction and long-term monitoring (2) within instruments for temperature compensation	High	Yes	4-terminal resistance measurement or compensating quarter-bridge
Carlson instrument	Concrete dams for construction and long-term monitoring	Medium	No	4-terminal resistance measurement or compensating half-bridge
Thermocouple	RCC dams for construction monitoring	Low	Yes	Low-level voltage and a suitable thermocouple reference junction

Fig. 5-40. Thermistor

Fig. 5-41. Thermistor embedded in concrete dam

resistance. This high sensitivity makes the output from a thermistor easy to measure using a standard ohmmeter circuit with a 2-terminal connection. Because the nominal resistance of a thermistor at ambient temperatures is high compared with the lead wire resistance and the change in resistance with a change in temperature is also high, the error contribution from the initial resistance and the temperature coefficient of the lead wires is usually not significant. Another advantage of thermistors is that they operate without special signal-conditioning circuitry, which adds to their robustness and economy.

Table 5-2 indicates that interchangeable thermistors of the same part number will yield medium accuracy. Basically, this means that there is an accuracy specification given for a particular thermistor model (0.18, 0.9, or 1.8°F, or 0.1, 0.5, 1°C), and devices do not have to be individually calibrated to achieve the accuracy stated by the manufacturer. Manufacturers of vibrating-wire piezometers often embed a thermistor in their instruments to measure the temperature of the surrounding ground water or to correct temperature-related errors of the instrument readings. The thermistor is an ideal device for this situation because it is very low cost and the piezometer manufacturer does not have to perform a special temperature calibration and can state the temperature accuracy specified by the thermistor manufacturer.

Multiple thermistors can also be assembled on a multiconductor cable, as illustrated in Fig. 5-42. Such multipoint thermistor cables are also called thermistor strings. Recently, digital thermistor strings have been developed that can bus up to 256 thermistor nodes on a single 4-conductor cable.

5.3.6.2 Resistance Temperature Sensors. An RTD is a general term for any device that senses temperature by measuring the change in the resistance of a material. RTD can be a term of confusion because it is logical to assume that any device that exhibits a stable temperature coefficient of resistance can be referred to as an RTD. However, in industry practice, RTD generally refers to a metal that is wound onto a ceramic mandrel or else laser-trimmed on a ceramic substrate. The RTD is precision trimmed in manufacture to a specific resistance at a reference temperature. The purity of the metal affects the characteristics and the long-term stability.

RTDs provide the highest accuracy among the four devices described here. The interchangeability error for platinum RTDs is lower than that of any other temperature sensor, and the usable range is appropriate for any dam instrumentation application. For stable long-term temperature mea-

Fig. 5-42. Multipoint thermistor string
Source: RST Instruments, reproduced with permission.

surements, platinum RTDs are the best performers. The main drawback is that the signal-conditioning requirements imposed on automated measurement systems makes the total instrumentation system more expensive. In the case of RTDs, unlike thermistors, the RTD resistance is relatively low, and the errors due to lead wire resistance must be eliminated. Lead-compensating bridge completion circuits or 4-terminal resistance measurements are required. Either of these signal-conditioning approaches requires more than one 2-wire data logger measurement channel for each RTD or signal-conditioning electronics at the sensor.

5.3.6.3 Carlson Instruments.   Carlson instruments have been used for many years in concrete dams to measure temperature as well as strain. The primary purpose of the Carlson instrument is to measure strain over ranges that are larger than those that can be measured with bonded resistance-type strain gauges. Most contemporary concrete dam construction tends to use vibrating-wire strain gauges where Carlson instruments were formerly used. However, Carlson instruments are still used on many newly constructed concrete dams in developing countries, particularly where there is a local industry that supplies the instruments at lower cost.

The nickel–chromium alloy wire used in Carlson instruments has a fairly stable temperature coefficient of resistance, like the RTDs discussed previously. However, it is not as stable as platinum. Furthermore, Carlson instruments are not trimmed to a specific resistance at a reference temperature. To use Carlson instruments as temperature sensors, the calibration data for temperature are provided with each new instrument. Therefore, they are not interchangeable on measurement channels on a data recording system without also modifying the calibration constants used in data processing.

Carlson instruments have similar complexity in signal conditioning, as previously described for RTDs. The result is a more expensive implementation requirement for automated data acquisition.

5.3.6.4 Thermocouples.   Unlike the other three devices, which sense temperature based on a temperature coefficient of resistance, thermocouples generate a voltage difference at the junction of two dissimilar metals that is a function of the junction temperature. Full-scale voltages generated by thermocouples are bipolar in the millivolt range. To achieve the specified accuracy, the measurement circuit must be able to measure accurately below the 10 mV level. In addition, the measurement circuit must include a thermocouple reference junction of known temperature, along with the circuitry and software to compute the compensation. Although thermocouples are relatively inexpensive and interchangeable, the signal-conditioning requirements include a thermocouple reference junction, and higher than usual demands are placed on precision measurement performance, particularly if the measuring unit must operate in outdoor temperatures.

The predominant use of thermocouples in dam instrumentation applications is to monitor temperatures of roller-compacted concrete (RCC) dams during construction. Thermocouples are the most widely used temperature sensors because of the wide selection of temperature ranges for different thermocouple types, and more specifically because many thermocouples can be used for very high process temperatures. However, this is not a reason to use them for monitoring RCC curing temperatures. The thermocouple for RCC monitoring is a Type T (copper-constant). The interchangeability accuracy for a Type T thermocouple is typically ±1.8°F (±1°C). The largest error contribution usually comes from the electronic reference junction simulation, which is a performance specification of the measurement system.

5.3.6.5 Other Devices.   Other vibrating-wire devices and integrated circuit (IC) electronic temperature sensors are both used to a limited extent in dam monitoring. The characteristics of these devices are explained next.

Vibrating-Wire Devices.   Some manufacturers have used vibrating-wire strain gauges and piezometers to measure the temperature of their surroundings by using the temperature coefficient of resistance of the excitation and/or pickup coil winding as an "RTD," which is not to be confused with the trimmed platinum RTD described previously. This method of temperature measurement is separate from the frequency measurement of the vibrating-wire gauge and requires an accurate resistance measurement to calculate temperature.

Although uncommon in terms of usage, an entirely different technique is to employ the thermal expansion of a metal cylinder, isolated within an external housing, to modify the tension of the vibrating wire, and therefore its resonant frequency, in a predictable way. The manufacturer provides a calibration of frequency as a function of temperature. Such a device obviously cannot be used to measure pressure or strain because the sensing element is isolated from external forces.

Whether the vibrating-wire device is sensing temperature by the RTD method or the vibrating-wire method, each transducer is usually provided by the manufacturer with unique calibration coefficients or a calibration curve. The user must obtain the temperature accuracy statement for the device from the manufacturer.

Integrated Circuit Temperature Sensors.   IC microchip-based sensors use microcircuitry and the temperature coefficient of silicon semiconductors to produce an output voltage that is directly proportional to temperature. Some IC temperature sensors contain an even higher level of circuit integration, producing the temperature as a digital signal that can be connected directly to microprocessor input/output ports.

These sensors are used in modern "intelligent instruments," particularly those with a requirement to automatically correct for temperature effects that would otherwise cause errors in the output of the instrument. In addition to temperature corrections, the temperatures of remote instruments and their surroundings can also be reported by an instrument that contains an IC temperature sensor. Therefore, these temperature sensors typically serve a dual purpose in geotechnical or structural instrumentation.

There are many variations of these devices, and their performance is somewhat dependent on how they are employed in the instrument circuit design. Generally, however, their accuracy is similar to that of a thermistor. However, the user does not usually interface directly with these devices because they are "embedded" components used by the manufacturer. Accuracy performance for temperature should therefore be obtained from the instrument maker and not from the IC temperature sensor data sheet.

5.3.7 Mass Movement (Nonstrain)

Movement occurs in all structures in response to internal and external loads. The movement discussed in this section refers to relatively large movements, which can be as high as several feet. The strain that was discussed previously in this chapter refers to small deformations. Monitoring movement may be required to answer a question about a dam's performance. Instruments for monitoring movement such as inclinometers and tiltmeters measure changes in angles. Relative vertical movements (settlement or heave) can also be measured by a variety of instruments, including leveling devices and extensometers, and relative horizontal movements can be measured by convergence and crack meters. Absolute horizontal and vertical movements are generally measured and monitored using the geodetic (survey) methods discussed in Chapter 6.

5.3.7.1 Inclinometers.
There are two basic types of inclinometers: manual and in-place. Both require installation of specialized inclinometer casing. The casing is constructed with two pairs of longitudinal grooves at right angles along the interior surface of the pipe. The grooves prevent a tiltmeter (the inclinometer probe) from rotating when lowered into the pipe. The orientation of the grooves defines the azimuth (bearing) of measured deflections. Pipe sections are typically 10 ft (3 m) long, and sections of pipe are joined together using a coupling. Several coupling designs have been developed, ranging from pop-riveted slip couplings to proprietary self-locking connections.

Casing Installation. Inclinometers used in geotechnical engineering, shown in Figs. 5-43 and 5-44, are composed of tiltmeters with a fixed gauge length. They are used to monitor the lateral deflections of an inclinometer

Fig. 5-43. Inclinometer casing with self-locking connection
Source: RST Instruments, reproduced with permission.

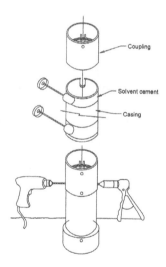

Fig. 5-44. Inclinometer casing with external coupling
Source: ASCE Task Committee on Instrumentation and Monitoring Dam Performance (2000).

casing in a borehole. Although a casing may be installed vertically, horizontally, or inclined, the majority of inclinometer surveys are conducted in vertical casings. Horizontal casings are used to measure settlement. Typically, a baseline survey is performed once the casing is installed. Subsequent surveys allow a determination of the rate of progressive lateral deflection of a borehole. Inclinometers have been used in slopes, embankments, sides of excavations, and concrete dams. Inclinometers are often used to detect sliding movement along weak planes in natural slopes and embankment dams. Installed in the slopes around the reservoir, they can be used to monitor landslides.

Inclinometer casings are generally installed in a vertically drilled borehole with one set of diametrically-opposed grooves oriented downslope. However, an inclinometer casing may be installed in inclined boreholes to measure lateral displacement or in shallow level trenches to measure settlement. It can also be installed on the face of a dam, cast into concrete, or built up by adding sections as an earth fill is raised. Vertically installed casings are usually extended below the foundation of a dam into ground that is unlikely to move. The portion of the casing in nondeforming ground will serve to provide a fixed reference for subsequent surveys, eliminating the need for continued resurveying of the top of the casing.

To prevent buckling of the casing where settlement in excess of 1% is anticipated, telescoping couplings, shown in Fig. 5-45, are used to allow the inclinometer to move with the surrounding soil. Normal telescoping couplings between sections of casing can accommodate from 5 to 10% settlement. Specially designed telescoping casings can accommodate from 10 to 30% settlement.

Vertical inclinometers subjected to large horizontal strain often shear and can no longer be measured. Typical inclinometer casings allow a deflection curvature with a radius of approximately 6.5 to 10.5 ft, depending on the casing design, before they begin to shear. The corresponding horizontal deflection depends on the length of the deflection zone.

Inclinometer casings installed horizontally in trenches must be backfilled carefully with select fill to prevent damage. Horizontal installations are generally limited in length to 500 ft (about 150 m) if a manual inclinometer probe is used because the probe has to be pulled through the casing.

Inclinometer casings are sometimes installed on the upstream faces of concrete-faced rockfill dams and RCC dams to measure the deflection of the face. In this case, the casing is attached to the face of the dam or is embedded in the concrete face, sometimes with expansion joints or other provisions to accommodate thermal expansion and contraction.

Manual Inclinometers. An inclinometer probe, shown in Fig. 5-46, contains two gravity-referenced tilt sensors, oriented 90° apart, with each measuring the angle of the borehole casing with respect to the vertical. The probe has two sets of wheels that ride in grooves on the interior of the

Fig. 5-45. Telescopic inclinometer casing section
Source: RST Instruments, reproduced with permission.

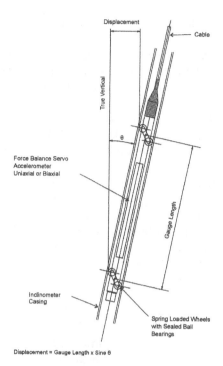

Fig. 5-46. Manual probe inclinometer
Source: ASCE Task Committee on Instrumentation and Monitoring Dam Performance (2000).

inclinometer casings. A wheel spacing of 2 ft (0.5 m) is standard and is referred to as the "gauge length."

Tilt sensors within a manual inclinometer probe are typically MEMS or force balance accelerometers; magnetostrictive and electrolytic sensors have also been used. The sensors are delicate and susceptible to damage and thus require regular recalibration. The typical recommendation from a manufacturer is recalibration every 12 months under moderate usage. Because they are solid-state components, MEMS sensors, shown in Fig. 5-47, are less susceptible to calibration degradation owing to shocks on the probe than other types of sensors.

The inclinometer probe is attached to a control cable that is used to excite the tilt sensors and to transmit the signal back to the readout unit. The cable is marked every 2 ft (0.5 m) to allow the probe to be quickly and repeatedly raised or lowered by its gauge length at successive readings during the survey. The other end of the cable is plugged into a readout unit. Older units display the readings only, requiring the user to manually record each reading. Modern readout units store recorded data in memory to eliminate written records and incorporate a display to visualize the inclinometer profile on-site and identify possible bad readings. This speeds up the survey pro-

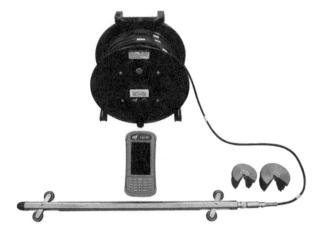

Fig. 5-47. Digital MEMS inclinometer probe
Source: RST Instruments, reproduced with permission.

cess, eliminates errors in reading handwritten notes, and allows the survey data to be automatically transferred to a computer.

Manual inclinometer surveys are labor intensive, slow, and produce data that require additional processing. This method does not lend itself to real-time monitoring with the ability to automatically activate alarm systems if thresholds are exceeded. Manual surveys may be adequate in situations where little or no movement is expected and reading intervals are long. For projects with questions related to movement of a performance indicator, manual instruments may not be the best choice.

In-Place Inclinometers. An in-place inclinometer, shown in Fig. 5-48, consists of a series of individual inclinometer sensors installed at successive depths in an existing inclinometer casing in a borehole and connected by rods, which are typically 1.65, 3.3, 6.5, or 10 ft (0.5, 1, 2, or 3 m) long, thereby covering the entire length of the borehole and providing a continuous inclinometer profile. A string of in-place inclinometer sensors can also be suspended in an inclinometer casing at a depth of interest, (e.g., a depth at which a shear movement has already been identified by a traditional inclinometer survey using a manual inclinometer probe). Uniaxial in-place inclinometers (measuring tilt in one plane) and biaxial in-place inclinometers (measuring tilt in two orthogonal vertical planes) are available. In-place inclinometers can be installed horizontally, vertically, or inclined.

In-place inclinometers, consisting either of strings covering the full length of an inclinometer casing or strings suspended from the collar, can easily be removed periodically (e.g., yearly) to conduct complete inclinometer profiles with a manual inclinometer to confirm measurements. In-place inclinometer sensors can also be reused elsewhere after removal.

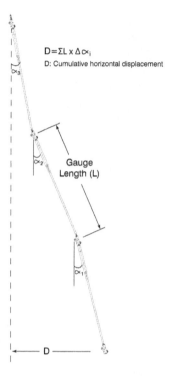

$D = \Sigma L \times \Delta \alpha_i$

D: Cumulative horizontal displacement

Gauge
Length (L)

D

Fig. 5-48. In-place inclinometer
Source: RST Instruments, reproduced with permission.

MEMs, force balance accelerometers, electrolytic tilt sensors, and vibrating-wire sensors have been used in in-place inclinometers. Today, MEMS sensors are the most common as they are available in digital versions that allow them to bus many sensors in a borehole on a single signal cable, making installation and removal much easier.

In addition to greatly reducing the labor required to perform a survey, in-place inclinometers also allow the incorporation of ADASs, thus reducing data-acquisition errors and increasing the likelihood of capturing reliable profiles. ADASs may also be used to retrieve data remotely, and the system can be programmed to automatically and immediately send an alarm if a Threshold or Action Level is exceeded.

5.3.7.2 Shape Accelerometer Arrays. A shape accelerometer array (SAA), shown in Fig. 5-49, is similar to an in-place inclinometer. It consists of an assembly of long, typically 1.65 ft (0.5 m), slender rigid elements connected by special joints that allow bending in any direction. The members are fitted with triaxial MEMS sensors to measure tilt and acceleration. The

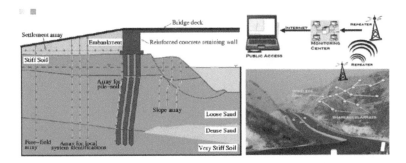

Fig. 5-49. Shape accelerometer array
Source: Bennett et al. (2007), reproduced with permission.

tilt data gained from SAAs are similar to data from a biaxial inclinometer, but SAAs are capable of deforming over much larger deformations than inclinometers. Typical arrays are installed in 1 in. (25 mm) PVC electrical conduits. The conduit can be installed in a borehole similar to an inclinometer casing or it may be fixed to a structural component. Provided that deformations are within allowed limits, the SAA assembly can be removed from the conduit and reused in another location. SAAs can readily be interfaced to a data logger and transmit measurements wirelessly.

SAAs can be installed horizontally, vertically, or inclined. The SAA does not extend or compress and is oriented perpendicular to the anticipated displacement direction.

There are several advantages to this system, which is starting to replace the use of an array of in-place inclinometers for monitoring the movement of an embankment slope. The SAA provides an adaptable, cost-effective means to monitor movement; however, installing several arrays of in-place inclinometers can become extremely expensive. Additional advantages over manual or in-place inclinometers include higher spatial resolution, larger tolerance to shearing, autonomous data collection and ease of calibration, reusable inexpensive casing, low installation and maintenance costs, and full automation.

5.3.7.3 Time-Domain Reflectometry. Time-domain reflectometry (TDR), shown in Fig. 5-50, has been used increasingly for monitoring settlement and slope stability in recent years, but dam-monitoring applications have been few. Originally developed to locate faults in communication and power lines, data collection consists of simply attaching an electronic cable tester to a coaxial cable grouted into a borehole and then taking a reading. The basic principle is similar to that of radar. An electrical pulse is sent down the coaxial cable. When the pulse encounters a break or necking, which causes an electrical impedance change, it is reflected. The reflection

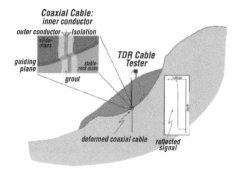

Fig. 5-50. Basic setup of TDR measuring site
Source: Thuro et al. (2007), reproduced with permission.

shows as a "spike" on a plot of reflected signal versus time (Kane 1998). Because the pulse velocity in the cable is known, the depth to the zone of deformation is easily determined, as illustrated in Fig. 5-50 (Thuro et al. 2007).

TDR monitoring is often used on critical slopes to determine the locations of shear planes and often has warning or alarm systems configured with the monitoring program. Cable placements are also oriented to monitor vertical deflections, settlement, or heaving. At present, TDR cannot easily identify the direction or magnitude of deformation, only its location along the cable. There are many factors affecting the correlation of measured impedance to shear deformation: cable strength, composition and diameter, soil and grout interaction, and signal attenuation. There are ongoing studies to develop reliable correlations. A good recommendation is to use TDR cables of the same type and same length in all boreholes of a given instrumentation program, if possible, so that a comparison between reflected signal magnitudes will at least allow qualitative ranking of the magnitude of the deformations in different boreholes.

A TDR cable can exhibit several reflected signals, which would be indicative of a number of deformation zones, as long as the external jacket of the cable does not become damaged and allow water ingress in the cable. The reflected signal becomes very high and unusable past the zone of water ingress. A cut coaxial cable provides a high reflected signal as well.

The advantages of TDR are simplicity, low cost, and rapid data collection. It takes only a few minutes to read any TDR cable, regardless of its length. Small-diameter boreholes and coaxial cable can be used. Half-inch (12.7 mm) coaxial cable is highly recommended because of its relatively low signal attenuation. The borehole is then grouted with no casing required. Used in conjunction with other monitoring instruments, TDR ground deformation monitoring systems can be a cost-effective way to supplement the complete monitoring program for dams and slopes.

5.3.7.4 Tiltmeters and Beam Sensors. Tiltmeters measure rotational movement using gravity as the reference datum. Tiltmeters, shown in Fig. 5-51, consist of a mounting bracket or plate and a tilt sensor in a housing that is attached to the mounting bracket. The tilt sensor and housing assembly are referred to as the "tiltmeter." When rigidly attached to a dam, a tiltmeter measures the rotation of the dam in the vicinity of the attachment point. Rotational movements occur when a dam is subjected to applied stresses, bending moments, or shears. These loads arise from reservoir pressures, thermoelastic expansion and contraction, volume changes from alkali–aggregate reactions, and other factors.

Tiltmeters are available in both uniaxial and biaxial versions. Uniaxial tiltmeters measure rotation in one vertical plane that is typically aligned with a principal direction in the dam, such as the upstream–downstream direction. Biaxial tiltmeters measure rotations in two orthogonal vertical planes and can therefore measure longitudinal and transverse rotations simultaneously.

Technologies used in the tiltmeter sensors have included force balance, electrolytic, ceramic, MEMS, and normal and inverted pendulum systems. Variations of the tiltmeter include portable tiltmeters, surface-mount tiltmeters, borehole tiltmeters, and submersible tiltmeters. Each of these is discussed in the following sections.

The rotation data provided by tiltmeters may be used directly to monitor bending and angular movements owing to settlement and to estimate

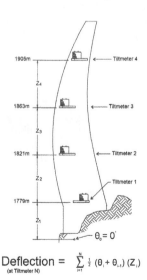

$$\text{Deflection} \atop \text{(at Tiltmeter N)} = \sum_{i=1}^{N} \tfrac{1}{2}\,(\theta_i + \theta_{i\text{-}1})\,(Z_i)$$

Fig. 5-51. Dam deflections from tiltmeter measurements
Source: ASCE Task Committee on Instrumentation and Monitoring Dam Performance (2000).

overturning forces. Displacements are calculated from tiltmeter data in a manner similar to that used in processing the data from inclinometers. Readings from a vertical array of tiltmeters installed in galleries or on a dam face may be incrementally summed to obtain a lateral displacement (deflection) profile (Dienum 1987). The same process may be used with an array of tiltmeters in a horizontal line to obtain vertical displacements resulting from settlement or other causes.

Because they are capable of very high precision, tiltmeters can detect and track small movements that are below the resolution threshold of some geodetic surveying techniques. Tiltmeters used in dams typically resolve movements of 1 μrad (0.2 arcsec) or smaller.

Surface-Mount Tiltmeters. Surface-mount tiltmeters, shown in Fig. 5-52, consist of a base plate, a sensor, and a readout device. They are installed on structural members or surfaces to measure the vertical rotation of the surface.

Most surface-mount tiltmeters incorporate a MEMS sensor, electrolytic tilt sensor, force balance accelerometer, or vibrating wire as the sensing element. These four types of sensing elements are described next.

In recent years, MEMS tiltmeters have gained considerable acceptance because of their low thermal sensitivity, excellent repeatability (7 arcsec), wide measuring range (±15°), which allows installation without the need to level the tiltmeter, and digital output, which makes it possible to bus several tiltmeters on a single signal cable.

Electrolytic tiltmeters, which have no mechanical moving parts, are the most rugged, offer the greatest precision (<0.1 arcsec = 0.5 μrad), and have good long-term stability.

Fig. 5-52. Biaxial surface-mount MEMS tiltmeter (with vertical mounting bracket) Source: RST Instruments, reproduced with permission.

Force balance accelerometers output a voltage signal proportional to the sine of the tilt angle. High-gain units have a resolution of about 0.5 arcsec, a precision of 5–8 arcsec, and a range of ±1°. They have the highest power consumption of the four types, which is important if the tiltmeters are to be operated from batteries. They are also much more expensive than their counterparts.

Sensitive tiltmeters installed on structural members easily measure structural deformation caused by thermal expansion and contraction. If the expected thermoelastic strain and tilt of the structural concrete or steel are of the same order of magnitude as the measured tilt, thermoelastic movement is an important contributor to the measurement.

Several steps can be taken to minimize the temperature-induced output change of the tiltmeter itself. First, a tiltmeter with a temperature coefficient that is appropriate for the temperature changes that will occur at the installation point should be selected. Temperature fluctuations can be minimized by installing tiltmeters in the shade or providing a light-colored sun shield over each instrument. Mounting studs should be short and of equal length to minimize differential thermal expansion and contraction. If mounting brackets are used, they should be made as rigid and compact as possible. These considerations are especially important if the tiltmeters will be installed on the face of a dam, where temperature changes can be extreme.

Submersible Tiltmeters. Submersible tiltmeters, shown in Fig. 5-53, perform functions similar to those of dry surface-mount tiltmeters, but they are designed and constructed for use in submerged applications and under extreme hydrostatic pressures. Submersible tiltmeters gain ruggedness from

Fig. 5-53. Biaxial submersible MEMS tiltmeter (with horizontal mounting plate)
Source: RST Instruments, reproduced with permission.

the design and construction of their housings and connections. Instrument housings are typically machined from a solid block of metal such as aluminum, stainless steel, or titanium. Connections are typically constructed of neoprene and are rated up to 10,000 psi (70 MPa).

Tilt measurements may be via electrolytic tilt sensors, ceramic tilt sensors, or MEMS. Uniaxial and biaxial versions are available. The dimensions of the housings vary, and variations are available for downhole installation and for use on pipelines. The instruments typically include temperature sensors.

Applications particular to submersible tiltmeters may include measuring tilt in deep layers of an earthen dam, the upstream face of a concrete dam, permanent or temporary cofferdams, foundations or substructures of offshore structures, or submerged pipelines.

Borehole Tiltmeters. Borehole tiltmeters, shown in Fig. 5-54, are installed in holes drilled in concrete, rock, and soils. They may be installed in dams, their abutments, and reservoir slopes. Borehole tiltmeters have long cylindrical bodies and are designed to be semipermanently sanded or permanently grouted into a borehole. Sanding consists of pouring and

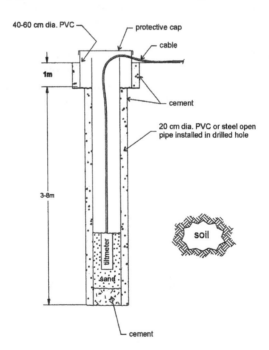

Fig. 5-54. Borehole tiltmeter installation in soil
Source: ASCE Task Committee on Instrumentation and Monitoring Dam Performance (2000).

compacting sand around the tiltmeter. This method is simple and fast and allows the tiltmeter to be retrieved at a later time. After an initial settling-in period of a few days to a few weeks, the installation becomes stable. Disturbance to the sand is always a possibility and may result from water-level changes in the hole, vibration (earthquake), or human intrusion. If disturbance is a concern, grouting the tiltmeter into the hole could be considered. Borehole tiltmeters must contain downhole signal-conditioning electronics for maximum output stability and signal quality. Because downhole temperatures are relatively constant, thermal effects are generally less important with borehole tiltmeters than with surface-mount tiltmeters. Most borehole tiltmeters in production today use electrolytic tilt sensors.

Portable Tiltmeters. A portable tiltmeter, shown in Fig. 5-55, is not fixed in place. Instead, an array of tilt plates is attached to the structure by epoxy cement or screws. The portable tiltmeter is then positioned on each of the tilt plates, one plate at a time, to take a set of readings. Currently, force balance accelerometer sensors and electrolytic tilt sensors are used in portable tiltmeters. This method has the advantage of reducing equipment costs, but it requires more labor to collect the data. It also is less precise than fixed-in-place tiltmeters because of errors associated with the repeated manual repositioning of the sensors.

Beam Sensors. Beam sensors, shown in Fig. 5-56, are a special type of tiltmeter in which a uniaxial sensor is mounted on a long beam, typically a square aluminum tube. The two ends of the beam are attached to the structure to be monitored. The beam sensor measures the rotation that

Fig. 5-55. Portable MEMS tiltmeter and tilt plate
Source: RST Instruments, reproduced with permission.

Beam Sensors

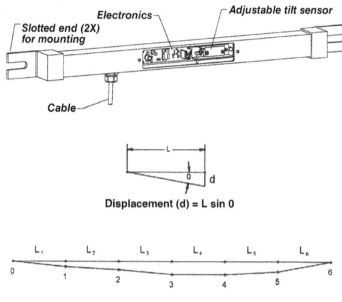

Fig. 5-56. *Beam sensor instrument and interpretation technique for a chain of sensors*
Source: *ASCE Task Committee on Instrumentation and Monitoring Dam Performance (2000).*

occurs in a vertical plane between the two ends of the beam. The beams are typically installed horizontally to measure the vertical displacement of one end with respect to the other or vertically to measure the horizontal displacement of one end with respect to the other. A series of beams attached end-to-end can be used to compute deflections along a horizontal or vertical profile.

Most beam sensors in use today incorporate electrolytic tilt or MEMS sensors because of their relatively low cost compared to force balance accelerometers.

Plumblines and Inverted Pendulums. A plumbline is a gravity-referenced instrument. It consists of a few simple components: a stainless-steel wire attached to a fixed point, a weight, and a damping bucket containing a damping fluid, usually oil, to damp movements of the weight owing to wind, vibration, and air circulation. Plumblines, shown in Fig. 5-57, and

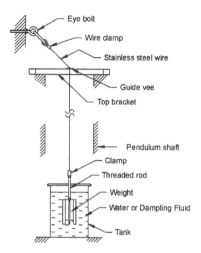

Fig. 5-57. Schematic diagram of a plumbline
Source: ASCE Task Committee on Instrumentation and Monitoring Dam Performance (2000).

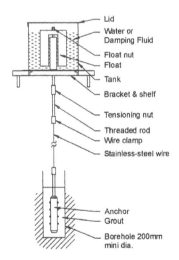

Fig. 5-58. Schematic diagram of an inverted pendulum
Source: ASCE Task Committee on Instrumentation and Monitoring Dam Performance (2000).

inverted pendulums are commonly used in concrete dams to measure changes in the verticality of the structure.

An inverted pendulum, shown in Fig. 5-58, differs from a plumbline in that it is anchored in a borehole in the foundation of the dam, which makes it necessary to locate a float and a damping bucket above the anchoring point.

Plumblines are usually installed in formed shafts extending vertically in the concrete of the dam, for example, from one service gallery level to the next one. Plumbline lengths usually do not exceed 165–260 ft (50–80 m) in a concrete dam. Longer plumblines have been used but are not recommended because of excessive vibration of the wire caused by wind and air circulation in the pendulum shaft. For taller dams, several plumblines can be installed in vertical alignment, one above the other, starting with an inverted pendulum anchored in the foundation, to obtain a complete profile of any change in verticality (Fig. 5-59).

Horizontal movements of the dam with respect to the plumbline wire, both in the upstream and downstream directions and in left bank–right bank directions, have traditionally been measured with mechanical measuring tables or by reading tables. Recently, electromechanical pendulum readout devices that connect plumblines and inverted pendulum systems to ADASs have become available, allowing real-time remote readings.

Mechanical measuring tables are available in several designs. One of them, illustrated in Fig. 5-60, consists of a horizontal metal console attached to the concrete wall of a small chamber through which the pendulum wire runs. The console is equipped with two sets of 3-point reference seats for seating a removable microscope frame, one set for each of the two horizontal directions of movement to be measured. The microscope, mounted on a traveling slide, is used to locate the wire with respect to a graduated scale. A resolution of 0.0008 in. (0.02 mm) and an accuracy of 0.004 in. (0.1 mm)

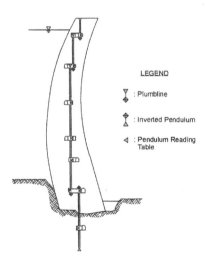

LEGEND

Y : Plumbline

☧ : Inverted Pendulum

◁ : Pendulum Reading Table

Fig. 5-59. String of several plumblines and one inverted pendulum installed in a concrete arch dam
Source: ASCE Task Committee on Instrumentation and Monitoring Dam Performance (2000).

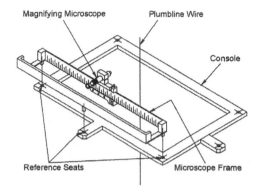

Fig. 5-60. Mechanical measuring table for plumbline wire with removable micro-scope frame
Source: ASCE Task Committee on Instrumentation and Monitoring Dam Perfor-mance (2000).

can normally be achieved with this type of measurement, although measure-ments to one-tenth of an inch are generally adequate for dam-performance monitoring.

Remote reading tables, often called telependulums, are of the noncon-tacting type, that is, the reading principle does not involve contact of the pendulum wire to a sensing plunger of any type that could interfere with the wire movement. Most remote reading tables today, examples of which are shown in Fig. 5-61, are based on induction with frequency output, capac-itive, or on optical principles, which convert the pendulum wire displace-ment with respect to a frame into an electrical or a digital signal proportional to the displacement in two horizontal directions. Some reading tables also allow for reading the vertical displacement of the pendulum wire.

Remote reading tables based on optical principles incorporate a light source on one side of the wire, such that the shadow of the wire is projected onto a linear photodiode array attached to the table frame on the opposite side. Averaging techniques are then used to identify the photodiode pixels that receive the maximum light or, conversely, the minimum light so that the wire position with respect to the array can be known. Optical reading tables may be accurate to 0.0004–0.002 in. (0.01–0.05 mm). This type of reading table requires careful use because humidity or condensation on the wire or the photodiode array may compromise the accuracy of the readings. A heated cabinet is very often recommended for proper use of this table.

5.3.7.5 Settlement Gauges.

Borehole Extensometers. In its simplest configuration, a borehole extensometer comprises several basic components: an anchor, a rod, a flexible sleeve, a reference plate, and grout. After boring a hole, the

Fig. 5-61. Direct and inverted pendulums with optical telependulums

anchor/rod/sleeve assembly is constructed outside the hole. The assembly is then lowered into the hole and the anchor is grouted in place. After this is accomplished, the reference plate is set and the distance between the top (head) of the rod and the reference plate is measured. As the anchor settles with the soil or rock, the rod will be pulled down within the tube, and the distance between the reference plate and the head of the rod will change. For manual readings, the distance can be read with a dial indicator. More sophisticated designs add a vibrating-wire displacement transducer or linear potentiometer to automate readings and allow use of a real-time ADAS. Placing multiple rods in a single borehole with anchors at various elevations allows determination of the settlement of multiple points at a single location.

Several types of anchors are available for use, including grouted anchors, hydraulic anchors, and Borros anchors. A hydraulic anchor does not require grout and thus may also be useful for installation in an upward-directed borehole, such as at the roof of a tunnel in rock. The anchor is set in place by hydraulically inflating a bladder, which secures the anchor against the walls of the borehole.

Borehole extensometers are used in dam foundations and abutments to measure compression and extension movements. Such movements may be caused by the weight of the dam, stresses induced by reservoir filling, slippage along specific geological features, and other factors.

Borehole extensometers may be used to assess the compression of the foundation due to the load of the dam or to monitor extension or compression movement of specific joint, bedding plane, or geological features. In special cases, they may be installed in holes drilled in existing concrete dams. They are also used to monitor settlement of embankment dams.

Borehole extensometers are either single-point, measuring the change of distance between a single anchor and a reference head, or multipoint, which incorporate several anchoring points within the borehole. In multipoint borehole extensometers (MPBX), each anchor point is separately connected to the reference head, allowing measurements of the change of distance between each anchoring point and the reference head. Some MPBX models, referred to as single-rod, incremental, or in-line MPBXs, feature a single rod along which the displacements of several anchoring points can be measured, whereas multirod MPBXs have a single rod connected to each anchoring point. These two types of MPBX are described further in the next section.

Multirod MPBX. Multirod-type MPBXs, shown in Fig. 5-62, are read either mechanically or electronically. A typical mechanical readout consists of a dial gauge to measure the distances between the ends of the extension rods and a stainless-steel reference plate.

For digital reading, electronic displacement transducers are attached to the rods and to the reference head at the collar of the hole. It is also possible to design the MPBX for both mechanical and electronic readings, as shown in Fig. 5-63. This is achieved by adding a reference tip to the displacement transducer and a reference surface at the top of the measurement head. A dial gauge set up on this surface is used to sense the reference tip.

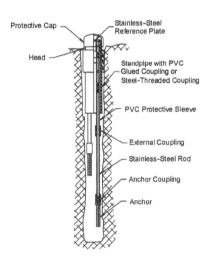

Fig. 5-62. Multirod borehole extensometer, mechanical version
Source: ASCE Task Committee on Instrumentation and Monitoring Dam Performance (2000).

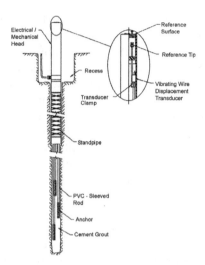

Fig. 5-63. Multirod borehole extensometer, combined electromechanical version
Source: ASCE Task Committee on Instrumentation and Monitoring Dam Performance (2000).

Single-Rod (Incremental) MPBX.   Single-rod MPBXs are alternative designs in which several displacement transducers are located in a single borehole, each separated from the next one by a length of rod or tubing. The main advantage of this design is that there is no electrical measuring head protruding out of the borehole, as is generally the case with a multirod MPBX. In cramped areas, such as small-diameter galleries, or for aesthetic or functional reasons, this can be an advantage. Single-rod MPBXs are often referred to as incremental or in-line extensometers because total displacement between the bottom anchor and the collar of the hole is equal to the sum of the relative displacements between the individual anchoring points.

Several single-rod MPBX designs are available. One design, illustrated in Fig. 5-64, uses submersible displacement transducers assembled in line with a stainless-steel rod covered by PVC tubing. Many transducers can be pre-assembled in line, inserted, and then grouted in a borehole. Alternatively, in case of highly fractured ground or when grouting is not the preferred anchoring method, hydraulic anchors can be used, as shown in Fig. 5-65. This type of extensometer can use a central steel rod with a 3/8 in. (9.5 mm) diameter or more and is therefore suitable to read both compression and extension movements. The total measurement range of the extensometer is the sum of the ranges of the individual transducers.

Another incremental MPBX model for use in concrete or rock is illustrated in Fig. 5-66. It consists of a series of mechanical anchors, connecting tubes, and centralizers installed one in front of the other in a borehole. The

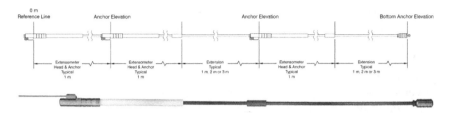

Fig. 5-64. Single-rod (incremental) multipoint borehole extensometer with grouted anchors
Source: RST Instruments, reproduced with permission.

Fig. 5-65. Hydraulic bladder anchor for highly fractured ground
Source: RST Instruments, reproduced with permission.

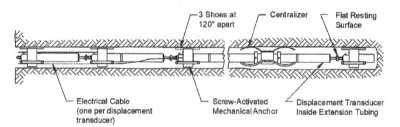

Fig. 5-66. Retrievable incremental multipoint borehole extensometer
Source: ASCE Task Committee on Instrumentation and Monitoring Dam Performance (2000).

anchors are screw-activated using an installation tool that is inserted into the borehole at the time of installation. Three anchor shoes expand in the radial direction to secure the anchor to the borehole wall. This anchoring mechanism allows the complete MPBX to be retrieved at a later time for reinstallation elsewhere. Each connecting tube incorporates a displacement transducer (LVDT or potentiometer), which is mounted longitudinally and has a flat surface at the tip of its spring-loaded feeler arm. During installation, this surface is placed in contact with the mechanical anchor. The feeler arm is then compressed backward slightly to allow the transducer to measure both compression and extension movements in the borehole (Thompson et al. 1990).

Probe Extensometers.   Probe extensometers consist of an array of targets positioned longitudinally along an access tube into which a probe can be lowered to sense the positions of the targets. The vertical access tube can either be located in a borehole or built into an embankment or fill as it is constructed. Depth readings from probe extensometer surveys are recorded by hand. The reading method is not compatible with monitoring by an ADAS.

Although the rod extensometers described earlier are intended to be permanently installed in boreholes, the sensor probe of a probe extensometer can be used for many boreholes. Also, although the measurement range of fixed extensometers is generally about 2–4 in. (5–10 cm), with some systems having a range of 8 in. (20 cm), probe extensometers can measure settlement greater than 3 ft (1 m).

Magnetic and inductive probe extensometers are not commonly used in the rock foundations and abutments of dams or in concrete dams because of the low resolution.

The working principle of probe extensometers is typically magnetic or inductive. Magnetic and inductive probes are similar in the sense that both types are located at the end of a graduated survey tape and are lowered into the access pipe from a reel. When the sensing element is adjacent to a target, an audible tone is emitted at the reel, which enables the operator to note and record the depth of the target. For magnetic systems, the targets consist of magnets in plate targets and datum targets that activate a reed switch located in the probe, as shown in Fig. 5-67. For inductive systems,

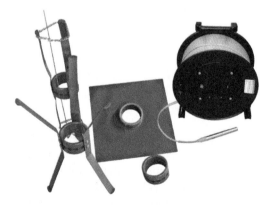

Fig. 5-67. Magnetic settlement system components: Spider targets, plate target, datum target, and reed switch probe
Source: RST Instruments, reproduced with permission.

stainless-steel wire rings or plates are used as targets, and the induction coil is contained in the probe.

Borros Heave/Settlement Points. Borros heave/settlement points, shown in Fig. 5-68, are mechanical assemblies used to measure vertical movements in soils and fills. The heave/settlement point consists of a three-prong anchor, a .25 in. (6 mm) inner riser pipe, and a 1 in. (25 mm) outer pipe.

The Borros point can be installed in a drilled hole or during fill construction by setting the anchor and adding extension lengths of inner riser pipe and outer pipe. Similar to a borehole extensometer, settlements are measured by observing or measuring the change of distance between the top ends of the inner and outer pipes.

USBR Settlement System. A USBR settlement system, shown in Fig. 5-69, consists of telescoping steel pipe sections with cross arms to anchor the system to the soil. Readings are taken by lowering a probe with spring-loaded fins into the pipe. The fins catch on the bottom end of the smaller-diameter pipe sections, and a tape attached to the tip of the probe gives the depth from the top of the installation. When the nose of the probe hits the

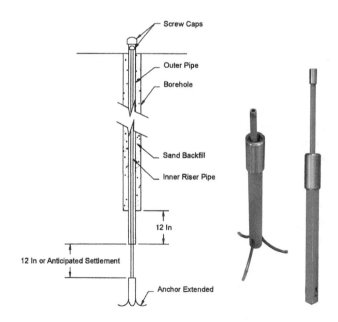

Fig. 5-68. Borros heave/settlement point. Schematic (left) and extended, retracted anchor (right)
Source: RST Instruments, reproduced with permission.

Fig. 5-69. U.S. Bureau of Reclamation settlement probe
Source: ASCE Task Committee on Instrumentation and Monitoring Dam Performance (2000).

bottom of the hole, the fins are retracted and the probe removed. This system is not in common use today.

Embedment Soil Strain Meter or Soil Extensometer. Long-baseline strain meters or extensometers, shown in Figs. 5-70A and 5-70B, are used in embankment and RCC dams to measure large deformations and cracking. The device incorporates a vibrating-wire displacement transducer, linear potentiometer, or LVDT inside telescoping plastic pipes designed to measure the relative movement of the two flanges at the ends of the pipes. Extensometers have been made up to 30 ft (10 m) long. Chains of them connected end-to-end have been positioned to obtain displacement profiles near the abutments of earth and rockfill dams and along their longitudinal and transversal axes. Chains of extensometers can also obtain profiles of cracking along the axes of RCC dams. These are also described under the strain section of this chapter.

Differential Settlement and Liquid-Level Gauges. Dam embankment and foundation settlement relative to a stable benchmark and differential settlement within the dam can be measured by a variety of hydraulic sensors. Differential settlements can also be measured using horizontal inclinometers and chains of beam sensors. The three principal hydraulic settlement measurement techniques are the

- Overflow technique, in which the elevation of a remote sensor is measured knowing that the water level in the two arms of a manometer are equal;
- Water-pressure technique, where the depth of the remote sensor below a known water level is calculated by measuring the water pressure at the sensor; and
- Water-level technique, in which a series of remote sensors is located in a more or less horizontal plane and their elevations are measured with respect to a common water level.

Overflow Technique. To use the overflow technique, shown in Figs. 5-71A and 5-71B, a sealed container is installed in the fill or concrete and connected to a readout station by three nylon tubes. One tube serves as a drain, another as a vent line, and a third is filled with de-aired water or antifreeze solution and connected at the readout location to a transparent, graduated standpipe. In operation, water is added to the standpipe until the water

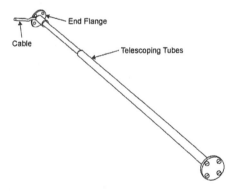

Fig. 5-70A. Soil extensometer schematic
Source: ASCE Task Committee on Instrumentation and Monitoring Dam Performance (2000).

Fig. 5-70B. Soil extensometer
Source: RST Instruments, reproduced with permission.

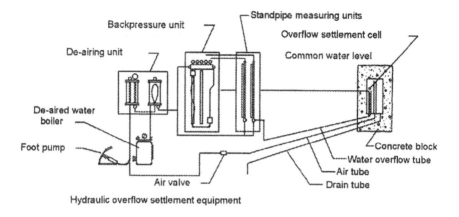

Hydraulic overflow settlement equipment

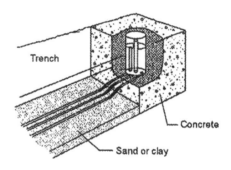

Installation of the overflow cell

Fig. 5-71A. Hydraulic overflow settlement system
Source: ASCE Task Committee on Instrumentation and Monitoring Dam Performance (2000).

Fig. 5-71B. Hydraulic overflow settlement system installation in a CFRD dam
Source: RST Instruments, reproduced with permission.

flows over the lip of the overflow weir inside the sealed container. Excess water drains out of the container through the drain line, and the vent line ensures an equal air pressure on both ends of the liquid column. At overflow, the water level is read on the graduated standpipe. As the overflow cell settles, there will be a corresponding drop of the water level in the standpipe.

Accuracies in the range of 0.02 in. (5 mm) are attainable with this method. The advantage of the system is its inherent simplicity. However, there may be complications:

• The cell needs to be at about the same elevation as the readout location, and the range of settlement is generally restricted to the length of the standpipe, although the range can be extended using manometers and back-pressuring equipment;
• Care must be taken to maintain a constant grade on the tubing to provide free drainage and to prevent the formation of kinks or siphons where air bubbles can become trapped;
• The use of a de-aired fluid is usually necessary, although filtered water (reverse osmosis or other) or antifreeze has also been used successfully. Regular flushing may be required if the water becomes fouled;
• Temperature can affect the readings; and
• The system is difficult to automate.

Liquid Pressure Technique. A typical pressure settlement system is shown in Fig. 5-72. A reservoir is located on stable ground or at a point where its elevation can be easily surveyed. It is connected via two plastic tubes to a pressure sensor (either electrical or pneumatic) located at the point of settlement. The tubes are filled with de-aired water or antifreeze solution. As the sensor settles, the height of the liquid column increases, and the change in pressure is measured by the transducer. Reading this type of system is easily automated when an electrical pressure sensor is used. Twin tubes

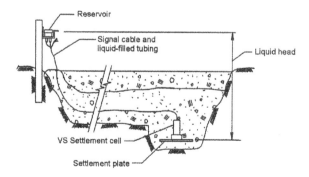

Fig. 5-72. Liquid pressure settlement system—standard type
Source: ASCE Task Committee on Instrumentation and Monitoring Dam Performance (2000).

permit the system to be flushed to purge any air bubbles that may have accumulated.

One advantage of the pressure settlement system is the potential for in situ transducer calibration; this is done by raising and lowering the reservoir. The zero-pressure reading can be calibrated by blowing the liquid out of the lines. With proper care, accuracies of ±0.2 in. (5 mm) are obtainable, although accuracies of ±0.4–0.8 in. (10–20 mm) are more typical. A disadvantage is the long horizontal run of tubing, which must be laid at a relatively constant grade to avoid air traps and which must be back-pressured or flushed periodically to remove air bubbles. Temperature changes of the fluid in the vertical tubing exiting the ground also affect the readings. If water is used, measures must be taken to prevent the growth of algae in the liquid lines.

A variation of the standard water pressure settlement system locates the pressure transducer either directly in bedrock or on a pipe attached to bedrock (Fig. 5-73). It is connected via a liquid-filled tube to the reservoir, which is directly over the transducer and is located at the point of interest. The advantage of this system is that it avoids the need for long horizontal tubing runs and does not require flushing. The disadvantage is that stable bedrock is required and must often be reached by drilling boreholes. Reading this type of system can be automated when an electrical pressure sensor is used.

Hydraulic Settlement Profilers - Standard Types. Another variation of the standard water pressure settlement system is the hydraulic settlement profiler, shown in Fig. 5-74. A pressure transducer inside a torpedo-shaped device is connected by a fluid-filled tube to a liquid reservoir. In use, the torpedo is pulled through an access pipe buried in the foundation or embankment.

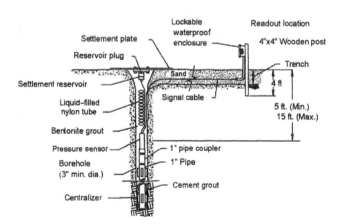

Fig. 5-73. Liquid pressure settlement system—bedrock type
Source: ASCE Task Committee on Instrumentation and Monitoring Dam Performance (2000).

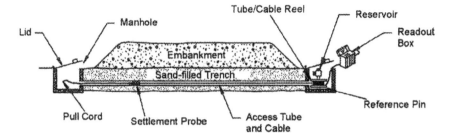

Fig. 5-74. *Hydraulic settlement profiler*
Source: *ASCE Task Committee on Instrumentation and Monitoring Dam Performance (2000).*

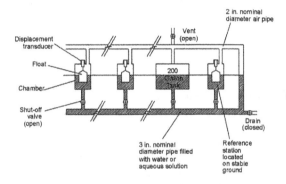

Fig. 5-75. *Liquid-level settlement system*
Source: *ASCE Task Committee on Instrumentation and Monitoring Dam Performance (2000).*

The transducer records the height of the water column between the transducer level and the reservoir level and thus provides a measure of the elevation of the access tube at any point along its length.

The access pipe is open at both ends. If the access pipe dead-ends in the embankment, it is connected at its far end via a U-shaped joint to a second parallel access tube containing the pull-in cable.

A variation of this system is to fill the access pipe with water and pull the transducer through it, thus eliminating the fluid-filled tube. The hardware for this second type of system is somewhat simpler, but if the access pipe develops a leak, it may be impossible to fill with water.

Liquid-Level Technique. In comparison with the other techniques described in this section, greater accuracy and precision can be achieved using a liquid-level settlement system (Fig. 5-75). A series of chambers, all at the same elevation, is connected by a liquid-filled tube so that the water inside each chamber is at the same level. Ideally, one of the chambers should be located

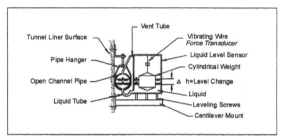

Fig. 5-76. Precision settlement system—water-level type
Source: ASCE Task Committee on Instrumentation and Monitoring Dam Performance (2000).

on solid ground that does not settle. Settlement of any chamber causes an apparent rise in the liquid level in that chamber, which is measured by an electronic displacement transducer attached to a float. A vent line connects the upper portions of all the chambers to equalize the air pressure above the liquid. Systems of this kind can have accuracies of ±0.04 in. (1 mm) and can be read by an ADAS.

An even more precise settlement monitoring system can be made using another water-level technique illustrated in Fig. 5-76. This system uses a 3 in. (76.2 mm) diameter horizontal pipe that is half-filled with water or anti-freeze solution. This pipe can be fixed to the wall of an underground gallery running in the dam embankment. At intervals along the pipe are a series of chambers connected hydraulically to the pipe, so that the water level in the chamber is at the same level as the water inside the pipe. The water level in the chamber is measured by a vibrating-wire force transducer, which consists of a weight hanging from the vibrating-wire and is partially submerged in the water so that changes in water level alter the buoyancy force on the weight and the tension in the wire. Settlement of any chamber produces an apparent rise of water level in that chamber. Systems of this kind can measure settlements of as little as 0.001 in. (0.025 mm).

Maintenance of Liquid-Filled Settlement Systems. Hydraulic systems suffer mainly from air bubbles in the liquid-filled tubes and to a lesser extent from blockages in the tubes caused by the growth of algae and bacteria. To minimize these problems, the following measures are recommended:

- Use de-aired water or antifreeze.
- Use distilled water rather than tap water.
- Dissolve chlorine bleach or small amounts of copper sulfate (one 6-mm crystal per 10 L) in the water to retard bacteria and algae growth (antifreeze solutions are already toxic to bacteria and algae).
- Use a wetting agent (nonfoaming) to reduce surface tension.
- Use twin tubes rather than a single tube so that the system can be flushed.
- Apply backpressure to water-pressure settlement systems to keep air bubbles in suspension, reducing the need for de-airing. Backpressure must then be measured and deducted from the readings.

5.3.7.6 Convergence Meters. Convergence meters are used to measure the change of distance between two points located on the walls of underground openings or at the surface of concrete or rock structures. They can be removable or nonremovable. Removable convergence meters are usually called tape extensometers, shown in Fig. 5-77, and comprise the following components: an extensometer head that applies a constant tension to the tape and includes a dial indicator graduated to 0.0004 in. (0.01 mm), a measuring tape attached to the extensometer, and anchored reference points. The overall repeatability of tape extensometer measurements, however, is reduced in practice to about 0.01 in. because of variations in tensioning the tape each time measurements are taken and in connecting and removing the tape to the eye-bolt anchors.

Recent advances in laser distance measurement technology have been used to develop laser extensometers, shown in Fig. 5-78, which can measure the distance between a reference anchor point to which the laser is temporarily mounted and a number of reflective monitoring points around the perimeter of the underground opening with high repeatability [typically 0.0125 in. (0.3 mm)]. The main advantage of a laser extensometer is that manual manipulation, which poses a difficulty for tape extensometer measurements, is not required.

Fixed convergence meters, shown in Fig. 5-79, are prop-up instruments that rest against the two surfaces between which convergence is to be measured.

5.3.7.7 Joint Meters and Crack Meters. Cracking in concrete dams may result from many factors: differential settlements of the foundation, elevated temperatures during concrete curing, shrinkage and swelling with changes in moisture content, alkali–aggregate reaction (AAR), and changes in reservoir levels. Attempts are made to control this cracking by inducing it to occur along construction joints, which may be sealed using water stops or grouted shut at a later date. In earth dams, as a result of differential settlements,

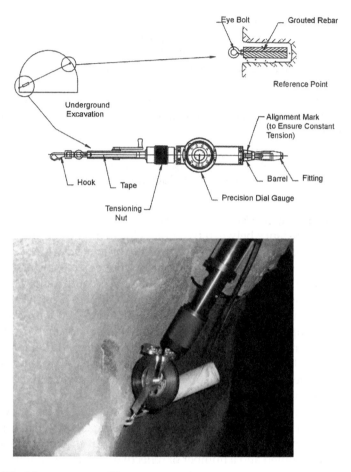

Fig. 5-77. Measurement with tape extensometer

zones of high shear or tension can cause cracking, especially at the abutments. Many types of crack meters exist for manual or remote monitoring of both surface cracks and interior cracks in one, two, or three dimensions. Instruments known as joint meters are used to monitor the opening or closing of construction joints.

Manual Crack Meters.   A simple crack meter is made by Avongard, shown in Fig. 5-80. Two plates overlap for part of their length. The bottom plate is fixed to one side of the crack and bears a grid pattern calibrated in millimeters. The overlapping upper plate, fixed to the other side of the crack, is transparent and is marked with a crosshair. As the crack opens and closes or shears in the plane of the surface, the amount of movement is shown by the position of the crosshair above the underlying grid.

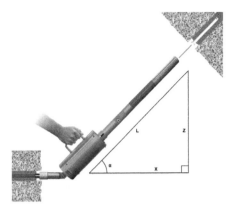

Fig. 5-78. Laser extensometer for convergence measurement
Source: RST Instruments, reproduced with permission.

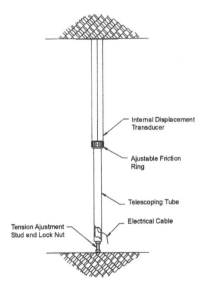

Fig. 5-79. Prop-up convergence meter
Source: ASCE Task Committee on Instrumentation and Monitoring Dam Performance (2000).

Movement of surface cracks or joints can be monitored in three dimensions using a dial indicator and a block system. A portable dial gauge is inserted into the openings to measure the gap between the two blocks in three orthogonal directions. The advantages of manual crack meters are their simplicity, low cost, and ease of use; disadvantages include the lack of real-time, continuous, automatic readings. A manual 3D crack meter,

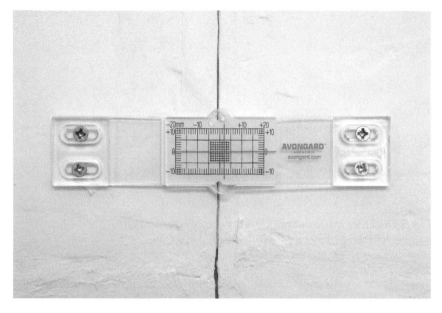

Fig. 5-80. Telltale crack meter
Source: Avongard.com, reproduced with permission.

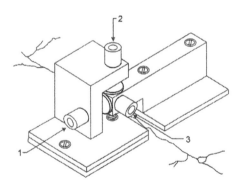

1, 2, & 3: Dial Gauge Ports

Fig. 5-81. Manual 3D crack meter
Source: ASCE Task Committee on Instrumentation and Monitoring Dam Performance (2000).

shown in Fig. 5-81, can be automated by the addition of displacement sensors, overcoming the disadvantages of simple manual crack meters. The sensors can be of the vibrating-wire, linear potentiometer, or LVDT type. The major advantage of implementing electronics is the potential for continuous, real-time, remote sensing and recording.

Electronic Crack Meters. A simple one-dimensional electronic crack meter is shown in Fig. 5-82. The meter is designed to be held by lag bolts or anchors drilled into the concrete or masonry on either side of the crack. The crack meter shown works on the vibrating-wire principle, but similar crack meters are available using LVDT transducers (Fig. 5-83).

Submersible Crack Meters. Submersible crack meters, shown in Fig. 5-84, are designed to monitor movement at submerged joints and cracks in concrete dams, tunnels, and tanks. Measurements are made by vibrating-wire sensors or potentiometer-based sensors. Uniaxial, biaxial, and triaxial

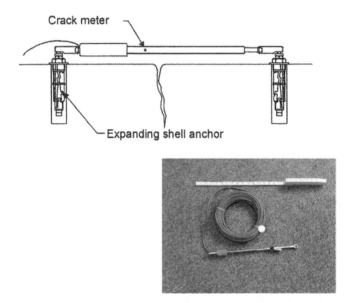

Fig. 5-82. Electronic crack meter with vibrating-wire transducer

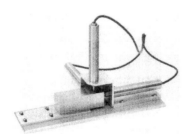

Fig. 5-83. 3D electronic crack meter with linear potentiometer transducers
Source: ASCE Task Committee on Instrumentation and Monitoring Dam Performance (2000).

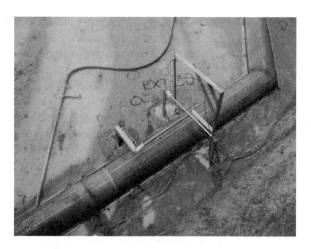

Fig. 5-84. 3D vibrating-wire submersible crack meter
Source: RST Instruments, reproduced with permission.

versions are available. The instruments are rated for use in up to 820 ft (250 m) of hydrostatic water pressure. Readings may be made manually or automatically. Data loggers capable of providing output for an array of crack meters are available.

Applications particular to submersible crack meters may include measuring crack progression in submerged concrete or rock, such as on the face of a dam, at dam abutments, or at the dam/abutment interface.

Embedment Joint Meters. Contraction joints are created between adjacent monoliths of concrete dams. Joints are also formed at each concrete lift. To monitor the opening and closing of these joints at points within the dam, an embedment-type joint meter may be used. Joint meters most commonly use vibrating-wire transducers, but Carlson transducers, based on the electrical resistance principle, also are used. The method of installation is in Figs. 5-85 and 5-86.

5.3.8 Seismic Response (Movement, Pressure, and Load)

Ground-motion measurements are not used as a predictive monitoring tool because they record earthquakes, without providing any warning prior to the event. There are a variety of devices that measure ground motion and can also record acceleration, velocity, and/or displacement over time. Records from these devices may trigger the need for other surveillance or measurements at the site to evaluate responses to ground shaking. Data from ground-shaking measurements at or near a site are used to analyze how the dam performed during an earthquake compared to its predicted

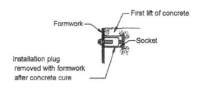

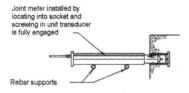

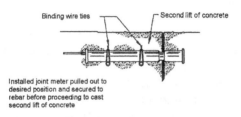

Fig. 5-85. Embedment joint meter
Source: ASCE Task Committee on Instrumentation and Monitoring Dam Performance (2000).

Fig. 5-86. Joint meter prior to and after installation

performance. Seismic response measurements described in this section are limited to strong-motion accelerographs. Mechanical seismic recorders are not discussed.

A strong-motion accelerograph combines one or more accelerometers and data-recording capabilities. It is used to measure and record dynamic ground accelerations produced by large earthquakes. The data are used to evaluate the response of an existing dam and to provide time histories to aid in the design of future dams (USCOLD 1989).

Ideally, seismic instruments would be placed in the free field (a location not influenced by the dam or other large structures), as well as on the dam crest or other pertinent structures. The response of the free-field instrument is representative of the baseline earthquake motion, and the crest instrument shows how the dam responds to the shaking.

An accelerograph, shown in Fig. 5-87, continually monitors accelerometer data until a user-set threshold is reached or exceeded. When this threshold is exceeded, the data is stored in 2–3-min data record increments, including the 30 s prior to the event. Standard accelerographs record data at 200 samples/s or more. It is standard practice to attach a GPS receiver to the accelerograph to obtain an accurate and reliable time stamp. One seismogeodetic system available is now adding GNSS (Global Navigation Satellite System) position recordings to significantly enhance the information recorded from the accelerograph. This generates more accurate histories of velocity and position data than is possible by integrating the acceleration data alone.

Strong-motion accelerographs typically come equipped with at least three channels and include an internal triaxial force balance accelerometer (two orthogonal horizontal accelerometers plus one vertical accelerometer). Instruments can be stand-alone, and recorded events can either be downloaded automatically by modem or manually retrieved by a laptop computer or by collecting the removable memory card. Alternatively, instruments can be connected in a network, typically fiber optic, and interfaced to an external LAN. Most instruments are supplied with software for making time history plots and generating ASCII file formats for data analysis. Instruments usually have a relay output to trigger actions when a user-set threshold is reached. One seismogeodetic system combines an accelerograph with a geodetic positioning GNSS receiver, as illustrated in Fig. 5-88, and relay data using the same GNSS network.

Seismic measurements require high rates of data sampling and recording to accurately capture ground shaking. Because most general-purpose data-acquisition systems do not support the required high data rates and high resolution, accelerographs are typically self-contained units with their own data-recording capabilities. For accurate recording of earthquake events, accelerographs must be securely anchored (1) to rock to capture foundation base motion and (2) to a point(s) of interest on the dam—typically

Fig. 5-87. Digital accelerographs

the crest. Installation is accomplished by attaching the accelerograph to a concrete floor or prepared rock surface using anchor bolts.

Temporary seismographs may be installed during construction to monitor vibrations induced by blasting, dynamic compaction, or other activities that induce large vibrations. Seismographs measure pressure-wave velocities, and the velocities are evaluated to assess the potential for damaging structures or initiating a dam construction failure sequence.

Fig. 5-88. Real-time fully digital accelerometer, which can relay data via GNSS
Source: Courtesy of Trimble, reproduced with permission.

There are two basic categories of strong-motion accelerographs widely used: analog and digital accelerographs. These are described in the next section (Shakal and Huang 1996).

5.3.8.1 Analog Accelerographs. These pioneering devices, developed in the 1930s, are triggered by earthquakes and record the time versus acceleration on film. These devices were not always reliable, and it proved difficult to interpret results due to inconsistency in the x, y, and z components. An analog record may also need to be digitized, which is expensive and time-consuming. The instrument's current applicability is in regions of low likelihood of activity, where few records are expected over the lifetime of the instrument.

Rapid use of the data is not possible because of the need to develop the film and then digitize it. Analog accelerometers are legacy devices and are not in general use today.

5.3.8.2 Early Digital Accelerographs. First introduced in the 1970s, these accelerographs digitally record earthquakes on magnetic tape. Early digital accelerographs required significant maintenance to keep them functioning, and the magnetic tape was difficult to utilize with developing computer interfaces. These instruments are being replaced by improved accelerographs.

5.3.8.3 Intermediate Digital Accelerographs. These instruments were introduced in the 1980s and are much more reliable and easier to use than the early digital accelerographs. The drawback is that electrical demands preclude the use of batteries, and so these devices are only operable at dams where power is available. Electrical service is often interrupted during earthquakes, and standby generators require a few seconds to transfer power. The consequence is that there may be gaps in the data when power is lost. Intermediate digital accelerographs have a greater dynamic range than the early film recorders because they record in a 12-bit format.

5.3.8.4 Modern High-Resolution Digital Accelerographs. The newest accelerographs record in a 24-bit format, and they have a much higher resolution than the intermediate digital accelerographs. Power consumption is still a problem for remote operation. For a comparison of instruments, see Table 5-3 at the end of this chapter.

5.3.9 Weather (Rainfall and Wind)

Total amounts of precipitation and precipitation intensity (i.e., hourly rate of precipitation) are important hydrological parameters to consider during the planning, design, and operational phases of a dam, its spillway, and its outlet works. Rainfall measurements can be useful in understanding the regional hydrologic factors that help to determine the appropriate design flood loading that could be expected at the dam. Rainfall measurements are also useful when evaluating instrumentation data from piezometers and weirs, reservoir-level data, etc., as rainfall affects spikes and trends. Wind speed and direction monitoring can be important to understand wave run-up affecting dams. These parameters may also be important in the evaluation of slope stability or avalanche risk on the reservoir side slopes.

5.3.9.1 Design Considerations. Not every dam requires a weather station. Often, meteorological data can be obtained from nearby sources. If a weather station is required, then the following considerations are applicable.

Placement of a weather station is important to get a representative sampling of measurements. The National Oceanic and Atmospheric Administration (NOAA) and weather station manufacturers provide guidelines and specifications for locating weather monitoring instruments; however, variations from that guidance may be justified to meet the objectives of the user, cable length limits, radio transmission distance limits, or other factors. Wind causes errors in precipitation measurements, and the errors become more significant for snow and frozen precipitation. Trees, buildings, or other tall objects located close to gauges may deflect precipitation because of wind turbulence. To avoid wind and the resulting turbulence problems, rain gauges should be located in wide open spaces or on elevated sites, such as the tops of buildings. The best site for a gauge is one where it is protected in all directions, such as in an opening in a grove of trees. The height of the protection should not exceed twice its distance from the gauge. Research to optimize manufactured windscreens for snow is ongoing, but accepted designs such as the Alter windscreen and others have existed for many years.

Helpful advice is available from NOAA's guidance for its Cooperative Observer Program: http://www.nws.noaa.gov/om/coop/standard.htm.

Snow depth and water content may be helpful to estimate seasonal runoff to inform flood management. Snow surveys can provide these data.

Anemometers for wind magnitude and direction are best mounted on a rooftop or on a high pole. The anemometer should be mounted at least 6 ft above a roof line for best results. For ground installations, the anemometer should be placed on a pole in an open area, unobstructed to the wind, and at least 5 ft above the ground.

The distance between a measurement and a readout point may have important implications when deciding whether to use a precipitation gauge compatible with an ADAS.

Access also may be an issue for service and maintenance.

Manual gauges require no power. However, gauges need power if they use ADAS, motorized equipment, or require heating. If power is impractical to provide at a gauge location, electric heat and motorized equipment are not possible. However, the low power required to energize a gauge's transducer can still be provided through signal cables, from batteries, or from solar cells at the gauge location.

5.3.9.2 Rain Gauges. Although precipitation gauge manufacturing dates back over a century in the United States, important design improvements were made in the 1980s in the sensing and data-collection functions of precipitation gauges.

Mechanically, the two principal types of precipitation gauges are pan-type gauges and bucket gauges. Pan-type gauges are usually placed in an array or grid with a large surface area and are used principally for calibration and research; therefore, such gauges are not further discussed in this chapter. Bucket gauges are further subdivided into three classes: weighing gauges, tipping gauges, and manual gauges. Typical bucket capacities are 12 in. (300 mm), 24 in. (600 mm), and 30 in. (750 mm) of precipitation. The buckets have a cylindrical chimney or stack. The orifice of the stack is normally a specified height above the ground, for example, 60 in. (1.5 m), to minimize ground effects. Orifice diameters range from 6 to 12 in. (150 to 300 mm).

If snow is collected, the outside surface of the stack may be painted dark to absorb the maximum heat and aid in melting. To minimize the tendency for snow to adhere to the inside of the stack, nonstick coatings may be applied. Buckets are made of a material that tolerates cold weather without cracking and has low friction to avoid snow sticking. Certain types of plastics meet this criterion and are inexpensive. When temperatures are below freezing, adding antifreeze to the bucket melts the snow that collects.

Measurement units in the United States are inches (in.), with a customary resolution of ±0.01 in. Outside the United States, the SI system of measurement prevails and precipitation is measured in millimeters (mm), with a typical resolution of ±0.1 mm.

Weighing Bucket Gauges.   Weighing bucket gauges measure, as their name implies, by weighing the precipitation in a collection bucket. Resolution is better than 0.004 in. (0.1 mm) versus a tipping bucket gauge resolution of 0.01 in. (0.25 mm). Better resolution improves intensity measurements. Because they can operate without heating, weighing bucket gauges are the most practical type for measuring snowfall. The weighing sensor may be a load cell on one of the chains that supports a platform in which the bucket rests. Other methods of weighing include weighing the support platform of the bucket and measuring the pressure of the liquid precipitation with a pressure transducer at the bottom of the bucket. The weighing mechanism is typically a vibrating-wire or resistance strain gauge load cell. In the gauge shown in Fig. 5-89, a vibrating-wire load cell is filled on one of three internal chains supporting the bucket platform. Accurate measurements require the bucket to be level. Vibrating-wire load cells used in weighing buckets have the same advantages as other types of vibrating-wire instruments. These include simple design, minimal long-term error due to drift, minimal temperature effects, and error-free transmission over long signal cables (one mile or more).

Weighing bucket gauges use antifreeze (ethylene glycol) to melt frozen precipitation. A drawback is that this reduces capacity and requires more frequent maintenance to empty the bucket. Typically, the capacity of a 24-in. (600-mm) bucket would be reduced to about 60% capacity when enough antifreeze is used to protect down to −13°F (−25°C). With the use of antifreeze comes the need for environmentally proper disposal of the collected precipitation/antifreeze mix in waste containers. To minimize evaporation, a thin layer of aircraft hydraulic oil can be added to the bucket. The precipitation drops through the oil without emulsifying when oil of the correct viscosity is used.

For precipitation measurements at dams and reservoirs where automated data acquisition is used, weighing bucket gauges may be superior to pan or tipping-type gauges.

Tipping Bucket Gauges.   Tipping bucket gauges are designed to tip upon collection of a fixed amount of precipitation, typically 0.01 in. (.25 mm) of water. The collected water is then discharged to the ground or a collection

Fig. 5-89. Weighing bucket precipitation gauge with double-alter shield
Source: Geonor, reproduced with permission.

bucket. Hence, they have unlimited capacities if the precipitation is discharged to the ground. The tipping action causes a reed switch or a mercury switch to trigger a pulse counter. Pulses may be easily interfaced to an ADAS. A disadvantage for remote operation in freezing temperatures is the need for heating, usually from an AC power source. A disadvantage of using electrical heating is the need for temperature controls; in other words, if the heater is continuously powered, melted precipitation may evaporate. Another disadvantage is that debris can foul a tipping bucket. For these reasons, tipping buckets are generally not suitable for measurement of snowfall or for use in remote locations. The presence of moving parts also creates a need for maintenance and cleaning.

Manual Precipitation Gauges. Manually read gauges are low in initial cost, but they are labor intensive. They are not suitable for remote, hard-to-access locations, or where intensity measurements are needed.

5.4 FACTORS AFFECTING INSTRUMENT PERFORMANCE

Long-term performance of dam-monitoring instrumentation is sensitive to the maintenance of devices, connections, and cabling. Exposed cabling is susceptible to damage. Connections are susceptible to damage if cables can be easily pulled or snagged. Moisture entering connections and causing corrosion is a common maintenance problem. Dry connections allow the instrumentation system to work properly.

5.4.1 Cabling, Tubing, Junction Boxes, Terminal Boxes

Because they are easily damaged during handling and installation, cables and tubing are often the weak link in instrument installations. Cables need to be sufficiently robust to survive handling during installation and the long-term effects of water and chemicals in the ground. As discussed in this section, lightning protection is also essential to monitoring systems with potential lightning exposure.

A cable, shown schematically in Fig. 5-90, consists of insulated conductors, a shield, a drain wire, and an outer jacket. The conductors are usually stranded copper (AWG 20–24) with polypropylene insulation. Stranded conductors are preferred because they are less likely to fatigue as the cable is bent. For sensor signals that can be affected by noise, cables with twisted-pair construction are used. An overall shield and drain wire also help prevent interference from electrical noise. The shield is terminated at the sensor without being connected to the sensor. At the surface, the shield is connected to ground.

The cable jacket is usually polyvinyl chloride (PVC), polyurethane, or polyethylene. The jacket provides overall protection to the components of the cable and electrical insulation from transient ground currents. The jacket is required to provide a minimum insulation of 300 V. PVC is the most commonly used jacket material as it can be sealed in the sensor with potting.

Polyurethane is more expensive than PVC, but it provides better long-term waterproofing and better resistance to abrasion. Polyethylene has an even better resistance to water absorption, but it requires an expensive mechanical seal because potting will not adhere to it. The most common jacket is black, which contains ultraviolet (UV) inhibitors to protect the jacket from UV light. Black is used if the cable is to be exposed for any length of

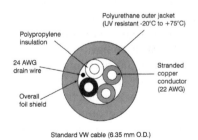

Fig. 5-90. Typical construction of 4-conductor cable for vibrating-wire instruments
Source: RST Instruments, reproduced with permission.

time. Water-blocked cable may be specified where tearing of the cable jacket is a possibility: armored and reinforced cable is also available.

For individual sensor installations in boreholes, cables are usually terminated at the top of the borehole in a waterproof enclosure. Large installations usually require a central readout point with cables routed from the borehole to the readout point in trenches. Whenever cable is exposed aboveground, it must be enclosed in conduit to protect it from UV light, physical damage, and animals (rodents like to take bites out of it).

Cables routed through earth embankments must be flexible enough (stretch) to accommodate settlement that may occur. Cables are typically laid in a trench, backfilled with 6 in. to 1 ft (150–300 mm) of granular fill above and below the cable, the fill having been passed through a #40 sieve. The fill in the trench is compacted to the same relative compaction as the embankment using hand-operated equipment. Trenches do not pass through the impermeable zone of a dam because they form a potential leakage path. Bentonite seals are often placed at regular intervals in trenches to serve as waterstops.

Cables passing from one zone of the dam to another or passing through coarse fill are protected in grease-filled conduits. Burying cable splices below ground level is not advised for long-term projects. If buried, splices are required, and splices need to be strong enough to withstand burial and maintain a waterproof connection.

Junction boxes and terminal boxes are commonly metal or fiberglass. If they are sealed, they contain a silica-gel desiccant to prevent condensation. The desiccant must be changed on a regular basis. The cable entry point can be sealed with grommets or putty.

5.4.2 Lightning and Transient Protection

As mentioned previously, instruments with electronic or electrical components are highly susceptible to damage from electrical transients caused by lightning discharges. The degree of susceptibility depends on the type of components employed in the instrument, the instrument location, and the manner in which signal cables are connected to and extended from the instrument. Instruments that are installed outdoors on embankment dams are especially vulnerable to lightning-induced transients because this environment lacks the usual first line of defense against lightning effects, a grounded building or structure. Destructive surges can also be caused by other mechanisms such as electrostatic discharge (ESD), which is typically a handling issue, and ground potential surges, which can be violent and frequent in the vicinity of switchyards. In the case of permanent installations, the protection strategies are essentially the same for all types of surges. Several extensive monitoring arrays have been destroyed by lightning. Instrument manufacturers and suppliers are the best source of site- and instrument-specific advice regarding lightning and transient protection.

5.4.3 Calibration and Maintenance

Instrument maintenance requires accessibility. Buried instruments are normally not recoverable and should be selected with high long-term, low-maintenance reliability in mind. Buried instruments intended to last the design life of the dam require careful research before selection. Existing instrument performance data over periods of several decades may not be available to aid in instrument selection. Judgment must be exercised to select an appropriate instrument with a view to maximizing needed data, maintenance requirements, and cost. It may be advisable to locate as many instruments as possible in recoverable installations so that they can be replaced if necessary.

Instruments should be calibrated at the factory prior to shipment, and calibration data must be furnished with each instrument. Whenever possible, the calibration should be checked again after it is received to make sure the instrument was not damaged in shipment. Items such as pressure gauges used to measure uplift and weirs used to monitor seepage should be calibrated regularly. Calibration of the readout devices is also very important and should be done on a scheduled basis, at least annually.

Maintenance is generally the responsibility of the technicians charged with taking the instrument readings. It is important that an engineer or representative responsible for dam safety periodically inspect the condition of the instrumentation to ensure that it is being properly maintained. Readout boxes and terminals are often subject to wet or damp conditions. Keeping them as dry as possible will help prevent corrosion to the terminals. A record of each instrument on a dam including its location, the date of its last inspection, its condition, and a photograph of the instrument, if possible, add value to a performance-monitoring plan.

Maintenance of an instrument should also be a key consideration when selecting the proper instrument. An example of the importance of this can be seen in the manifolds where piezometer tubes at a dam are encrusted with calcium deposits. The very-small-diameter plastic tubes are brittle and cannot be cleaned out, nor can the calcium buildup be removed because more damage to the tubes would occur. This has led to the concern that as the bundles of tubes pass through the core of the dam, piping of core material could occur.

Table 5-3. Instrument Comparison.

		Instrument Name				
	Manual inclinometers	In-place inclinometers (MEM Sensors)		Inclinometer casing	Shape accelerometer arrays (SAA)	Time-domain reflectometry (TDR), Section 5.3.7.3
		Multiple-interval	Multiple-point			
DEFORMATION AND MOVEMENT						
Inclinometers, Section 5.3.7.1						
Parameter measured	Lateral deflections of inclinometer casing in a borehole.	Lateral deflections of inclinometer casing in a borehole.	Lateral deflections of inclinometer casing in a borehole.	Required for manual inclinometer and in-place inclinometers.	Lateral deflections of inclinometer casing or PVC access pipe in a borehole.	Determine locations of active shear plane in a borehole.
Common applications	Slope stability in dam embankment and abutments, reservoir slopes.	Slope stability in dam embankment and abutments, reservoir slopes.	Slope stability in dam embankment and abutments, reservoir slopes.	Access casing for inclinometer surveys.	Slope stability in dam embankment and abutments, reservoir slopes.	Slope stability in dam abutments and reservoir slopes.
Current practice or legacy instrument?[a]	Current	Current	Legacy/current	Current	Current	Current
Installation specifics/ease of installation	Requires inclinometer casing. Inclinometer probe, 2 ft long, lowered in casing using marked signal cable and raised to take readings at fixed intervals (typically every 2 ft).	Requires inclinometer casing. String of sensors with rigid 2 to 10 ft extension rods between wheels.	Requires inclinometer casing. String of point sensors separated by flexible spacer rods, 2 to 10 ft long.	Casing lowered and grouted in borehole. Buoyancy must be controlled in holes with high water table, or casing raised in embankment during construction of new dam. May break at slide plan.	Preassembled tubular string of 1.6 ft long sensors ready for insertion in casing or access pipe.	Coaxial cable grouted in borehole or inclinometer casing or access pipe.

General considerations	Requires operator dexterity for repeatable readings.	Sensor string can be lowered to depth of anticipated movement. Can be removed and reused in another borehole.	ABS casing, 2.75 or 3.34 in. diameter. Groove spiral must be minimal. Maximum settlement accommodated without telescopic sections is 1%.	Sensor string can be lowered to depth of anticipated movement. Can be removed and reused in another borehole.	Coaxial cable, 12.7 mm (0.5 in.) diameter preferred for stronger signal.
General methods of reading output[b]	Portable data logging readout to collect readings as probe is raised manually in the casing.	Portable readout, ADAS.	N/A	Portable readout, ADAS with dedicated conditioning interface.	Portable TDR readout unit or ADAS with dedicated TDR conditioning interface given.
Sensor options	MEMS accelerometer, servo-accelerometer.	MEMS accelerometer, electrolytic tilt sensors.	MEMS accelerometer, electrolytic tilt sensors.	MEMS accelerometer.	Time-domain reflectometry.
Sensitivity	0.0006 deg. (MEMS digital) typical.	0.0006 deg. (MEMS digital) typical.	0.0006 deg. (MEMS digital) typical.	N/A	Provides only location of shear movement, not magnitude.
Accuracy	±0.002 deg. (MEMS digital) typical.	±0.002 deg. (MEMS digital) typical.	±0.002 deg. (MEMS digital) typical.	±0.029 deg.	Shear location along cable within 2 to 5 ft depending on cable length.
Repeatability	±1 to 3 mm per 30 m of profile (system accuracy for MEMS digital).	±0.002 deg. (MEMS digital) typical.	±0.002 deg. (MEMS digital) typical.	±1.5 mm per 32 m of profile.	2 to 5 ft.

Table 5-3. (*Continued*) Instrument Comparison.

	Instrument Name					
	In-place inclinometers (MEM Sensors)					
	Manual inclinometers	Multiple-interval	Multiple-point	Inclinometer casing	Shape accelerometer arrays (SAA)	Time-domain reflectometry (TDR), Section 5.3.7.3
Accuracy/Repeatability[c]	Probe requires calibration annually, or when checksums become too high.	Ensure stable positioning of the string in the casing. Connectors between sensors must be waterproof.	Ensure stable positioning of the string in the casing. Connectors between sensors must be waterproof.	Protect collar end of casing from damage.	Ensure stable positioning of the string in the access pipe.	If coaxial cable jacket gets damaged, water ingress will corrupt the TDR signal.
Longevity[d]	Probe: YYYY Cable: YYY	YYYY	YYYY	YYYYY	YYYY	YYYY
Maintenance requirements	Periodic probe calibration. Periodic cable inspection.	Verify surface cable integrity. Removing or lifting string can be considered.	Verify surface cable integrity. Removing or lifting string can be considered.	Verify casing integrity. Silt and dirt can accumulate in casing grooves.	Verify surface cable integrity. Removing or lifting string can be considered.	Verify surface cable integrity.
Relative expense[e]	Instrument: $$$$ Readout: $$$ ADAS: N/A	Instrument: $$ Readout: $$$ ADAS: $$$ to $$$$	Instrument: $$ Readout: $$$ ADAS: $$$ to $$$$	Instrument: $ Readout: N/A ADAS: N/A	Instrument: $$$$ Readout: N/A ADAS: $$$$	Instrument: $ Readout: $$$ ADAS: $$$$

[a]"Current practice or legacy" refers to common practice in current U.S. applications. However, legacy instruments may still be in use and available and may have some advantages in particular applications.

[b]General methods of reading output: can include manual, on-demand, and output that can accommodate ADAS (Automated Data Acquisition Systems).

[c]Accuracy/repeatability considerations: include thermal exposure and long-term stability.

[d]Longevity: Y, 1 year or less; YY, 1–5 years; YYY, 5–10 years; YYYY, 10–15 years; and YYYYY, more than 15 years.

[e]Relative expense: $, $1–$199; $$, $200–$999; $$$, $1,000–$2,999; $$$$, $3,000–$9,999; and $$$$$, more than $10,000.

	Instrument Name				
	Surface-mount tiltmeters	Submersible tiltmeters	Borehole tiltmeters	Portable tiltmeters	Beam sensors
Tiltmeters and Beam Sensors, Section 5.3.7.4					
Parameter measured	Rotational movement using gravity as the reference datum.	Rotational movement using gravity as the reference datum.	Rotational movement using gravity as the reference datum.	Rotational movement using gravity as the reference datum.	Rotational movement using gravity as the reference datum.
Common applications	Tilt in gravity dams, appurtenant concrete structures.	Settlement-induced tilt of inclined concrete face in CFRD dams, other submerged applications (typically up to 700 ft depth).	Localized tilt in dam embankments, superficial overburden layers.	Tilt in gravity dams.	Horizontal settlement profile in inspection gallery, dam crest.
Current practice or legacy instrument?[a]	Current	Current	Current	Current	Current
Installation specifics/ease of installation	Vertical or horizontal mounting bracket.	Vertical or horizontal mounting bracket, protective cover recommended.	Grouted in borehole.	Fixed horizontal or vertical reference tilt plates read with a portable tiltmeter.	Series of beams attached end to end.
General considerations	Concrete or steel structure on which tiltmeter is mounted can also display temperature-dependent tilt movements. Temperature measurement to tilt can assist in assessing temperature effects.	Concrete or steel structure on which tiltmeter is mounted can also display temperature-dependent tilt movements. Temperature measurement to tilt can assist in assessing temperature effects.	Concrete or steel structure on which tiltmeter is mounted can also display temperature-dependent tilt movements. Temperature measurement to tilt can assist in assessing temperature effects.	Concrete or steel structure on which tiltmeter is mounted can also display temperature-dependent tilt movements. Temperature measurement to tilt can assist in assessing temperature effects.	Concrete or steel structure on which tiltmeter is mounted can also display temperature-dependent tilt movements. Temperature measurement to tilt can assist in assessing temperature effects.
General methods of reading output[b]	Portable readout, ADAS.	Portable readout, ADAS.	Portable readout, ADAS.	Portable readout.	Portable readout, ADAS.

Table 5-3. (*Continued*) Instrument Comparison.

	Instrument Name				
	Surface-mount tiltmeters	Submersible tiltmeters	Borehole tiltmeters	Portable tiltmeters	Beam sensors
Sensor options	MEMS accelerometer, electrolytic tilt sensors.	MEMS accelerometer, electrolytic tilt sensors.	MEMS accelerometer, electrolytic tilt sensors.	MEMS accelerometer, electrolytic tilt sensors.	MEMS accelerometer, electrolytic tilt sensors.
Sensitivity	0.0006 deg. (MEMS digital) typical.	0.0006 deg. (MEMS digital) typical.	0.0006 deg. (MEMS digital) typical.	0.0006 deg. (MEMS digital) typical.	0.0006 deg. (MEMS digital) typical.
Accuracy	±0.002 deg. (MEMS digital) typical.	±0.002 deg. (MEMS digital) typical.	±0.002 deg. (MEMS digital) typical.	±0.002 deg. (MEMS digital) typical.	±0.002 deg. (MEMS digital) typical.
Repeatability	±0.002 deg. (MEMS digital) typical.	±0.002 deg. (MEMS digital) typical.	±0.002 deg. (MEMS digital) typical.	±0.01 deg. (system accuracy for MEMS digital).	±0.01 deg. (system accuracy for MEMS digital).
Accuracy/ Repeatability[c]	Must be protected from direct sun exposure.	Protect from vibrations and external shocks.	Protect from vibrations and external shocks.	Requires operator proficiency for repeatable readings.	Protect from vibrations and external shocks.
Longevity[d]	YYYY	YYYYY	YYYYY	YYY	YYY
Maintenance requirements	Protect from any contact or shock. Consider secondary enclosure.	Protect from any contact or shock. Consider secondary enclosure.	Verify surface cable integrity.	Periodic tiltmeter calibration.	Protect from any contact or shock. Periodic inspection.
Relative expense[e]	Instrument: $$ Readout: $$ ADAS: $$$ to $$$$	Instrument: $$$ Readout: $$ ADAS: $$$ to $$$$	Instrument: $$$ Readout: $$ ADAS: $$$ to $$$$	Instrument: $$$ Readout: $$ to $$$ ADAS: N/A	Instrument: $$$ Readout: $$ ADAS: $$$ to $$$$

[a] "Current practice or legacy" refers to common practice in current U.S. applications. However, legacy instruments may still be in use and available and may have some advantages in particular applications.

[b] General methods of reading output: can include manual, on-demand, and output that can accommodate ADAS (Automated Data Acquisition Systems).

[c] Accuracy/Repeatability considerations: include thermal exposure and long-term stability.

[d] Longevity: Y, 1 year or less; YY, 1–5 years; YYY, 5–10 years; YYYY, 10–15 years; and YYYYY, more than 15 years.

[e] Relative expense: $, $1–$199; $$, $200–$999; $$$, $1,000–$2,999; $$$$, $3,000–$9,999; and $$$$$, more than $10,000.

Plumblines and Inverted Pendulums, under Section 5.3.7.4 (Tiltmeters and Beam Sensors)

	Instrument Name	
	Plumblines	Inverted pendulums
Parameter measured	Rotational movement using gravity as the reference datum.	Rotational movement using gravity as the reference datum.
Common applications	Overall inclination in concrete gravity and arch dams.	Overall inclination in concrete gravity and arch dams relative to foundation. Used when access to bottom of well will be inconvenient after installed.
Current practice or legacy instrument?[a]	Current	Current
Installation specifics/ease of installation	Requires vertical shaft, typically 8–12 in. diameter extending from dam crest to foundation.	Requires vertical 8–12 in. diameter borehole in foundation.
General considerations	Suspension cable should not exceed 200 ft in length. Protection against water dripping in shaft is required. Simpler than inverted pendulum.	Suspension cable should not exceed 200 ft in length. Protection against water dripping in shaft is required. Simpler than inverted pendulum.
General methods of reading output[b]	Mechanical reading table, or electromechanical or optical automatic readout interfaced to ADAS.	Mechanical reading table, or electromechanical or optical automatic readout interfaced to ADAS.
Sensor options	Mechanical, electromechanical, or optical.	Mechanical, electromechanical, or optical.
Sensitivity	0.01 mm (0.0004 in.) typical with optical readout.	0.01 mm (0.0004 in.) typical with optical readout.

Table 5-3. (*Continued*) Instrument Comparison.

	Instrument Name	
	Plumblines	Inverted pendulums
Accuracy	0.05 mm (0.002 in.) typical with optical readout.	0.05 mm (0.002 in.) typical with optical readout.
Repeatability	0.05 mm (0.002 in.) typical with optical readout.	0.05 mm (0.002 in.) typical with optical readout.
Accuracy/Repeatability[c]	Damping fluid must be checked periodically. No friction in tank and in shaft. No air draft in shaft.	Damping fluid must be checked periodically. No friction in tank and in shaft. No air draft in shaft.
Longevity[d]	YYYY	YYYY
Maintenance requirements	Change damping fluid and clean all components periodically. Verify obstructions.	Change damping fluid and clean all components periodically. Verify obstructions.
Relative expense[e]	Instrument: $$$ Readout: $$$$ ADAS: $$$$$	Instrument: $$$$ Readout: $$$$ ADAS: $$$$$

[a]"Current practice or legacy" refers to common practice in current U.S. applications. However, legacy instruments may still be in use and available and may have some advantages in particular applications.

[b]General methods of reading output: can include manual, on-demand, and output that can accommodate ADAS (Automated Data Acquisition Systems).

[c]Accuracy/Repeatability considerations: include thermal exposure and long-term stability.

[d]Longevity: Y, 1 year or less; YY, 1–5 years; YYY, 5–10 years; YYYY, 10–15 years; and YYYYY, more than 15 years.

[e]Relative expense: $, $1–$199; $$, $200–$999; $$$, $1,000–$2,999; $$$$, $3,000–$9,999; and $$$$$, more than $10,000.

Borehole Extensometers, under Section 5.3.7.5 (Settlement Gauges)

	Instrument Name					
	Multirod MPBX	Single-rod (Incremental) MPBX	Probe extensometers	Borros heave/ settlement points (Geonor settlement probes)	Bureau of Reclamation Settlement System	Convergence meters, Section 5.3.7.6
Parameter measured	Compression and extension movements along axis of borehole.	Compression and extension movements along axis of borehole.	Compression and extension movements along axis of borehole.	Compression and extension movements along axis of borehole.	Compression and extension movements along axis of borehole.	Change of distance between two points located on the walls of underground openings.
Common applications	Compression in concrete dam foundations. Movements of geological features in dam abutments.	Compression in concrete dam foundations. Movements of geological features in dam abutments.	Settlement in earth dams.	Settlement in earth dams.	Settlement in earth dams.	Underground powerhouse, diversion tunnel.
Current practice or legacy instrument?[a]	Current	Current	Current	Current	Legacy	Current
Installation specifics/ease of installation	Requires 75–100 mm diameter borehole, grouted anchors or other types (hydraulic bladder . . .).	Requires 75–100 mm diameter borehole, grouted anchors or other types (hydraulic bladder . . .).	Requires vertical access pipe (25–35 mm diameter) or inclinometer casing, and magnetic settlement plates or targets along pipe.	Riser pipe and concentric inner rigid rod connected to bottom anchor.	Telescoping steel pipe sections with cross arms.	Measuring tape extended between reference anchors or laser measurement.

Table 5-3. (*Continued*) Instrument Comparison.

	Multirod MPBX	Single-rod (Incremental) MPBX	Probe extensometers	Borros heave/ settlement points (Geonor settlement probes)	Bureau of Reclamation Settlement System	Convergence meters, Section 5.3.7.6
			Instrument Name			
General considerations	If compression is anticipated, rods must be rigid enough to not bend.	Electrical head is recessed in borehole.	Requires operator dexterity for repeatable readings. Survey from bottom to top of hole must be done in a similar way every time.	Requires operator dexterity for repeatable readings. Survey from bottom to top of hole must be done in a similar way every time.	Requires operator dexterity for repeatable readings. Survey from bottom to top of hole must be done in a similar way every time.	Requires operator dexterity for repeatable readings.
General methods of reading output[b]	Mechanical reading (depth gauge) or electrical (VW displacement transducers, potentiometers, CVDT).	Electrical (VW displacement transducers, potentiometers . . .).	Manual reading only: probe lowered in access pipe. Depth read on graduated electric tape.	Manual reading: inner rod displacement relative to riser pipe.	Manual reading only: probe lowered in pipe.	Dial gauge (mechanical or digital) or laser.
Sensor options	VW, potentiometer, or many other options.	VW or potentiometer.	Reed switch probe.	Manual reading scale. Can also be automated with electric displacement transducer.	Cannot be automated.	Laser distance sensor accuracy varies between 0.1 and 1 mm.
Sensitivity	0.001–0.05 mm (0.00004–0.002 in.) typical.	0.001–0.05 mm (0.00004–0.002 in.) typical.	0.1–0.5 mm (0.04–0.2 in.).	0.5 mm (0.02 in.) (for manual reading) typical.	0.5 mm (0.02 in.) typical.	0.01 mm (0.0004 in.) for dial gauge.
Accuracy	0.01–0.05 mm (0.0004–0.002 in.) typical.	0.01–0.05 mm (0.0004–0.002 in.) typical.	0.1–0.5 mm (0.04–0.2 in.).	0.5 mm (0.02 in.) (for manual reading) typical.	0.5 mm (0.02 inch) typical.	0.01–0.05mm (0.0004–0.002 in.) for dial gauge.

Repeatability	0.01–0.05 mm (0.0004 to 0.002 in.) typical.	0.01–0.05 mm (0.0004–0.002 in.) typical.	0.1–0.5 mm (0.04–0.2 in.).	0.5 mm (0.02 in.) (for manual reading) typical.	0.5 mm (0.02 in.) typical.	0.1–0.3 mm (0.004–0.01 in.) for dial gauge, 0.1–1 mm for laser.
Accuracy/Repeatability[c]	Rigid rods provide better reading accuracy and repeatability, especially when compression movements are anticipated.	Rigid rods provide better reading accuracy and repeatability, especially when compression movements are anticipated.		Friction or shear movements in borehole can affect readings.	Survey from bottom to top of hole must be done in a similar way every time.	Temperature dilation of measuring tape can be significant source of error.
Longevity[d]	YYYYY	YYYYY	YYYY	YYYY	YYYY	YYY
Maintenance requirements	Periodic check of electrical head. Verify surface cable integrity.	Verify surface cable integrity.	Periodic inspection of readout and graduated electric tape.	Periodic inspection.	Periodic instrument inspection.	Periodic instrument and anchor points inspection.
Relative expense[e]	Instrument: $$$ to $$$$ Readout: $$$ ADAS: $$$ to $$$$	Instrument: $$$ to $$$$ Readout: $$$ ADAS: $$$ to $$$$	Instrument: $$ Readout: $$ ADAS: N/A	Instrument: $$ to $$$ (if electric transducer) Readout: $$ ADAS: N/A or $$$ to $$$$ (if electric transducer)	Instrument: $$$ Readout: $ ADAS: N/A	Instrument: $ Readout: $$$$ ADAS: N/A

[a]"Current practice or legacy" refers to common practice in current U.S. applications. However, legacy instruments may still be in use and available and may have some advantages in particular applications.
[b]General methods of reading output: can include manual, on-demand, and output that can accommodate ADAS (Automated Data Acquisition Systems).
[c]Accuracy/Repeatability considerations: include thermal exposure and long-term stability.
[d]Longevity: Y, 1 year or less; YY, 1–5 years; YYY, 5–10 years; YYYY, 10–15 years; and YYYYY, more than 15 years.
[e]Relative expense: $, $1–$199; $$, $200–$999; $$$, $1,000–$2,999; $$$$, $3,000–$9,999; and $$$$$, more than $10,000.

Table 5-3. (*Continued*) Instrument Comparison.

Joint Meters and Crack Meters, Section 5.3.7.7

	Instrument Name				
	Manual crack meters	Electric crack meters	Submersible crack meters	Embedment joint meters	Embedment soil strain meter or soil extensometer
Parameter measured	1D, 2D, and 3D movement of surface cracks or joints.	1D, 2D, and 3D movement of surface cracks or joints.	1D, 2D, and 3D movement of surface cracks or joints.	1D extension/contraction of joints between concrete blocks.	1D horizontal extension/compression in earth embankments.
Common applications	Concrete surfaces, rock joints.	Concrete joints and cracks in dams, tension cracks in slopes.	Upstream face of concrete dams and appurtenant structures.	RCC dams, CFRD dams.	Extension/compression in earth core and shells of embankment dams, or rockfill of CFRD dams.
Current practice or legacy instrument?[a]	Current	Current	Current	Current	Current
Installation specifics/ease of installation	Surface mounting, generally with grouted or mechanical anchors.	Surface mounting, generally with grouted or surface anchors.	Surface mounting, generally with grouted anchors.	Embedded in concrete between joints during construction.	Extensometer with length of 2 to 10 ft embedded during embankment construction.

General considerations	Requires operator dexterity for repeatable readings.	Concrete or steel structure on which crack meter is mounted can also display temperature-dependent tilt movements.	Concrete or steel structure on which crack meter is mounted can also display temperature-dependent tilt movements.	Concrete or steel structure on which crack meter is mounted can also display temperature-dependent tilt movements.	String of multiple extensometers can be installed in series.
General methods of reading output[b]	Mechanical reading; micrometer or depth gauge.	Portable readout, ADAS.	Portable readout, ADAS.	Portable readout, ADAS.	Portable readout, ADAS.
Sensor options	Removable spring-loaded potentiometer or LVDT.	VW displacement transducers, potentiometers, LVDT.	VW displacement transducers, potentiometers, LVDT.	VW displacement transducers, potentiometers, LVDT.	VW displacement transducers, potentiometers, LVDT.
Sensitivity	0.01 mm (0.0004 in.) for dial gauge.	0.01–0.04 mm (0.0004–0.0016 in.) typical for VW, depending on measuring range.	0.01–0.04 mm (0.0004–0.0016 in.) typical for VW, depending on measuring range.	0.01–0.04 mm (0.0004–0.0016 in.) typical for VW, depending on measuring range.	0.01–0.04 mm (0.0004–0.0016 in.) typical for VW, depending on measuring range.
Accuracy	0.02–0.05 mm (0.0008–0.002 in.).	0.01–0.1 mm (0.0004–0.004 in.) (±0.2% F.S. for VW typical).	0.01–0.1 mm (0.0004–0.004 in.) (±0.2% F.S. for VW typical).	0.01–0.1 mm (0.0004–0.004 in.) (±0.2% F.S. for VW typical).	0.01–0.1 mm (0.0004–0.004 in.) (±0.2% F.S. for VW typical).
Repeatability	0.02–0.05 mm (0.001–0.002 in.).	0.01–0.1 mm (0.0004–0.004 in.) depending on transducer type and measuring range.	0.01–0.1 mm (0.0004–0.004 in.) depending on transducer type and measuring range.	0.01–0.1 mm (0.0004–0.004 in.) depending on transducer type and measuring range.	0.01–0.1 mm (0.0004–0.004 in.) depending on transducer type and measuring range.

Table 5-3. (*Continued*) Instrument Comparison.

	Instrument Name				
	Manual crack meters	Electric crack meters	Submersible crack meters	Embedment joint meters	Embedment soil strain meter or soil extensometer
Accuracy/ Repeatability[c]	Ensure repeatable positioning of micrometer. Use reference measuring bar.	VW displacement transducers have temperature correction factor.	VW displacement transducers have temperature correction factor.	VW displacement transducers have temperature correction factor.	VW displacement transducers have temperature correction factor.
Longevity[d]	YYY	YYYY	YYYYY	YYYY	YYYY
Maintenance requirements	Periodic instrument and anchor points inspection.	Protect from any contact or shock. Periodic inspection.	Protect from any contact or shock.	Verify surface cable integrity.	Verify surface cable integrity.
Relative expense[e]	Instrument: $ Readout: $$$ ADAS: N/A	Instrument: $$ Readout: $$$ ADAS: $$$ to $$$$	Instrument: $$ Readout: $$$ ADAS: $$$ to $$$$	Instrument: $$ Readout: $$$ ADAS: $$$ to $$$$	Instrument: $$ Readout: $$$ ADAS: $$$ to $$$$

[a]"Current practice or legacy" refers to common practice in current U.S. applications. However, legacy instruments may still be in use and available and may have some advantages in particular applications.

[b]General methods of reading output: can include manual, on-demand, and output that can accommodate ADAS (Automated Data Acquisition Systems).

[c]Accuracy/Repeatability considerations: include thermal exposure and long-term stability.

[d]Longevity: Y, 1 year or less; YY, 1–5 years; YYY, 5–10 years; YYYY, 10–15 years; and YYYYY, more than 15 years.

[e]Relative expense: $, $1–$199; $$, $200–$999; $$$, $1,000–$2,999; $$$$, $3,000–$9,999; and $$$$$, more than $10,000.

			Instrument Name			
	Overflow technique	Liquid pressure technique	Hydraulic settlement profilers - standard types	Hydraulic settlement profiler - double fluid type	Liquid-level settlement system	Precision settlement system - water-level type
Differential Settlement Gauges and Liquid-Level Gauges, under Section 5.3.7.5 (Settlement Gauges)						
Parameter measured	Settlement of earth core and shoulders, settlement of concrete face.	Settlement of earth core and shoulders, settlement of concrete face.	Continuous settlement profile.	Continuous settlement profile.	Precision settlement.	Precision settlement.
Common applications	Embankment dams, CFRD dams.	Embankment dams, CFRD dams.	Small embankments.	Embankment dams.	Tunnels, powerhouse.	Tunnels, powerhouse.
Current practice or legacy instrument?[a]	Current	Current	Current	Legacy	Current	Current
Installation specifics/ease of installation	Overflow cells embedded during construction, connected by hydraulic tubing to a graduated vertical-level indicator glass tube in downstream station. Vent and drain tubing also required.	Settlement plates incorporating a pressure transducer embedded during construction, connected by twin hydraulic tubing to a reservoir in downstream station.	Torpedo-mounted pressure transducer connected to fluid-filled tube embedded during construction.	Loop of tubing filled with de-aired water embedded during construction.	Fluid-filled chambers connected along horizontal 2–3 in. diameter tubing.	Fluid-filled chambers connected laterally along horizontal 4–8 in. diameter tubing.

Table 5-3. (Continued) Instrument Comparison.

	Instrument Name					
	Overflow technique	Liquid pressure technique	Hydraulic settlement profilers - standard types	Hydraulic settlement profiler - double fluid type	Liquid-level settlement system	Precision settlement system-water-level type
General considerations	Smooth slope (1–2% approx.) of tubing downward to cells is required. De-aired or filtered water must be used.	Smooth slope of tubing downward to cells is required. De-aired water or a water–glycol solution must be used. Subsequent de-airing is needed periodically to remove air bubbles. Diameter of tubing must be larger to remove bubbles.	Requires operator dexterity for repeatable readings.	Requires mercury.	Horizontal tubing must be reasonably level to avoid air bubble traps.	Horizontal tubing must be level within tight tolerance (less than half the diameter of the tubing).
General methods of reading output[b]	Manual reading on vertical scale or low-range pressure transducer at bottom of indicator tube.	Change of fluid head read on low-range pressure transducer.	Pressure transducer pulled along tube and pressure read at intervals.	Mercury pumped from one end. Pumping pressure read continuously to provide profile.	Displacement transducer connected to a float in chamber.	VW force transducer connected to a float positioned laterally to the horizontal tubing.
Sensor options	Manual (graduated vertical scale) and/or VW or 4–20 mA pressure transducer.	VW or 4–20 mA pressure transducer.	VW or 4–20 mA pressure transducer.	VW or 4–20 mA pressure transducer.	VW or 4–20 mA pressure transducer.	VW or 4–20 mA pressure transducer.
Sensitivity	1 mm (0.04 in.) or 0.25 mm (0.01 in.) typical if electric transducer.	0.25–1 mm (0.01–0.04 in.) depending on measuring range of transducer.	0.25–1 mm (0.01–0.04 in.) depending on measuring range of transducer.	10 mm (0.4 in.) typical, depending on measuring range of transducer.	0.1 mm (0.004 in.) typical.	0.025 mm (0.001 in.) typical.

Accuracy	1–2 mm (0.04–0.08 in.) for manual and 0.5 mm (0.02 in.) typical for electric.	0.5–2 mm (0.02–0.08 in.) typical.	0.5–2 mm (0.02–0.08 in.) typical.	10 mm (0.4 in.) typical.	0.25–1 mm (0.01–0.04 in.) typical.	0.025 mm (0.001 in.) typical.
Repeatability	1–2 mm (0.04–0.08 in.) for manual and 0.5 mm (0.02 in.) typical for electric.	0.5–2 mm (0.02–0.08 in.) typical.	0.5–2 mm (0.02–0.08 in.) typical.	10 mm (0.4 in.) typical.	0.25–1 mm (0.01–0.04 in.) typical.	0.025 mm (0.001 in.) typical.
Accuracy/ Repeatability[c]	Air bubbles can develop in tubing and will cause considerable reading scatter if not purged.	Air bubbles can develop in tubing and will cause considerable reading scatter if not purged. One reference cell can be used to correct for barometric pressure changes (if unvented transducer) and for fluid density changes with temperature.	Ensure repeatable positioning of the pressure transducer at each interval.	Air bubbles can develop in tubing and will cause considerable reading scatter if not purged.	One reference chamber mounted on a stable location can be used to improve accuracy.	One reference chamber mounted on a stable location can be used to improve accuracy.
Longevity[d]	YYYYY	YYYYY	YYY	YYYYY	YYYYY	YYYYY
Maintenance requirements	Periodic overall flushing.	Periodic flushing or de-airing of fluid with suitable apparatus.	Periodic readout inspection.	Periodic flushing with suitable apparatus.	Periodic flushing and inspection.	Periodic flushing and inspection.
Relative expense[e]	Instrument: $$ Readout: $$$ ADAS: $$$$	Instrument: $$ Readout: $$$ ADAS: $$$ to $$$$	Instrument: $$$ Readout: $$$$ ADAS: N/A	Instrument: $$ Readout: $$$ ADAS: $$$$	Instrument: $$$ Readout: $$$ ADAS: $$$$	Instrument: $$$ Readout: $$$ ADAS: $$$$

a"Current practice or legacy" refers to common practice in current U.S. applications. However, legacy instruments may still be in use and available and may have some advantages in particular applications.

bGeneral methods of reading output: can include manual, on-demand, and output that can accommodate ADAS (Automated Data Acquisition Systems).

cAccuracy/Repeatability considerations: include thermal exposure and long-term stability.

dLongevity: Y, 1 year or less; YY, 1–5 years; YYY, 5–10 years; YYYY, 10–15 years; and YYYYY, more than 15 years.

eRelative expense: $, $1–$199; $$, $200–$999; $$$, $1000–$2,999; $$$$, $3,000–$9,999; and $$$$$, more than $10,000.

Table 5-3. (*Continued*) Instrument Comparison.

	Instrument Name	
	Open standpipe piezometers	Observation well
GROUNDWATER LEVEL AND PORE WATER PRESSURES		
Open Standpipe Piezometers (Casagrande Piezometers) and Observation Wells, Section 5.3.1.3		
Parameter measured	Pore water pressure and water level in embankment dams.	Water level in embankment dams.
Common applications	Pore water pressure or water level in downstream filter, chimney and blanket drain, downstream shell.	Water level in downstream shell or foundation.
Current practice or legacy instrument?[a]	Current	Current
Installation specifics/ease of installation	Vertical PVC or steel pipe installed during construction or in borehole, with filter tip at bottom end.	Vertical PVC or steel pipe installed during construction with slotted section at bottom end or along full length.
General considerations	Filter tip can become clogged with time. Needs to be cleaned or purged if instrument becomes unresponsive to water column.	Filter tip can become clogged with time. Needs to be cleaned or purged if instrument becomes unresponsive to water column.
General methods of reading output[b]	Manual: water-level indicator.	Manual: water-level indicator.
Sensor options	VW or strain gauge pressure transducer hanging from collar of standpipe.	VW or strain gauge pressure transducer hanging from collar of standpipe.

Sensitivity	1 mm or 0.01 ft tape markings.	1 mm or 0.01 ft tape markings.
Accuracy	2 mm (0.08 in.) typical with water-level indicator.	2 mm (0.08 in.) typical with water-level indicator.
Repeatability	2 mm (0.08 in.) typical with water-level indicator.	2 mm (0.08 in.) typical with water-level indicator.
Accuracy/Repeatability[c]	Water drops on wall of deep standpipes can cause erroneous readings.	Water drops on wall of deep standpipes can cause erroneous readings.
Longevity[d]	YYYY to YYYYY	YYYY to YYYYY
Maintenance requirements	Periodic response testing by adding water and confirming return to previous level. Flushing may be required.	Periodic response testing by adding water and confirming return to previous level. Flushing may be required.
Relative expense[e]	Instrument: $$ Readout: $$ ADAS: N/A	Instrument: $$ Readout: $$ ADAS: N/A

[a]"Current practice or legacy" refers to common practice in current U.S. applications. However, legacy instruments may still be in use and available and may have some advantages in particular applications.
[b]General methods of reading output: can include manual, on-demand, and output that can accommodate ADAS (Automated Data Acquisition Systems).
[c]Accuracy/Repeatability considerations: include thermal exposure and long-term stability.
[d]Longevity: Y, 1 year or less; YY, 1–5 years; YYY, 5–10 years; YYYY, 10–15 years; and YYYYY, more than 15 years.
[e]Relative expense: $, $1–$199; $$, $200–$999; $$$, $1000–$2,999; $$$$, $3,000–$9,999; and $$$$$, more than $10,000.

Table 5-3. (Continued) Instrument Comparison.

	Instrument Name					
	Pneumatic and Hydraulic Piezometers, Section 5.3.1.4		Resistance strain gauge piezometers, Section 5.3.1.5	Vibrating-wire piezometers, Section 5.3.1.6	Fiber optic piezometers, Section 5.3.1.7	Quartz pressure sensors, Section 5.3.1.8
	Pneumatic piezometers	Twin-tube hydraulic piezometers				
Parameter measured	Pore water pressure and water level in embankment dams.	Pore water pressure and water level in embankment dams.	Pore water pressure and water level in embankment dams.	Pore water pressure and water level in embankment dams.	Pore water pressure and water level in embankment dams.	High-accuracy water pressure.
Common applications	Pore water pressure or water level in downstream filter, chimney and blanket drain, downstream shell.	Pore water pressure or water level in downstream filter, chimney and blanket drain, downstream shell.	Pore water pressure or water level in downstream filter, chimney and blanket drain, downstream shell.	Pore water pressure or water level in earth core, down-stream filter and blanket drain, downstream shell, at foundation contact, in foundation behind grout curtain.	Pore water pressure or water level in earth core, downstream filter and blanket drain, down-stream shell, at foundation contact, in foundation behind grout curtain.	Specialized applications (upper and lower reservoir level, seepage weirs, others. . . .). Fairly uncommon for dams.
Current practice or legacy instrument?[a]	Legacy / limited current	Legacy	Current	Current	Current	Current
Installation specifics/ease of installation	Direct embedment in zone of interest or in open standpipe piezometer.	Direct embedment in zone of interest.	Lowered in open standpipe piezometer or direct embedment in zone of interest (less common).	Direct embedment in zone of interest or lowered in open standpipe piezometer. Can be installed in boreholes (rock or soil) by the fully-grouted method.	Direct embedment in zone of interest or lowered in open standpipe piezometer. Can be installed in boreholes (rock or soil) by the fully-grouted method.	N/A

General considerations	Twin tubing for nitrogen circulation can be subject to kinking or crushing. Must be suitably protected.	De-aired water or a water–glycol solution required. Subsequent de-airing often needed to remove air bubbles.	More sensitive to lightning and transient damage than VW piezometers.	May require lightning and transient damage protection if surface cable runs extend more than approx. 150–300 ft. Signal cable is 4-conductors plus shield. Heavy-duty cable recommended if permanently embedded in earth dams subject to differential movements.	Immune to lightning and transient damage. Cable splicing in case of damage requires specialized equipment.	N/A
General methods of reading output[b]	Portable readout. Automation is possible but requires elaborate hardware and electronics.	Manual readout (Bourdon gauge). Automation is possible with electric pressure transducers.	Manual readout and ADAS.	Manual readout and ADAS.	Manual readout and ADAS.	Manual readout and ADAS.
Sensor options	Portable readout can be designed for low or high water pressure.	VW or 4-20 mA pressure transducer.	Analog (4-20 mA or 0-5 VDC) or digital (RS485) outputs.	Frequency output (typically between 2,000 and 3500 Hz). Digital is possible but uncommon.	Fabry–Perot or Bragg grating interferometry.	Frequency or digital (RS232 and RS485) outputs.
Sensitivity	0.1% F.S. typical	10 mm (0.4 in.) typical, depending on measuring range of transducer.	0.025% F.S. typical.	0.025% F.S. typical.	0.025% F.S. typical.	0.000001% F.S. typical.
Accuracy	±0.25% F.S. typical.	10 mm (0.4 in.) typical.	±0.1% F.S. typical.	±0.1% F.S. typical.	±0.1% F.S. typical.	±0.01% F.S. typical.
Repeatability	0.1% F.S. typical.	10 mm (0.4 in.) typical.	0.025% F.S. typical.	0.025% F.S. typical.	0.025% F.S. typical.	±0.01% F.S. typical.

Table 5-3. (Continued) Instrument Comparison.

	Instrument Name					
	Pneumatic piezometers	Twin-tube hydraulic piezometers	Resistance strain gauge piezometers, Section 5.3.1.5	Vibrating-wire piezometers, Section 5.3.1.6	Fiber optic piezometers, Section 5.3.1.7	Quartz pressure sensors, Section 5.3.1.8
Accuracy/ Repeatability[c]	Very long tubing can reduce accuracy of readings.	Air bubbles can develop in tubing and will cause considerable reading scatter if not purged.	Long lead cables can degrade the signal (for 0–5 VDC especially) and increase risk of lightning damage.	Some VW piezo-meters have exhibited zero drift in the past. Issue is now controlled by manufacturers but should be spot-checked if piezo-meters will be permanently embedded.	Zero drift should be spot-checked if piezometers will be permanently embedded.	Digital version is fully characterized and exhibits better accuracy and thermal properties.
Longevity[d]	YYYY to YYYYY	YYYY to YYYYY	YYY to YYYY	YYYYY	Unknown	YYY to YYYY
Maintenance requirements	Verify surface tubing integrity.	Verify surface tubing integrity. Calibrate Bourdon gauge.	Verify surface tubing integrity. If possible, periodic (5 year?) transducer recalibration.	Verify surface tubing integrity. If possible, periodic (5 year?) transducer recalibration.	Verify surface tubing integrity. If possible, periodic (5 year?) transducer recalibration.	Verify surface tubing integrity. If possible, periodic (5 year?) transducer recalibration.
Relative expense[e]	Instrument: $ to $$ Readout: $$$$ ADAS: N/A	Instrument: $$ Readout: $$ or $$$ (if electric) ADAS: N/A or $$$$ (if electric)	Instrument: $$ Readout: $$$ ADAS: $$$ to $$$$	Instrument: $$ Readout: $$$ ADAS: $$$ to $$$$	Instrument: $$$ Readout: $$$$ ADAS: $$$$ to $$$$$	Instrument: $$$ Readout: $$$ ADAS: $$$ to $$$$

[a]"Current practice or legacy" refers to common practice in current U.S. applications. However, legacy instruments may still be in use and available and may have some advantages in particular applications.

[b]General methods of reading output: can include manual, on-demand, and output that can accommodate ADAS (Automated Data Acquisition Systems).

[c]Accuracy/Repeatability considerations: include thermal exposure and long-term stability.

[d]Longevity: Y, 1 year or less; YY, 1–5 years; YYY, 5–10 years; YYYY, 10–15 years; and YYYYY, more than 15 years.

[e]Relative expense: $, $1–$199; $$, $200–$999; $$$, $1,000–$2,999; $$$$, $3,000–$9,999; more than $10,000.

	Instrument Name		
	Fully grouted piezometers	The Westbay System	The Waterloo System

Multiple Piezometers in a Single Borehole, under Section 5.3.7.5 (Settlement Gauges)

	Fully grouted piezometers	The Westbay System	The Waterloo System
Parameter measured	Multilevel pore water pressure in soils and interstitial pressure in rocks.	Multilevel pore water pressure in soils and interstitial pressure in rocks.	Multilevel pore water pressure in soils and interstitial pressure in rocks.
Common applications	Multilevel water pressure in embankment dams and dam foundations (rock or soil).	Multilevel water pressure in embankment dams and dam foundations (rock or soil).	Multilevel water pressure in embankment dams and dam foundations (rock or soil).
Current practice or legacy instrument?[a]	Current	Current	Current
Installation specifics/ease of installation	Grouted with bentonite–cement mix in a borehole. Grout mix proportions of cement, bentonite, and water must be adjusted according to surrounding soil or rock type.	Casing couplers incorporating valve measurement port and inflatable packers permanently installed in a borehole.	Series of special double packers with sample entry ports in-between, connected by watertight casing.
General considerations	Individual piezometers can be grouted in, or strings of piezometers pre-assembled on a multiconductor cable can be grouted at once.	Specialized installation procedure. System allows other purposes: fluid sampling or pumping, chemical sensing.	Specialized installation procedure. System can be customized: open tube or dedicated pressure measurement, sampling, purging.
General methods of reading output[b]	Portable readout and ADAS.	Water pressure measured by probe lowered in casing.	Water pressure measured by pressure transducers in sample entry ports. Can be connected to ADAS.
Sensor options	Generally, VW piezometers or other types with low displacement diaphragms.	4–20 mA pressure transducer.	4–20 mA pressure transducer.

Table 5-3. (*Continued*) Instrument Comparison.

| | Instrument Name | | |
	Fully grouted piezometers	The Westbay System	The Waterloo System
Sensitivity	Depending on piezometer type used.	0.025% F.S. typical.	0.025% F.S. typical.
Accuracy	Depending on piezometer type used.	±0.1% F.S. typical.	±0.1% F.S. typical.
Repeatability	Depending on piezometer type used.	0.025% F.S. typical.	0.025% F.S. typical.
Accuracy/ Repeatability[c]	Permeability of grout mix must be similar, within a tolerance range of 2–3 orders of magnitude, to the permeability of surrounding rock or soil.	Proper installation and sealing with packers is key to multilevel measurements.	Proper installation and sealing with packers is key to multilevel measurements.
Longevity[d]	YYYYY	YYYY to YYYYY	YYYY to YYYYY
Maintenance requirements	Verify surface cable integrity.	Periodic inspection.	Periodic inspection.
Relative expense[e]	Instrument: $$ Readout: $$$ ADAS: $$$ to $$$$	Instrument: $$$ Readout: $$$ ADAS: $$$$	Instrument: $$$ Readout: $$$ ADAS: $$$$

[a]"Current practice or legacy" refers to common practice in current U.S. applications. However, legacy instruments may still be in use and available and may have some advantages in particular applications.

[b]General methods of reading output: can include manual, on-demand, and output that can accommodate ADAS (Automated Data Acquisition Systems).

[c]Accuracy/Repeatability considerations: include thermal exposure and long-term stability.

[d]Longevity: Y, 1 year or less; YY, 1–5 years; YYY, 5–10 years; YYYY, 10–15 years; and YYYYY, more than 15 years.

[e]Relative expense: $, $1–$199; $$, $200–$999; $$$, $1,000–$2,999; $$$$, $3,000–$9,999; and $$$$$, more than $10,000.

SEEPAGE, FLOW, AND TURBIDITY, under Section 5.3.2.1 (Flow Measurement)

	Instrument Name		
	Calibrated containers and stopwatch	Parshall flumes	V-Notch weirs
Parameter measured	Water flow rate (gal./min).	Water flow rate (gal./min).	Water flow rate (gal./min).
Common applications	Downstream seepage flow.	Downstream seepage flow.	Downstream seepage flow.
Current practice or legacy instrument?[a]	Current	Current	Current
Installation specifics/ease of installation	Must be sized according to expected flow rate. Requires channeling seepage waters to measuring point.	Must be sized according to expected flow rate. Requires channeling seepage waters to measuring point.	Must be sized according to expected flow rate. Requires channeling seepage waters to measuring point.
General considerations	Manual method. No continuous data.	Can accommodate variable seepage flows and high solids content?	More suitable for lower flow rates. Requires periodic maintenance to ensure proper flow tends to trap sediment.
General methods of reading output[b]	Time to fill calibrated container is converted to flow rate.	Many options: staff gauge, pressure sensor, VW weir monitor, ultrasonic level transducer (see following).	Many options: staff gauge, pressure sensor, VW weir monitor, ultrasonic level transducer (see following).
Sensor options	N/A	Vented strain gauge or vented VW level pressure transducer, 1 or 2 m (3.28 to 6.56 ft) range typical.	Vented strain gauge or vented VW level pressure transducer, 1 or 2 m (3.28 to 6.56 ft) range typical.
Sensitivity	N/A	Depending on level measurement sensor used (see following).	Depending on level measurement sensor used (see following).

Table 5-3. (*Continued*) Instrument Comparison.

	Instrument Name		
	Calibrated containers and stopwatch	Parshall flumes	V-Notch weirs
Accuracy	0.1 gal./min, depending on container size.	Depending on level measurement sensor used (see following).	Depending on level measurement sensor used (see following).
Repeatability	0.1 gal./min, depending on container size.	Depending on level measurement sensor used (see following).	Depending on level measurement sensor used (see following).
Accuracy/Repeatability[c]	Operator dependent.	Water level in flume converts to flow rate using conversion equation. System is less accurate at lower end of flow capacity.	Water level in flume converts to flow rate using conversion equation. System is less accurate at lower end of flow capacity.
Longevity[d]	YY		YYYYY
Maintenance requirements	No maintenance. Periodic replacement.	Periodic cleaning.	Periodic cleaning.
Relative expense[e]	Instrument: $ to $$ Readout: N/A ADAS: N/A	Instrument: $$$ Readout: N/A ADAS: N/A	Instrument: $$$ Readout: N/A ADAS: N/A

[a] "Current practice or legacy" refers to common practice in current U.S. applications. However, legacy instruments may still be in use and available and may have some advantages in particular applications.
[b] General methods of reading output: can include manual, on-demand, and output that can accommodate ADAS (Automated Data Acquisition Systems).
[c] Accuracy/Repeatability considerations: include thermal exposure and long-term stability.
[d] Longevity: Y, 1 year or less; YY, 1–5 years; YYY, 5–10 years; YYYY, 10–15 years; and YYYYY, more than 15 years.
[e] Relative expense: $, $1–$199; $$, $200–$999; $$$, $1,000–$2,999; $$$$, $3,000–$9,999; and $$$$$, more than $10,000.

Water-Level Measurement Sensors for Flumes and Weirs, under Section 5.3.2.2 (Water-Level Measurement)

	Instrument Name				
	Low-range pressure sensors	Vibrating-wire (VW) weir monitor	Ultrasonic and radar level sensors	Flowmeters (velocity meters)	Turbidity meters
Parameter measured	Water level in flumes and weirs.	Water level in flumes and weirs.	Water level in flumes and weirs.	Water flow rate (gal./min).	Water turbidity.
Common applications	Downstream seepage flow.	Downstream seepage flow.	Downstream seepage flow.	Downstream seepage flow.	Suspended solids in downstream seepage waters.
Current practice or legacy instrument?[a]	Current	Current	Current	Current	Current
Installation specifics/ease of installation	Measurement range must correspond to expected water height in flume or weir.	Measurement range must correspond to expected water height in flume or weir.	Measurement range must correspond to expected water height in flume or weir.	Various design and working principles: generally fitted to a pipe in which water is channeled. Can also be used in open channels but requires separate level measurement.	Submerged in water.
General considerations	Freezing in winter can damage sensor. Vented pressure transducer requires periodic replacement of desiccant cartridge.	Vented VW transducer requires periodic replacement of desiccant cartridge.	Mounted on bracket above water surface. Instrument may be susceptible to lightning or transient damage.	Pipe must be full of water for proper measurement.	Recent models need no maintenance or recalibration, others require periodic wiper replacement.
General methods of reading output[b]	Portable readout or ADAS.	Portable readout or ADAS.	Portable readout or ADAS.	Portable readout or ADAS.	Portable readout or ADAS.
Sensor options	Vented strain gauge or vented VW level pressure transducer, 1 or 2 m (3.28–6.56 ft) range typical.	Vented VW force transducer, 0.15–1.5 m range (0.5–5 ft) range typical.	4–20 mA, 0–10 VDC or digital output, 0.1–3 m (0.3–10 ft) typical.	4–20 mA transmitter, others. 1–1,000 gal./min min–max flow typical.	4–20 mA, 0–10 VDC or digital output, 0.1–400 or 1,000 NTU range typical.

Table 5-3. (*Continued*) Instrument Comparison.

	Instrument Name				
	Low-range pressure sensors	Vibrating-wire (VW) weir monitor	Ultrasonic and radar level sensors	Flowmeters (velocity meters)	Turbidity meters
Sensitivity	0.025% F.S. typical.	0.025% F.S. typical.	0.025% F.S. typical.	0.025% F.S. typical.	0.025% F.S. typical.
Accuracy	±0.1% F.S. typical.	±0.1% F.S. typical.	±0.1–0.2% F.S. typical.	±0.25% F.S. typical.	±1.00% to ±3.00% F.S. typical.
Repeatability	0.025% F.S. typical.	0.025% F.S. typical.	±0.025% F.S. typical.	0.025% F.S. typical.	0.025% F.S. typical.
Accuracy/ Repeatability[c]	Protect from external shocks, freezing and water/humidity ingress in vent tube.	Protect from external shocks, freezing and water/humidity ingress in vent tube.	Repeatability can be affected by water surface and environment.	Minimum–maximum flow range vary with pipe diameter (1–4 in. typical, or bigger).	Fouling of optical surface can compromise repeatability.
Longevity[d]	YYY to YYYY	YYYY	YYY to YYYY	YYYY	YYYY
Maintenance requirements	Periodic inspection and cleaning. Periodic (5 year?) transducer recalibration.	Periodic inspection and cleaning. Periodic (5 year?) transducer recalibration.	Periodic inspection and cleaning. Periodic (5 year?) transducer recalibration.	Periodic inspection and cleaning. Periodic (5 year?) transducer recalibration.	Periodic inspection and cleaning. Periodic (5 year?) transducer recalibration.
Relative expense[e]	Instrument: $$$ Readout: $$$ ADAS: $$$ to $$$$	Instrument: $$$ Readout: $$$ ADAS: $$$ to $$$$	Instrument: $$$ to $$$$ Readout: $$$ ADAS: $$$ to $$$$	Instrument: $$$$ Readout: $$$ ADAS: $$$ to $$$$	Instrument: $$$ Readout: $$$ ADAS: $$$ to $$$$

[a]"Current practice or legacy" refers to common practice in current U.S. applications. However, legacy instruments may still be in use and available and may have some advantages in particular applications.

[b]General methods of reading output: can include manual, on-demand, and output that can accommodate ADAS (Automated Data Acquisition Systems).

[c]Accuracy/Repeatability considerations: include thermal exposure and long-term stability.

[d]Longevity: Y, 1 year or less; YY, 1–5 years; YYY, 5–10 years; YYYY, 10–15 years; and YYYYY, more than 15 years.

[e]Relative expense: $, $1–$199; $$, $200–$999; $$$, $1,000–$2,999; $$$$, $3,000–$9,999; and $$$$$, more than $10,000.

STRESS AND STRAIN

Strain Measurement Instruments, Section 5.3.4

Parameter		Instrument Name				
	Mechanical strain gauges	Electrical strain gauges — Resistance (Carlson strain meters)	Vibrating-wire gauges	Sister bar	Strain gauge rosettes in concrete, Section 5.3.4.2	Dummy gauges in concrete, Section 5.3.4.3
Parameter measured	Strain	Strain	Strain	Strain	Strain	Strain
Common applications	Strain in concrete, steel.	Strain in concrete, steel.	Strain in concrete, steel.	Strain in concrete.	2D and 3D strain.	Correction for hydration effects.
Current practice or legacy instrument?[a]	Current	Current/legacy	Current	Current	Current	Current
Installation specifics/ease of installation	Reference studs bonded or grouted to concrete/steel surface, approx. 10 cm (4 in.) apart.	Embedded in concrete.	Embedded in concrete or surface-mounted on steel.	Embedded in concrete.	2–6 strain gauges mounted in different orientations on a bracket.	Strain gauge installed in container with padded sides.
General considerations	Requires operator dexterity for repeatable readings.	Length of gauge should be 3–4 times size the maximum size of concrete aggregate.	Length of gauge should be 3–4 times size the maximum size of concrete aggregate.	Installed parallel to larger diameter rebar, or used in replacement of VW embedment strain gauge.	Considerable care required during concrete casting.	Considerable care required during concrete casting.

Table 5-3. (*Continued*) Instrument Comparison.

		Instrument Name				
		Electrical strain gauges				
	Mechanical strain gauges	Resistance (Carlson strain meters)	Vibrating-wire gauges	Sister bar	Strain gauge rosettes in concrete, Section 5.3.4.2	Dummy gauges in concrete, Section 5.3.4.3
General methods of reading output[b]	Mechanical reading: high-resolution micrometer.	Carlson portable readout or ADAS.	VW portable readout or ADAS.	VW portable readout or ADAS.	VW portable readout or ADAS.	VW portable readout or ADAS.
Sensor options	Micrometer or removable spring-loaded potentiometer or LVDT.	Carlson	VW	VW	VW	VW
Sensitivity	0.001 mm (0.0004 in.), e.g., 10 micro-strains.	2–5 micro-strains depending on base length of strain gauge.	0.4 micro-strain.	0.4 micro-strain.	Same as VW strain gauges.	Same as VW strain gauges.
Accuracy	0.001 mm (0.0004 in.), e.g., 10 micro-strains.	±0.5% F.S. typical (±0.1% F.S. if individually calibrated).	±0.5% F.S. for VW typical (±0.1% F.S. if individually calibrated).	±0.25% F.S. typical.	Same as VW strain gauges.	Same as VW strain gauges.
Repeatability	0.001 mm (0.0004 in.), e.g., 10 micro-strains.	2–5 micro-strains depending on base length of strain gauge.	1 micro-strain.	1 micro-strain.	Same as VW strain gauges.	Same as VW strain gauges.

Accuracy/ Repeatability[c]	Ensure repeatable positioning of micrometer. Use reference measuring bar.	Carlson strain gauges are typically not calibrated individually. Full scale is 1,000–4,000 micro-strains depending on base length of strain gauge.	VW strain gauges are typically not calibrated individually. Full scale is typically 3,000 micro-strains.	Sister bars are typically calibrated individually. Full scale is typically 3,000 micro-strains.	Same as VW strain gauges.	Same as VW strain gauges.
Longevity[d]	YYYY	YYYY	Surface-mounted: YYYY Embedded: YYYYY	YYYY	YYYY	YYYY
Maintenance requirements	Periodic instrument and reference stud inspection.	Verify surface cable integrity.	Verify surface cable integrity, inspect gauge if surface mounted.	Verify surface cable integrity.	Verify surface cable integrity.	Verify surface cable integrity.
Relative expense[e]	Instrument: $ Readout: $$$ ADAS: N/A	Instrument: $$ Readout: $$$ ADAS: $$$ to $$$$	Instrument: $ Readout: $$$ ADAS: $$$ to $$$$	Instrument: $$ Readout: $$$ ADAS: $$$ to $$$$	Instrument: $$ Readout: $$$ ADAS: $$$ to $$$$	Instrument: $ Readout: $$$ ADAS: $$$ to $$$$

[a]"Current practice or legacy" refers to common practice in current U.S. applications. However, legacy instruments may still be in use and available and may have some advantages in particular applications.

[b]General methods of reading output: can include manual, on-demand, and output that can accommodate ADAS (Automated Data Acquisition Systems).

[c]Accuracy/Repeatability considerations: include thermal exposure and long-term stability.

[d]Longevity: Y, 1 year or less; YY, 1–5 years; YYY, 5–10 years; YYYY, 10–15 years; and YYYYY, more than 15 years.

[e]Relative expense: $, $1–$199; $$, $200–$999; $$$, $1,000–$2,999; $$$$, $3,000–$9,999; and $$$$$, more than $10,000.

Table 5-3. (*Continued*) Instrument Comparison.

		Instrument Name		
		Total pressure cells and concrete stress cells		Distributed strain measurement using optical fiber, Section 5.3.4.4
	Stress inclusions	Diaphragm	Hydraulic	
Stress Measurements, Section 5.3.5				
Parameter measured	Stress	Stress	Stress	Strain
Common applications	Stress in concrete and rock.	Total stress.	Total stress acting perpendicular to the cell.	Distributed strain in soil embankments and concrete.
Current practice or legacy instrument?[a]	Current	Legacy	Current	Current
Installation specifics/ease of installation	Wedged in small diameter diamond-drilled borehole (EX, BX, or NX) with installation tool.	Direct embedment in concrete.	Direct embedment in soil or concrete.	Fiber optical cable embedded in soil or concrete, or mounted on concrete surface.
General considerations	Proper installation is critical. Can be a delicate process.	Cell stiffness must be similar to concrete stiffness.	If in soil, suitable compaction around cell required so that cell does not become a rigid or soft inclusion.	Specialized splicing in case of cable break.
General methods of reading output[b]	VW portable readout or ADAS.	Portable readout.	VW portable readout or ADAS.	ADAS with specialized optical interface.

Sensor options	VW	Carlson or resistive strain gauge.	VW or 4-20 mA pressure transducer.	Brillouin optical scattering.
Sensitivity	2–10 psi depending on borehole diameter.	0.1% F.S. typical.	0.025% F.S. typical.	10 micro-strains typical, 3 ft spatial resolution typical.
Accuracy	20%, full-scale range of 10,000 psi.	±0.5% F.S. typical.	±0.1 to 0.5% F.S. typical.	20 micro-strains typical, depends on cable length and measurement time.
Repeatability	N/A (0.5–5% estimated).	±0.5% F.S. typical.	±0.025% F.S. typical.	10–20 micro-strains typical.
Accuracy/Repeatability[c]	Protect from vibrations. Installation in highly fractured rock less accurate.	Proper contact of cell with concrete must be ensured.	Uniform soil compaction is important so that cell does not respond as soft inclusion.	Accuracy can be improved by increasing measurement time.
Longevity[d]	YYY	YYYYY	YYYY	Unknown (YYYYY?)
Maintenance requirements	Verify surface cable integrity.	Verify surface cable integrity.	Verify surface cable integrity.	Verify surface cable integrity.
Relative expense[e]	Instrument: $$ Readout: $$$ ADAS: $$$ to $$$$	Instrument: $$ Readout: $$$ ADAS: $$$ to $$$$	Instrument: $$ Readout: $$$ ADAS: $$$ to $$$$	Instrument: $ (price of fiber optic cable only) Readout and ADAS: $$$$$

[a]"Current practice or legacy" refers to common practice in current U.S. applications. However, legacy instruments may still be in use and available and may have some advantages in particular applications.

[b]General methods of reading output: can include manual, on-demand, and output that can accommodate ADAS (Automated Data Acquisition Systems).

[c]Accuracy/Repeatability considerations: include thermal exposure and long-term stability.

[d]Longevity: Y, 1 year or less; YY, 1–5 years; YYY, 5–10 years; YYYY, 10–15 years; and YYYYY, more than 15 years.

[e]Relative expense: $, $1–$199; $$, $200–$999; $$$, $1,000–$2,999; $$$$, $3,000–$9,999; and $$$$$, more than $10,000.

Table 5-3. (Continued) Instrument Comparison.

	Instrument Name		
	Strain gauge load cells (electrical-resistance-type)	Vibrating-wire load cells	Hydraulic load cells
LOAD, SECTION 5.3.5.2			
Parameter measured	Load	Load	Load
Common applications	Load in ground anchors and rockbolts.	Load in ground anchors and rockbolts.	Load in ground anchors and rockbolts.
Current practice or legacy instrument?[a]	Current	Current	Current
Installation specifics/ease of installation	Thick bearing plates for uniform load spreading are required.	Thick bearing plates for uniform load spreading are required.	Thick bearing plates for uniform load spreading are required.
General considerations	Eccentric or uneven loading should be avoided.	Eccentric or uneven loading should be avoided.	Eccentric or uneven loading should be avoided.
General methods of reading output[b]	Portable readout or ADAS.	Portable readout or ADAS.	Bourdon gauge, portable readout or ADAS.
Sensor options	Resistive strain gauge.	VW	Bourdon gauge, VW or 4–20 mA pressure transducer.
Sensitivity	0.025% F.S. typical.	0.025% F.S. typical.	0.25 to 1% F.S. typical with Bourdon gauge.

Accuracy	±0.5% F.S. typical.	±0.5% F.S. typical.	±0.25 to 1% F.S. typical with Bourdon gauge.
Repeatability	±0.025% F.S. typical.	±0.025% F.S. typical.	±0.25 to 1% F.S. typical with Bourdon gauge.
Accuracy/Repeatability[c]	Accuracy can vary depending on thickness of end bearing plates and if load is nonaxial.	Accuracy can vary depending on thickness of end bearing plates and if load is nonaxial.	Accuracy can vary depending on thickness of end bearing plates and if load is nonaxial.
Longevity[d]	YYYY	YYYY	YYYY
Maintenance requirements	Verify surface cable integrity, inspect load cell.	Verify surface cable integrity, inspect load cell.	Verify surface cable integrity, inspect load cell.
Relative expense[e]	Instrument: $$$ Readout: $$$ ADAS: $$$ to $$$$	Instrument: $$$ Readout: $$$ ADAS: $$$ to $$$$	Instrument: $$$ Readout: $$$ ADAS: $$$ to $$$$

[a]"Current practice or legacy" refers to common practice in current U.S. applications. However, legacy instruments may still be in use and available and may have some advantages in particular applications.

[b]General methods of reading output: can include manual, on-demand, and output that can accommodate ADAS (Automated Data Acquisition Systems).

[c]Accuracy/Repeatability considerations: include thermal exposure and long-term stability.

[d]Longevity: Y, 1 year or less; YY, 1–5 years; YYY, 5–10 years; YYYY, 10–15 years; and YYYYY, more than 15 years.

[e]Relative expense: $, $1–$199; $$, $200–$999; $$$, $1,000–$2,999; $$$$, $3,000–$9,999; and $$$$$, more than $10,000.

Table 5-3. (*Continued*) Instrument Comparison.

	Instrument Name			
	Thermistors, Section 5.3.6.1	RTDs (resistance temperature devices), Section 5.3.6.2	Carlson instruments Section 5.3.6.3	Thermocouples Section 5.3.6.4
TEMPERATURE				
Parameter measured	Temperature	Temperature	Temperature	Temperature
Common applications	Leakage detection in embankments, concrete and dam foundations, concrete curing temperature during construction to avoid cracking, temperature-induced deflections in dams.	Leakage detection in embankments, concrete and dam foundations, concrete curing temperature during construction to avoid cracking, temperature-induced deflections in dams.	Leakage detection in embankments, concrete and dam foundations, concrete curing temperature during construction to avoid cracking, temperature-induced deflections in dams.	Leakage detection in embankments, concrete and dam foundations, concrete curing temperature during construction to avoid cracking, temperature-induced deflections in dams.
Current practice or legacy instrument?[a]	Current	Current	Current/legacy	Current
Installation specifics/ease of installation	Installed in boreholes or direct embedment in soil or concrete. All vibrating-wire instruments incorporate a thermistor.	Installed in boreholes or direct embedment in soil or concrete. All vibrating-wire instruments incorporate a thermistor.	Installed in boreholes or direct embedment in soil or concrete. All vibrating-wire instruments incorporate a thermistor.	Installed in boreholes or direct embedment in soil or concrete. All vibrating-wire instruments incorporate a thermistor.
General considerations	Single thermistor at end of 2-conductor cable, or string of thermistors, up to 48, on multiconductor cable if analog or up to 256 on 4-conductor cable if digital.	Single sensor at end of 2-conductor cable.	Single sensor at end of 2-conductor cable.	Single sensor at end of 2-conductor cable.

	Portable readout or ADAS.	Portable readout or ADAS.	Portable readout or ADAS.	Portable readout or ADAS.
General methods of reading output[b]				
Sensor options	NTC (negative temperature coefficient) 2,252; 3K; 5K; 10K Ω.	100–1K Ω.	Carlson resistance.	J, K, T, and E types.
Sensitivity	±0.16°F typical.	±0.16°F typical.	0.1°F typical.	0.3°F typical.
Accuracy	±0.16°F typical.	±0.16°F typical.	±0.5°F typical.	±0.3°F typical.
Repeatability	±0.16°F typical.	±0.16°F typical.	±0.1°F typical.	±0.3°F typical.
Accuracy/Repeatability[c]	Higher-resistance thermistor suitable for long lead length, but less accurate. 3K Ω is a good compromise. Limited to 300°F.	Not suitable for long lead cable. Can measure higher temperatures than thermistors.	Long lead cables can degrade the signal.	Suitable for long leads. Reference required. Wide temperature range.
Longevity[d]	YYYY to YYYYY (if embedded)	YYYY to YYYYY	YYYYY	YYY to YYYY
Maintenance requirements	Verify surface cable integrity.	Verify surface cable integrity.	Verify surface cable integrity.	Verify surface cable integrity.
Relative expense[e]	Instrument: $ Readout: $$$ ADAS: $$$ to $$$$	Instrument: $ Readout: $$$ ADAS: $$$ to $$$$	Instrument: $ Readout: $$$ ADAS: $$$ to $$$$	Instrument: $ Readout: $$ ADAS: $$$ to $$$$

[a]"Current practice or legacy" refers to common practice in current U.S. applications. However, legacy instruments may still be in use and available and may have some advantages in particular applications.

[b]General methods of reading output: can include manual, on-demand, and output that can accommodate ADAS (Automated Data Acquisition Systems).

[c]Accuracy/Repeatability considerations: include thermal exposure and long-term stability.

[d]Longevity: Y, 1 year or less; YY, 1–5 years; YYY, 5–10 years; YYYY, 10–15 years; and YYYYY, more than 15 years.

[e]Relative expense: $, $1–$199; $$, $200–$999; $$$, $1,000–$2,999; $$$$, $3,000–$9,999; and $$$$$, more than $10,000.

Table 5-3. (*Continued*) Instrument Comparison.

Other Devices, Section 5.3.6.5

	Instrument Name		
	Vibrating-wire devices	Integrated circuit (IC) temperature sensors	Distributed temperature measurement using optical fiber, Section 5.3.4.4
Parameter measured	Temperature	Temperature	Temperature
Common applications	Leakage detection in embankments, concrete and dam foundations, concrete curing temperature during construction to avoid cracking, temperature-induced deflections in dams.	Leakage detection in embankments, concrete and dam foundations, concrete curing temperature during construction to avoid cracking, temperature-induced deflections in dams.	Temperature in embankments for leakage detection.
Current practice or legacy instrument?[a]	Current	Current	Current
Installation specifics/ease of installation	Installed in boreholes or direct embedment in soil or concrete. All vibrating-wire instruments incorporate a thermistor.	Incorporated in certain types of instruments, e.g., tiltmeters.	Fiber optical cable embedded in soil or installed in observation wells.
General considerations	Single sensor at end of 2-conductor cable.	Single sensor at end of 2-conductor cable.	Specialized splicing in case of cable break.
General methods of reading output[b]	Portable readout or ADAS.	Portable readout or ADAS.	ADAS with specialized optical interface.
Sensor options	VW	Analog or digital output.	Raman optical scattering.

Sensitivity	0.05°F typical.	0.3°F typical.	0.16°F to 0.8°F typical; 3 ft spatial resolution typical.
Accuracy	±0.5% F.S. typical.	±0.5°F typical.	±0.16°F to 0.8°F, depends on cable length and measurement time.
Repeatability	±0.025% F.S. typical.	±0.5°F typical.	±0.16°F typical.
Accuracy/Repeatability[c]	Suitable for long lead.	Can be damaged by transient.	Internal readout averaging typically a few seconds to 1 h, considerably increases accuracy and spatial resolution.
Longevity[d]	YYYYY	YYY	Unknown (YYYY?)
Maintenance requirements	Verify surface cable integrity.	Verify surface cable integrity.	Verify surface cable integrity.
Relative expense[e]	Instrument: $$ Readout: $$$ ADAS: $$$ to $$$$	Instrument: $ Readout: $$ ADAS: $$$ to $$$$	Instrument: $ (price of fiber optic cable only) Readout and ADAS: $$$$$

[a] "Current practice or legacy" refers to common practice in current U.S. applications. However, legacy instruments may still be in use and available and may have some advantages in particular applications.

[b] General methods of reading output: can include manual, on-demand, and output that can accommodate ADAS (Automated Data Acquisition Systems).

[c] Accuracy/Repeatability considerations: include thermal exposure and long-term stability.

[d] Longevity: Y, 1 year or less; YY, 1–5 years; YYY, 5–10 years; YYYY, 10–15 years; and YYYYY, more than 15 years.

[e] Relative expense: $, $1–$199; $$, $200–$999; $$$, $1,000–$2,999; $$$$, $3,000–$9,999; and $$$$$, more than $10,000.

Table 5-3. (*Continued*) Instrument Comparison.

	Instrument Name
	Pan-type gauges, Section 5.3.9.2
PRECIPITATION	
Parameter measured	Rainfall/snowfall
Common applications	Measure daily evaporation/precipitation.
Current practice or legacy instrument?[a]	Current
Installation specifics/ease of installation	Pan must be re-filled to known level if used for measuring evaporation.
General considerations	Steel cylinder 47.5 in. diameter, 10 in. height.
General methods of reading output[b]	Visual, portable readout or ADAS.
Sensor options	Graduated vertical scale and/or low-range VW or 4–20 mA pressure transducer.
Sensitivity	0.025% F.S. typical.
Accuracy	±0.1% F.S. typical.
Repeatability	±0.025% F.S. typical.
Accuracy/ Repeatability[c]	Transducer must be vented for barometric correction.
Longevity[d]	YYYY
Maintenance requirements	Visual inspection, periodic (5 year?) transducer re-calibration.
Relative expense[e]	Instrument: $$$ Readout: $$$ ADAS: $$$ to $$$$

[a]"Current practice or legacy" refers to common practice in current U.S. applications. However, legacy instruments may still be in use and available and may have some advantages in particular applications.
[b]General methods of reading output: can include manual, on-demand, and output that can accommodate ADAS (Automated Data Acquisition Systems).
[c]Accuracy/Repeatability considerations: include thermal exposure and long-term stability.
[d]Longevity: Y, 1 year or less; YY, 1–5 years; YYY, 5–10 years; YYYY, 10–15 years; and YYYYY, more than 15 years.
[e]Relative expense: $, $1–$199; $$, $200–$999; $$$, $1,000–$2,999; $$$$, $3,000–$9,999; and $$$$$, more than $10,000.

Bucket Gauges, under Section 5.3.9.2 (Rain Gauges)

	Instrument Name		
	Weighing bucket gauges	Tipping bucket gauges	Manual precipitation gauges
Parameter measured	Rainfall/snowfall	Rainfall/snowfall	Rainfall/snowfall
Common applications	Measure daily precipitation.	Measure daily precipitation.	Measure daily precipitation.
Current practice or legacy instrument?[a]	Current	Current	Current
Installation specifics/ ease of installation	Preferably 5 ft above ground, antifreeze added to bucket if snow is collected. Bucket must be emptied periodically.	Requires heating in freezing zones, unlimited measuring capacity.	Must be emptied.
General considerations	12, 24, or 30 in. capacity bucket with 6–12 in. orifice.	12, 24, or 30 in. capacity bucket with 6–12 in. orifice.	Graduated plastic cylinder.
General methods of reading output[b]	Portable readout or ADAS.	Portable readout or ADAS.	Visual scale.
Sensor options	VW or strain gauge load cell.	Pulse signal.	No sensor option.
Sensitivity	0.025% F.S. typical.	0.01 in. typical.	0.01 in. typical.

Table 5-3. (*Continued*) Instrument Comparison.

	Instrument Name		
	Weighing bucket gauges	Tipping bucket gauges	Manual precipitation gauges
Accuracy	±0.1% F.S. typical.	±1% up to 2 in./h typical.	0.05 in. typical.
Repeatability	±0.004 in. typical.	±1% up to 2 in./h typical.	±0.05 in. typical.
Accuracy/Repeatability[c]	Good temperature stability. Windshield improves accuracy.	Level installation is required. Inlet can become contaminated or blocked.	Freezing and wind-induced errors.
Longevity[d]	YYYYY	YYYY	YY
Maintenance requirements	Visual inspection, remove dirt and particles from weighing bucket.	Visual inspection, remove dirt and particles from filter protecting the tipping bucket.	Visual inspection.
Relative expense[e]	Instrument: $$$$ Readout: $$$ ADAS: $$$ to $$$$	Instrument: $$ Readout: $$$ ADAS: $$$ to $$$$	Instrument: $ Readout: N/A ADAS: N/A

[a]"Current practice or legacy" refers to common practice in current U.S. applications. However, legacy instruments may still be in use and available and may have some advantages in particular applications.

[b]General methods of reading output: can include manual, on-demand, and output that can accommodate ADAS (Automated Data Acquisition Systems).

[c]Accuracy/Repeatability considerations: include thermal exposure and long-term stability.

[d]Longevity: Y, 1 year or less; YY, 1–5 years; YYY, 5–10 years; YYYY, 10–15 years; and YYYYY, more than 15 years.

[e]Relative expense: $, $1–$199; $$, $200–$999; $$$, $1,000–$2,999; $$$$, $3,000–$9,999; and $$$$$, more than $10,000.

RESERVOIR AND TAILWATER LEVELS, under Section 5.3.2.2 (Water Level Measurements)

	Instrument Name				
	Head (staff) gauges	Float systems	Bubbler systems	Submerged pressure transducer	Ultrasonic and radar level sensing
Parameter measured	Water level	Water level	Water level	Water level	Water level
Common applications	Reservoir level	Reservoir level	Reservoir level	Reservoir level	Reservoir level
Current practice or legacy instrument?[a]	Current	Current	Current	SDI-12	Current
Installation specifics/ease of installation	Graduated scale mounted vertically. Must be readable from distance.	Float attached to a chain in a stilling well, activating a rotary encoder.	Tubing lowered in pipe mounted to vertical or inclined wall.	Transducer suspended in pipe mounted to vertical wall.	Transducer suspended in pipe mounted to vertical wall.
General considerations	Generally, 2 1/2 in. wide. Must be made with durable material and indelible graduations and markings.	Float must be installed in a stilling well.	System comprises air compressor and pressure transducer. Submerged portion is maintenance-free.	Vented pressure transducer requires periodic replacement of desiccant cartridge.	Mounted on holding bracket above water surface. Instrument maybe susceptible to lightning or transient damage.
General methods of reading output[b]	Visual scale.	Stand-alone unit or ADAS.	ADAS	Portable readout or ADAS.	Portable readout or ADAS.
Sensor options	No sensor option.	Rotary encoder (absolute or incremental/relative).	Analog (4–20 mA or 0–5 VDC) or digital (RS485) outputs.	VW or analog (4–20 mA or 0–5 VDC) or digital (RS485) outputs.	4–20 mA, 0–10 VDC or digital output, 0.3–20, 50 or 100 m (1 to 65, 164 or 328 ft) typical.

Table 5-3. (*Continued*) Instrument Comparison.

	Head (staff) gauges	Float systems	Bubbler systems	Submerged pressure transducer	Ultrasonic and radar level sensing
			Instrument Name		
Sensitivity	0.01 or 0.1 ft.	0.01 ft typical.	0.05% reading with minimum of 0.01 ft typical.	0.025% F.S. typical.	0.025% F.S. typical.
Accuracy	0.01 or 0.1 ft.	0.01 ft typical.	0.05% reading with minimum of 0.01 ft typical.	±0.1% F.S. typical.	±0.1 to 0.2% F.S. typical.
Repeatability	±0.01 or 0.1 ft.	±0.01 ft typical.	±0.05% reading with minimum of 0.01 ft typical.	±0.025% F.S. typical.	±0.025% F.S. typical.
Accuracy/ Repeatability[c]	Graduated marks every 0.01 or 0.1 ft, depending on distance of observation.	Good temperature stability.	Desiccant must be replaced periodically.	Transducer can be absolute or vented for barometric correction.	Repeatability can be affected by water surface and environment.
Longevity[d]	YYYY	YYYY to YYYYY	YYYY	YYYY	YYYY to YYYYY
Maintenance requirements	Visual inspection.	Visual inspection, cleaning, functional check.	Visual inspection, periodic (5 year?) transducer recalibration.	Visual inspection, periodic (5 year?) transducer recalibration.	Visual inspection, periodic (5 year?) transducer recalibration.
Relative expense[e]	Instrument: $$ Readout: N/A ADAS: N/A	Instrument: $$$ Readout: $$$ ADAS: $$$ to $$$$	Instrument: $$$ Readout: $$$ ADAS: $$$ to $$$$	Instrument: $$ Readout: $$$ ADAS: $$$ to $$$$	Instrument: $$ Readout: $$$ ADAS: $$$ to $$$$

[a]"Current practice or legacy" refers to common practice in current U.S. applications. However, legacy instruments may still be in use and available and may have some advantages in particular applications.
[b]General methods of reading output: Can include manual, on-demand, and output that can accommodate ADAS (Automated Data Acquisition Systems).
[c]Accuracy/Repeatability considerations: Include thermal exposure, and long-term stability.
[d]Longevity: Y, 1 year or less; YY, 1–5 years; YYY, 5–10 years; YYYY, 10–15 years; and YYYYY, more than 15 years.
[e]Relative expense: $, $1–$199; $$, $200–$999; $$$, $1,000–$2,999; $$$$, $3,000–$9,999; and $$$$$, more than $10,000.

	Instrument Name
	Multiprobe water-quality instruments, Section 5.3.2.2
TURBIDITY/WATER QUALITY	
Parameter measured	Temperature, conductivity, dissolved oxygen, pH, oxidation–reduction potential, turbidity, and water level.
Common applications	Characterize seepage waters.
Current practice or legacy instrument?[a]	Current
Installation specifics/ease of installation	Submerged in water.
General considerations	Recent models need no maintenance or recalibration, except periodic wiper replacement for the turbidity sensor.
General methods of reading output[b]	Portable readout or ADAS.
Sensor options	RS-232, SDI-12, RS-485, USB.
Sensitivity	Sensor-dependent.
Accuracy	Sensor-dependent.
Repeatability	Sensor-dependent.
Accuracy/Repeatability[c]	Fouling (oil, sediments, biofilms) of measuring surfaces can compromise accuracy. Individual sensors have different maintenance requirements.
Longevity[d]	YYYY
Maintenance requirements	Visual inspection, periodic (yearly?) sensor recalibration.
Relative expense[e]	Instrument: $$$$ Readout: $$$ ADAS: $$$ to $$$$

[a]"Current practice or legacy" refers to common practice in current U.S. applications. However, legacy instruments may still be in use and available and may have some advantages in particular applications.

[b]General methods of reading output: Can include manual, on-demand, and output that can accommodate ADAS (Automated Data Acquisition Systems).

[c]Accuracy/Repeatability considerations: Include thermal exposure, and long-term stability.

[d]Longevity: Y, 1 year or less; YY, 1–5 years; YYY, 5–10 years; YYYY, 10–15 years; and YYYYY, more than 15 years.

[e]Relative expense: $, $1–$199; $$, $200–$999; $$$, $1,000–$2,999; $$$$, $3,000–$9,999; and $$$$$, more than $10,000.

Table 5-3. (*Continued*) Instrument Comparison.

	Instrument Name
	Modern high-resolution digital accelerographs

SEISMIC MEAUSUREMENTS

Strong-Motion Accelerographs (High-Resolution Digital Accelerographs), Section 5.3.8.4

Parameter measured	Seismic response.
Common applications	Measure movement, pressure, and load.
Current practice or legacy instrument?[a]	Current
Installation specifics/ease of installation	Units can be stand-alone (combined recorder and accelerometer) or multichannel recorder with separate triaxial accelerometers. Units can be mounted vertically or horizontally.
General considerations	Flat response DC to 200–600 Hz typical. Rugged, solid-state MEMS accelerometers not subject to aging.
General methods of reading output[b]	Single-point network control. Remote communication possible (modem, ethernet, etc.). Automatic alerting. Distributed recording.
Sensor options	Traditional spring and coil force-balance accelerometer, or MEMS capacitive accelerometer. Measuring range: ±4g range typical.
Sensitivity	1.25 V/g (differential) typical for MEMS sensors. Similar for force-balance sensors.
Accuracy	±0.1 to 1.0% F.S. typical.
Repeatability	Usually not specified.
Accuracy/Repeatability[c]	MEMS sensors operating range from –40°C to 85°C.
Longevity[d]	YYYYY Rugged, solid-state MEMS accelerometers not subject to aging.
Maintenance requirements	Low-maintenance comprehensive system self-test MEMS sensors do not need recalibration.
Relative expense[e]	Instrument and ADAS: $$$$$

[a]"Current practice or legacy" refers to common practice in current U.S. applications. However, legacy instruments may still be in use and available and may have some advantages in particular applications.
[b]General methods of reading output: Can include manual, on-demand, and output that can accommodate ADAS (Automated Data Acquisition Systems)
[c]Accuracy/Repeatability considerations: Include thermal exposure, and long-term stability.
[d]Longevity: Y, 1 year or less; YY, 1–5 years; YYY, 5–10 years; YYYY, 10–15 years; and YYYYY, more than 15 years.
[e]Relative expense: $, $1–$199; $$, $200–$999; $$$, $1,000–$2,999; $$$$, $3,000–$9,999; and $$$$$, more than $10,000.

CHAPTER 6

GEODETIC MONITORING

Measure what is measurable, and make measurable what is not so.

Galileo Galilei

Geodetic monitoring methods for dam deformation monitoring consist of ground-surveying instruments and procedures used to detect horizontal and vertical movement over a series of fixed points forming a survey control network. The fixed points may be monuments along a dam crest, appurtenant structures, or abutments.

Geodetic engineers and land surveyors are challenged with the task of providing horizontal and vertical position information to the dam owner and engineer as performance indicators of movement. In this chapter, both geodetic engineers and land surveyors, although professionally distinct, are referred to as "surveyors," and geodetic surveys are referred to as "surveys." In practice, the geodetic engineer bears the responsibility for program management, and the land surveyor makes the measurements.

Surveys to a high order of accuracy may be geo-referenced, where possible, to a geodetic control framework as part of good practice. Not all geodetic monitoring surveys are tied to such a framework, but the survey procedures are similar. What do geodesy and geodetic surveys add to a dam performance plan?

A geodetic survey for dam monitoring advises the owner and engineer about movement. It must be repeatable and accurately related within a reliable geodetic system. The type of survey will depend on location and prior surveys. Surveys may be based on a local system in areas where they cannot be geo-referenced. Such surveys have been widely accepted and are still in use today. Where a reliable geodetic system is available, best practice is to survey within that system.

There are special cases. Dams in seismically active areas are more likely to have regulations or standard practice for relating surveys to a geodetic system such as a state plane coordinate system. These systems are resurveyed occasionally, and adjustments are made to coordinates of control stations that affect the coordinates of surveys tied to these networks.

6.1 GENERAL

"Geodetic surveying is a survey in which account is taken of the figure and size of the earth. Geodetic surveys are usually prescribed where the areas or distances involved are so great that the results of desired accuracy and precision can be obtained only by the processes of geodetic surveying" (ACSM 1989).

"Geodesy, also named geodetics, is a branch of applied mathematics and earth sciences. It is the scientific discipline that deals with the measurement and representation of the Earth, including its gravitational field, in a three-dimensional time-varying space. Geodesists also study geodynamical phenomena such as crustal motion, tides, and polar motion. For this they design global and national control networks using space and terrestrial techniques while relying on datums and coordinate systems" (ACSM 1989).

Surveys are typically needed for large projects spanning distances in excess of 7 to 9 mi (11.2–14.5 km) that require accurate techniques and precise survey instrumentation owing to the curvature of the earth. Large infrastructure projects such as dams and reservoirs require surveys appropriate for their size and connection to a larger utility distribution or flood control system.

Control surveys establish geodetic coordinates on monuments that are then remeasured time and time again to monitor horizontal and vertical movement of the monuments. Not every monitoring survey is as stringent as a control survey; however, the techniques, equipment, instrumentation, and care of work should match the survey requirements. Surveys require careful field and office procedures for error adjustment.

Surveys are classified by orders of accuracy and are dependent on the type of project and reporting requirements. This can best be answered by asking what is being monitored and the realistic tolerances. Dams will have different tolerances and expected accuracies compared to a railroad condition survey.

In some states, surveys are required by law to be referenced to a local/ regional/state geodetic system. These surveys are tied (by measurement) to the geodetic system such as a state plane coordinate system or a national geodetic framework. Fully understanding the difference in datums and coordinate systems is necessary. Surveys have specific guidelines for selection of equipment, execution, and type and amount of measurements to properly define and classify a survey.

Initial surveys tied to a geodetic network provide the framework for all future surveys. Each subsequent survey reports movements compared to the previous and initial surveys.

Repeatability is the cornerstone of dam surveys. Repeatability can only be achieved if the previous surveys are reliable and properly executed. A repeatable survey is one that uses the same measurement techniques and procedures for each survey in order to determine and report trends in horizontal and vertical movement of a dam or structure. Of all the errors in measurement that can be made, the easiest type of error to remove from a survey is the systematic error. Repeatability is the key factor in eliminating systematic errors.

In dam-monitoring surveys, a geodetic system improves the reliability of measurements. A local system not tied to a geodetic system may work well; however, it could limit the surveyor with newer equipment and technology. Regions of frequent earthquakes or subsidence require special care so one can analyze the positional location and accuracies of the network in which the dam is located.

6.2 SURVEY METHODS, OLD AND NEW

Historically, many conventional methods have been used to obtain survey measurements to monitor movement. Some of the most widely used methods are triangulation, trilateration, collimation, and other alignment techniques. Commonly used instruments for those methods, shown in Fig. 6-1, were transits, plane tables and alidades, and theodolites to measure horizontal and vertical angles, steel tapes to measure distance, and optical levels and graduated rods to measure elevations.

Older theodolites, which could only measure angles, are now largely being replaced by co-axial total stations (theodolites with built-in electronic distance measurement [EDM]) that can measure angles and distances at the same time and record the measurements in a digital file. Older levels that required manual readings off a graduated rod are now being replaced by electronic levels with digital readings off bar-coded rods that greatly reduce the errors in readings.

Regardless of the advances in surveying technology in the recent years, the historic methods must be understood by the surveyor so that consistency is maintained and data from previous surveys is not lost over the life span of the monitoring project. A surveyor must be able to retrace the previous surveys in order to understand the characteristics of a specific

Plane Table and Alidade   Transit

Fig. 6-1. Early instruments

measurement and the errors and reasonable accuracies to expect when comparing to current surveys. An example of this would be a modern surveyor using GPS to compare a measurement that was made in the initial survey using a steel tape. It must be understood that this is not a fair assessment of a distance because taping was a more common practice at the time of the original survey and likely more accurate than using GPS to measure the distance. The same could be said for using a total station to read the same distance that was taped. In most cases, the taped distance will be more accurate and preserved in such a way that any deviation from the original would not necessarily be mistaken as actual movement. The correct tool for the job must be used, and it must be fully understood that unrealistic movements may emerge when comparing new to historic surveys. Newer technology does not always equate to a more accurate measurement.

As software and instrumentation capabilities improve, it may be feasible and more efficient to replace older methods. Historic methods required more time and field crew members to accomplish.

An example is replacing a plumbline. At two concrete arch dams (Duffy 1999), retro-reflective prisms were installed on the face of each dam to measure the x, y, and z displacements on the face of the dam at various heights. This was done in lieu of plumbline measurements that had been taken historically but were decidedly hazardous for personnel to obtain. The new "plumbline" measurements are obtained at the same time as the horizontal and vertical measurements on the crest of the dam, thereby accomplishing both tasks simultaneously. The precision and accuracy of the dam face measurements improved with this method because the physical limitations of acquiring the plumbline readings often introduced error into the measurements.

Current methods of geodetic monitoring are commonly defined by the type of instrument used to complete the survey; however, the monitoring survey is usually a combination of multiple instrument types, such as total stations, levels, and GPS. Current monitoring projects have the benefit of great advances in technology. The advance in software and robotic instrumentation, combined with the internet and wireless communication, allows the surveyor the opportunity to automate most monitoring projects and remotely control the measurement scheme. It may be more costly up front, but automating a geodetic system greatly reduces labor costs of a field crew traveling to a project to conduct a monitoring survey. Automation not only reduces cost but also provides an expeditious and redundant measurement that can be conducted daily or continuously as needed. Compared to annual and biannual on-site surveys, an automated system does not sacrifice accuracy, but it pays for itself in a short period of time and is most useful in remote areas.

6.3 TOTAL STATIONS

A total station is the modern version of what previously was known as a transit or theodolite. This instrument is an electronic optical instrument used in surveying to obtain angle and distance measurements from the instrument to a specific point that typically has a prism attached to a rod or other device at the point that is being measured. Robotics technology is now the heart of the modern total station, and it allows the instrument person to operate the station by remote control from a distance away. This modernization has reduced the number of crew members needed to conduct typical survey work and allows one person alone to operate the instrument and the rod. Reflectorless technology is now available with limited use to measurement movement depending on surfaces being measured along with distance and angle of deflection. Reflectorless observations can reach locations on a dam face with limited accuracy and repeatability.

Automatic target recognition (ATR) is now a standard on most modern total stations. ATR is integrated into the optics of the total station and uses a charge-coupled device (CCD) video camera that when pointed near a target/prism determines the ATR offset and correctly points the instrument to the exact center of the target/prism. This allows for automated systems to be used because the operator can program predetermined angles for the instrument once programmed. Then, the instrument can turn to its target, and the correction will be made automatically by the ATR mode.

A typical application in dam monitoring allows a survey crew to measure vertical and horizontal angles and measure distances to reflectors or prisms at specific key locations on a dam. These measurements are collected in sets of angles and distances and measured multiple times in direct and reverse modes to reduce calibration and reading errors. The total station can be used to reach most locations on a dam to measure targets or prisms along the downstream face of a dam. Most targets are permanently mounted on the structure itself or can be placed on a monument to be measured. The total station can be used to check the alignment of a dam by sighting down straight rows of monuments or it can be used to provide coordinates of points on the dam and abutments.

Trigonometric leveling is done using a total station to derive elevations calculated from the vertical angles and slope distance measurements. The accuracy of the elevations will depend on how precisely the total station can measure vertical angles. This method of determining vertical positions can be very useful if applied correctly. In a situation where accessibility is restricted, such as when monitoring a heavily traveled road surface on a dam crest, monitoring points can be set along the edge of the road and measured from a reference point close to the site to keep personnel safe and allow measurements to be taken without disrupting traffic. In situations with large numbers of monitoring points that need to be measured frequently, it is useful and cost effective to combine horizontal and vertical measurements in one operation.

The advantages of using a total station is that it allows the surveyor to measure distances to remote locations that may be difficult to reach with conventional methods. The frequency and amount of measurements to be collected will be important in deciding whether to use a conventional total station or a robotic model. The robotic and automated capabilities of some total stations provide additional benefits when designing a monitoring system on a large or frequently monitored site.

The ability to measure both horizontal and vertical positions accurately in one field operation can be very cost effective. A project that might take a two-person crew 2 to 3 days to measure conventionally (using a total station for horizontal measurements and a level for vertical measurements) might only take one person an hour or two to measure using a robotic, semi-automated system. Depending on the frequency of the monitoring, the somewhat more expensive robotic system may pay for itself quickly compared to the labor costs occurred with other methods.

As mentioned above, the introduction of robotics to the total station allows a geodetic engineer/surveyor to set a total station at a permanent or semi-permanent location and the instrument can complete a measurement cycle of multiple points along the face of a dam. This is typically done with software applications either onboard the total station or with the attachment of a field computer/data collector. Raw measurements including horizontal and slope distances, vertical and zenith angles, and horizontal angles are collected and then calculated into a usable value using coordinate geometry in the software.

A typical reading output can be either the raw measurements or coordinates collected in the field. Further analysis is required to reduce errors and provide a true measurement and a coordinate of where the measurement was made. Modern total stations can be left in place and attached to an ADAS and can be operated and viewed in real time through software applications from remote locations using server applications and internet-based programs.

When designing a monitoring system, it is best to utilize equipment with digital data-collection systems. This eliminates substantial errors related to handwritten notes and provides a backup system for the data, possibly saving revisiting the site. Data-collection systems need to be reviewed for flexibility of methods of collection and output formats. A careful review of the process the data output goes through is needed to note whether the software performs automatic corrections to raw data that may not be obvious but unwanted.

Operation and maintenance will not be drastically different between various total station manufacturers. The success and reliability of a geodetic monitoring plan is having well-trained survey personnel who have high regard for their instrumentation and equipment and a clear understanding of what and why they are measuring. Most modern total stations are durable and weatherproof, but special care must be taken in operating, transporting, and storing such instruments. Regular calibration checks, along with proper care and cleaning, are done as often as possible or at the start of a monitoring survey. In the case of an ADAS, where the total station is left on a pillar or permanently mounted, weekly and monthly preventive maintenance is part of the geodetic monitoring plan.

Careful review of calibrations and errors in measurement cycles is part of the analysis for determining if an instrument is reading correctly. Equipment logs and a good routine for maintenance will promote the longevity of the total station. Site conditions and the maintenance plan will determine the longevity and the type of total station to be used. Total stations must be weatherproof and capable of operating in extreme temperature changes. The total station is an effective and efficient instrument if used correctly in dam monitoring.

6.3.1 Procurement

Consulting with established owners and case studies of different manufacturers is informative before purchasing a total station. Important considerations and questions are part of any informed procurement.

6.3.1.1 Precision.   Is EDM precision good enough to achieve the accuracy required for the distances to be measured? Is the angular precision good enough to measure both horizontal and vertical angles with sufficient accuracy so that horizontal and vertical displacements can be measured with one method of measurement (one visit per point)? Is the angular precision good enough for measuring the angles accurately enough so that few repetitions of measurements are required?

6.3.1.2 Frequency and Quantity of Measurements. Will it be used on a new site requiring monthly measurements or an existing site that is measured twice per year? How many monuments are there to measure? How many instrument setup positions will it require to measure all the monuments? How many work hours will it take to measure manually versus an automated or semi-automated method?

6.3.1.3 Software and Hardware Capabilities. Does the total station allow for electronic data collection? How flexible are the data-collection and data-output methods?

6.3.1.4 Site Conditions. Is the total station readily accessible or remote? In what type of climate is it located— will there be heat wave problems? Is it safe for personnel?

When designing a new monitoring system (or updating an old one), the choice of which total station to use is important. Newer models of total stations that are automated can be costly up front, but when cost is analyzed over the life of a monitoring project, a total station might be the best choice.

6.4 LEVELS

A level is an optical–electric instrument that is used to measure differences in height between two given points with respect to a reference elevation using the technique known as leveling. Leveling is typically done with a level set up over a tripod. The surveyor sights a staff or rod with a numbered scale or bar code to obtain a reading. There are several types of levels: spirit level, automatic level, and digital level. The two most commonly used in dam monitoring are the automatic level and the digital level.

Automatic levels make use of a compensator that ensures that the line of sight remains horizontal once the operator has roughly leveled the instrument (to within 0.05°). Three level screws are used to level the instrument. The surveyor sets the instrument up quickly and does not have to re-level the instrument after each sight on a rod at another point. The compensator also reduces the effect of minor settling of the tripod to the actual amount of motion instead of leveraging the tilt over the sight distance. Automatic levels have become standard in the last 25 years.

Digital levels electronically read a bar-coded scale on the staff. These instruments are the electronic version of the automatic level, but they include a data recording capability. This feature removes the requirement for the operator to read a scale and write down the value, which reduces errors. The measurements are made to bar-coded rods and horizontal distances are measured, along with difference in elevations. The digital level may also compute and apply refraction and curvature corrections. This has drastically improved the accuracy and reliability of more precise elevation measurements.

The vertical position (elevation) of a monitoring point can be measured using differential leveling methods or with a total station (trigonometric levels). The trigonometric leveling technique was previously discussed in Section 6.3 *Total Station*. A level is used estimate movement by measuring differences in elevation from a primary benchmark outside the area of influence. The elevations are recorded and later adjusted in the office and compared to previously measured elevations.

Typically, elevations from leveling are the most accurate geodetic measurements. Confidence can be placed on this method because of the accuracy specifications, procedure, and adjustment procedures used with a level. Elevations derived from leveling are more consistent and are typically within a few millimeters of the original elevation of a monument.

The advantages of using a level are greater accuracy as stated above. The elevations derived from total stations and GPS are not the same as a direct reading that is achieved from leveling. The cost per instrument is much less than that of a total station; however, this type of instrument requires at least two people to survey. The reliability of a level depends on the care, transportation, and storage of the instrument. Leveling has a very long track record of providing reliable measurements for dam monitoring. Technology gains have increased the measurement accuracy and data-collection ability. A disadvantage of digital levels is the time it takes to make a measurement.

A level used by a surveyor monitoring a dam needs to have several methods for importing and retrieving data. Multiple user formats should be available when selecting the right instrument, along with the ability to store measurements on an internal memory or type of data card reader. User-defined formats add flexibility to the leveling process and allow for efficient data exchange with field instruments to office computers.

Level operation and maintenance are much simpler. Simple calibration checks are made prior to each level run for a movement survey. It is helpful to include in the field notes and report the serial number of the level along with the collimation error adjusted for that particular level run.

6.4.1 Procurement

Consulting with established owners and case studies of different manufactures are informative before purchasing a level. Important considerations and questions are part of any informed procurement.

6.4.1.1 Precision. The precision of the level must be good enough to achieve the accuracy needed for the planned elevations to be measured. Will the instrument achieve the accuracy needed? Performance specifications are based on DIN 18723, which specifies standard deviation height measuring per one kilometer of double leveling.

The accuracy of differential leveling methods is usually expressed as a maximum loop closure statement. The National Geodetic Survey outlines the criteria to follow to achieve certain orders of accuracy when establishing new benchmarks (National Geodetic Survey, NGS 1994). First- and second-order criteria are used to generate elevations on the reference points that one would use to start and check in one's level circuit. Movement surveys on monitoring points are usually performed to the third-order criteria.

6.4.1.2 Frequency and Number of Measurements. The frequency and number of measurements depend on the answers to the following questions:

- Could the vertical position be measured at the same time that horizontal positions are measured (one visit per point)?
- Will it be used on a new site requiring monthly measurements or on an existing site that is measured less frequently?
- How many monuments are there to measure?
- What are the distances and terrain between monitoring points and reference points?
- How many work hours will it take to measure manually versus an automated or semi-automated method?

6.4.1.3 Software and Hardware Capabilities. Does the level allow for electronic data collection and processing? How flexible are the data-collection and data-output methods? Will the software allow for careful analysis and adjustment procedures required by the accuracy standards? When considering advanced computer systems, the software and firmware capabilities of the instrument require definition.

When designing a new monitoring system (or updating an old one), the choice of which level to use is important. Newer models of levels that are electronic are not a major cost compared to the overall value that can be achieved. Levels are not left in place as part of an ADAS, but they are a primary tool used by a survey crew for establishing elevations on projects and are recommended for use in movement monitoring.

6.5 GLOBAL POSITIONING

GPS, based on a satellite navigation system, is now widely used for dam monitoring. The system returns geo-referenced positions in any weather conditions at any location on the earth. GPS is maintained by the United States government and is freely accessible for many applications. The system provides for monitoring surveys to connect GPS receivers to Continuously Operating Reference Stations (CORS) that are permanently mounted to the ground with deep drill-braced monuments and provide precise reference frames to the postprocessing of GPS data files. These stations are maintained by local and scientific agencies, and some are tied into a national network by the NGS.

The advent of GPS has given the surveyor the ability to transport accurate horizontal control to a monitoring site from greater distances than traditional triangulation and trilateration networks. This control is almost always more accurate and less costly, and it is independent of weather conditions. GPS is especially effective in establishing and verifying geodetic control used at nearly all monitoring sites that have sufficient sky view. The initial investment in GPS is substantial, and agencies or companies performing monitoring surveys usually require several monitoring sites or other types of survey projects to justify this investment. This is especially so in the case of most dams and reservoirs, which will require a four-receiver system to be economical. Positional accuracy in this type of network ranges from 0.12 to 0.20 in. (3 to 5 mm) in the semi-major axis at two standard deviations.

To offset this cost investment, users can sometimes take advantage of existing CORS in many areas around the world where both a large population base and earthquake activity exist together. These stations have been established by large government agencies or scientific organizations to help solve regional problems related to seismic or subsidence activity. Most of these systems have automated the collection and processing aspects of the GPS data through phone modems or wireless radio and data-processing centers. The data can be retrieved via the internet at no charge to the user in many instances.

A second way to reduce costs with GPS is to establish on-site CORS by creating two reference points that surround the project, as shown in Fig. 6-2. By leaving two GPS units operating continuously (during the survey duration) at these reference points, only one or two survey personnel are needed to create a very dependable and accurate system for measuring horizontal displacements. Testing has shown that the horizontal positioning

Fig. 6-2. CORS setup   *Fig. 6-3. RTK setup*

using this technique is more accurate than the more traditional "bridged" network that uses the short vectors created between monitoring points. This method is also a more reliable adjustment because only adjustments of vectors between fixed or constrained reference points occur and not between vectors established between monitoring points themselves.

Another example of GPS related to structural monitoring involves concrete gravity dams. Static GPS systems have been designed to continuously monitor long-term movement in gravity dams from changes in reservoir level, thermal effects, and wind.

Real-time GPS systems capable of measuring small movement at up to 10-Hz sampling rates are available and are likely to become another tool in measuring earthquake effects in real time on embankment and concrete dams in future practice.

A real-time kinetic (RTK) system, shown in Fig. 6-3, requires a GPS base station that is broadcasting from a known reference location to a roving GPS unit(s) within radio range. The position of the rover is broadcast in real time at about a 1 s interval. These controlled conditions include

- Rover and base separated by less than two miles,
- Well-modeled tilt of the local geoid,
- Mean position of multiple readings, and
- Fixed-height tripods.

For any of these GPS methods, the use of the precise orbital ephemeris increases the horizontal positioning accuracy significantly to the level of 0.12 to 0.20 in. (2 to 3 mm) in the semi-major axis at two standard deviations. Other key considerations included in producing this level of GPS position accuracy are

- Horizon elevation mask (15 to 20°),
- Survey time at monument (minimum of 30 min), and
- Multipath error (reduced with choke-ring-style antennas).

GPS cannot produce vertical positions accurately over relatively short distances compared to using a level. However, some applications using RTK techniques show promise in the 0.4 to 0.6 in. (10 to 15 mm) range vertically under properly controlled conditions. RTK may prove useful during construction of large embankment dams when settlement from loading and compaction is expected to be fairly large.

6.6 LIGHT DETECTION AND RANGING

Light detection and ranging (LiDAR) is used in remote-sensing applications and measures distance to a surface by illuminating a target with a laser. Reflected light is analyzed to determine a distance to provide a high-resolution three-dimensional digital maps and terrain models. LiDAR has applications in geomatics, geology, seismology, forestry, and many other remote-sensing and contour-mapping needs. LiDAR can be utilized in terrestrial, airborne, and underwater applications. The benefits of LiDAR include detailed mapping of the character of the ground and structures from a remote application where access to an area is not feasible such as penstocks and tunnels, modeling of surface areas prone to earthquake damage, and terrain models that can be created over large areas and archived for future evaluation of movements after an earthquake.

LiDAR has been used in dam monitoring, along with sonar, optical, and hydrographic survey data, to provide digital terrain models of the above and below water surfaces of dams and their foundations. This technology is rapidly advancing and is a cost-effective addition to the standard dam-monitoring plan.

The terrestrial LiDAR technique uses a 3D scanner, shown in Figs. 6-4 and 6-5, to collect data on the shape and appearance of an object by scanning the object with millions of measurements to define the shape and size. The scanner collects distance information about a surface within its field of view. The image that is created by a 3D scanner allows for a 3D position of each point in the image. The scans are referenced to a common system, registered, and merged together to create a model. The scanner can be used for monitoring movement on any type of dam. Scan data can be measured, archived, and compared to future or past scan data.

Concrete dams can be scanned using existing movement survey control points and related to conventional measurements. In addition to analyzing the traditional points of interest, any portion of the point cloud may be analyzed. Scans of concrete dams can be analyzed several ways:

- Vertexes can be placed on points of interest that are clearly identifiable. The vertexes can be exported and saved in a text file for comparison to future scans.

Fig. 6-4. Terrestrial scanner
Source: Courtesy of Leica Geosystems HDS, reproduced with permission.

Fig. 6-5. Terrestrial scanner in use

- When a successive scan is done, the vertex file can be imported from the text file into the new scan, and any movement can be seen visually in the point cloud on each point of interest.
- A complex mesh can be created on the structure and archived. Once a successive scan is done and a new complex mesh is created, the original complex mesh can be overlaid to detect changes.
- The survey control lines being used with traditional equipment can be incorporated into the point cloud and used to report measurements using a cross-section method with a virtual reference plane created in the software.

Embankment dams can be scanned using existing movement control points and compared to measurements made by other techniques. In addition to analyzing the traditional points of interest, any portion of the point cloud can be analyzed. Scans of embankment dams can be analyzed several different ways:

- Vertexes can be placed on points of interest that are clearly identifiable, such as high points of rocks on the face of the dam.
- The vertexes can be exported and saved in a text file for comparison to future scans.
- When a second scan is done, the vertex file can be imported from the text file into the new scan and any movement can be seen visually in the point cloud on each point of interest, clearly showing whether the rock face settled between the two scan periods at the vertex location.

A complex mesh can be created and archived. Once a second scan is done and a second complex mesh is created, the original complex mesh can be loaded to identify changes, clearly showing whether any portion of the rock face settled between the two scan periods.

Accuracy of a single measurement is approximately 0.5 in. (6 mm) at a range of 3 to 150 ft (1 to 50 m). Higher accuracy may be achieved using software to average thousands of points on the object of interest.

There are limitations to this method. Water absorbs the laser light and yields no information. Consequently, the scanner will not measure points on a water surface. Also, the range of scanning distance varies by scanner. Most scanners will produce accurate scans to 1,000 ft (300 m). Scanners will not operate properly in heavy vibration conditions, including in high winds.

6.6.1 Scanner Position

The scanner operates like a traditional total station in respect to setting it up over a known survey point and backsighting a known survey point. The scanner can also calculate its setup position using resection from two or more survey control points. A twin target pole with 3 in. targets is used to target the backsight survey point or resection points.

6.6.2 Targeting

Targeting is used to join multiple scans together with higher accuracy than relying solely on the traverse points of the scanner. Additional targeting is required on high-accuracy jobs with multiple setups. The different types of targets available are

- 6 in. (0.15 m) hemispheres,
- 6 in. (0.15 m) spheres,
- 6 in. (0.15 m) planar targets (magnetic and adhesive),
- Black-and-white paper targets,
- 3 in. (0.07 m) planar targets (magnetic and adhesive), and
- 3 in. (0.07 m) twin target poles.

6.6.2.1 Target Distance Limits. Target choice depends on the distance of the scanner from the target. A 3 in. 0.07 m) target may be used when the target is 150 ft (45.7 m) or less from the scanner, but 6 in. (0.15) targets should not be placed more than 500 ft (152.4 m) from the scanner.

6.6.2.2 Cloud-to-Cloud Targets. Targets can be established virtually in clouds within the software after the field scanning is complete if there are unique points that can clearly be identified in multiple scans. This office process usually takes longer than the time it would have taken for the field surveyor to place targets in the scan area before the scan. For efficiency, it is always better to place the targets in the scans in the field; however, the requirements of a project may change after the field scans are complete. Cloud-to-cloud targeting is an option to increase accuracy in the office. Figs. 6-6 and 6-7 provide examples of dam LiDAR scans.

6.7 FUTURE TRENDS

Advances in remote sensing and electronics have opened numerous doors in geodetic monitoring. High-accuracy achievements and advances in spatial resolution have given the surveyor more tools to add to the monitoring plan. Two major remote-sensing applications are LiDAR and interferometric synthetic aperture radar (InSAR). Recent advances in electronics and computing processors make these techniques more affordable and readily available. Future geodetic technologies such as terrestrial 3D surveys with high-definition scanners and airborne LiDAR surveys will provide new options in the way surface monitoring is done. As a supplement to standard geodetic monitoring, 3D high-definition scans and detailed models of large land areas and structures can be made for monitoring movement.

Airborne LiDAR surveys tied to ground GPS stations are becoming a more cost-effective way to create digital terrain models (DTMs) and high-resolution ground models to identify ground features associated with slope movements. Color-coded LiDAR relief maps are being used to detect instability in terrain features or structures themselves. Another trend is InSAR that may further alter the way surface monitoring is done by using ground-based or satellite measurements to the ground over a time series to detect movement in regional land areas and some large dams.

As drone technology becomes more available and widely used for spatial applications, imagery resolution will become more effective for some ground-monitoring applications.

On a much larger scale, plate tectonics and volcanic movements are monitored using real-time GPS networks. These networks are based on a

Fig. 6-6. Arch dam scan
Source: Courtesy of Seattle City Light, reproduced with permission.

Fig. 6-7. LiDAR terrain model of Morris Sheppard Dam
Source: Texas Natural Resources Information System/Texas Water Development.

CORS, which is a GPS antenna permanently attached to a deep drill-braced monument that transmits very accurate positions of a station. Many of these networks are currently in place and have been used by the geodetic/scientific community for several years. Advances in geodetic software, internet protocols, and communication systems have provided a very dense and accurate GPS network.

This new trend in detailed 3D intelligence, LiDAR/InSAR, and active GPS networks, along with modern geodetic analysis software, will help develop interesting strategies for future monitoring planning.

6.7.1 Interferometric Synthetic Aperture Radar

An emerging technology with possible application in dam monitoring is synthetic aperture radar (SAR). SAR is an electromagnetic imaging sensor often used in remote-sensing applications. The SAR sensor is mounted on an aircraft or a satellite and is used to make a high-resolution image of the earth's surface. Images of several passes are enlarged to obtain changes in elevation (Peltzer 1994). SAR uses a wavelength of 2.5 in. to 3 ft (1 cm to 1 m), whereas optical sensors use wavelengths near that of visible light, or 1 μm. These longer wavelengths enable SAR to see through clouds and storms, whereas optical sensors cannot because a SAR sensor carries its own illumination source in the form of radio waves transmitted by an antenna. SAR can be used at any time of the day or night. Methods are currently being developed that will allow for a dam and the surrounding area to be monitored for vertical movements using SAR.

SAR is a radar-based measurement technique used in geodesy and remote sensing. This method uses two SAR images to generate maps of surface movement or digital elevation models. This is done by determining differences in the phase of waves returned from the satellite or aircraft. Centimeter-scale changes in movement over spans of days to years can be detected for a time series. SAR is currently used in geophysical monitoring of natural hazards, such as earthquake fault zones, landslide areas, volcanic, and subsidence areas.

SAR can be used to supplement current monitoring methods used on large dam projects or to provide a more regional view of the terrain surrounding a monitoring project.

Fig. 6-8 (Mazzanti et al. 2015) illustrates SAR identifying elevation changes at Three Gorges Dam in China.

Continuous measurement and movement detection can be done by the same technology using a terrestrial-based system, a device for which is shown in Fig. 6-9. Day-and-night, all-weather observations can be conducted utilizing radar signals toward the face of a dam or structure and

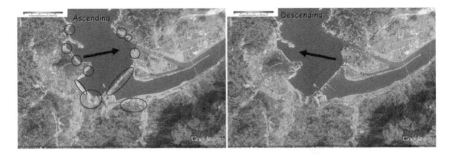

Fig. 6-8. Airborne SAR
Source: Courtesy of NAHZCA, Google Earth, reproduced with permission.

Fig. 6-9. Terrestrial SAR scanner
Source: Courtesy of Olson Instruments, reproduced with permission.

provide 2D displacement intensity maps that show back-scattered radar images.

6.7.2 Unmanned Aerial Vehicles

Unmanned aerial vehicles (UAVs), an example of which is shown in Fig. 6-10, are commonly referred to as drones and are remotely piloted aircraft. Other types of UAVs are unmanned air systems (UASs) and remotely piloted aircraft systems (RAPSs). The UAV is flown autonomously using an onboard computer system and flight software, or it is piloted using remote control. This technology is emerging, and payloads of special cameras and optics are bringing new monitoring opportunities as the accuracies and techniques are improved. The UAV can cover large areas that are inaccessible, unsafe, or too expensive for ground survey. Large dams and reservoirs can be measured quickly using a UAV. This capability is useful for emergency inspections and mapping and emerging technology to perform safer dam, spill gate, and transmission line condition inspections currently, requiring specialized high-angle access. UAVs provide high-resolution photographs and video for future comparison.

Fig. 6-10. Unmanned aerial vehicle (UAV)

6.7.3 Real-Time Networks

On a much larger scale, plate tectonics and volcanic movements are monitored using real-time GPS networks. These networks are based on CORS equipped with a GPS antenna permanently attached to a stable monument. CORS continuously transmit very accurate positions of the stations, as shown in Fig. 6-11. Many of these networks are currently in place and have been used by the geodetic and scientific communities for several years, and their utility is likely to improve with time.

6.8 PLANNING AND IMPLEMENTING

This section provides suggestions for creating a new monitoring plan or evaluating an existing plan. With aging dams and advances in technology, the demand to maintain and/or expand a monitoring plan requires thought. A monitoring program that includes surveying to measure performance indicators requires an easily understood plan that is portable to the next generation. Every dam is different, and there is no single answer or solution to planning and designing an effective program. Lessons learned from experience are helpful for planning and implementing a geodetic monitoring plan.

Selection of the survey method will be proposed by the surveyor to measure those performance indicators identified in the review of failure modes. The surveyor provides the dam owner and engineer with survey means and methods to accomplish the required measurements and assists in selecting a preferred alternative. The comparative accuracies, reliability, and costs of competing alternatives must be balanced within the overall dam-monitoring program. Allowing the surveyor to recommend a preferred survey alternative has benefits because it is the surveyor who must do the work and obtain the measurements.

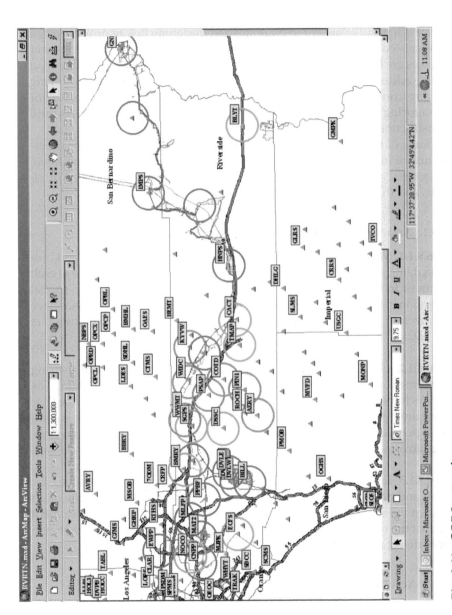

Fig. 6-11. CORS network

When designing and implementing a geodetic movement-monitoring plan for a new project or when evaluating an existing plan, there are several aspects to be considered. A monitoring plan for new projects should be done in coordination with the dam owner, engineers, and design staff. There is a positive benefit to planning and designing before project design is advanced. The locations of monuments, sensors, instruments, and power sources are better managed if they are part of the overall design rather than being added as an afterthought. Many projects will require surveying during construction, which becomes an important planning factor for monitoring movement because the surveyor must handle both tasks until construction is complete.

Some of the main factors to carefully consider and evaluate are (1) performance indicators to be measured, (2) type of structure to be monitored, (3) accuracy requirements, (4) frequency of monitoring, (5) expected products, and (6) the schedule for performance.

6.8.1. Arrangements for Different Types of Dams

6.8.1.1 Multiple-Arch and Buttress Dams. For multiple-arch and buttress dams, monitoring points are usually located at the nose and toe of each buttress. In the case of massive buttresses and large multiple arches, special attention is required for the foundations of the buttresses. If buttresses have contraction joints, joint movement requires measurement (Chrzanowski et al. 1992).

6.8.1.2 Gravity and Arch Dams. The requirements for measuring movement at the toe and along the abutments vary with the dam. The points to be measured are distributed as needed to address the need for measurements of performance indicators, as shown in Fig. 6-12. It is common practice to fit each monolith with monuments that allow for measuring movement of the crest.

6.8.1.3 Embankment Dams. Lines of monuments are located along the crest and on the downstream slope, as needed, to measure movement, as shown in Fig. 6-13. Monument spacing may consider loss of a monument so that the remaining spacing between monuments is small enough to reliably detect movement.

6.8.2 Accuracy Requirements

Required accuracy is a function of the expected behavior of a dam. The surveyor consults with the designer to learn the bounds of expected movements and then chooses the means and methods required to produce accurate measurements.

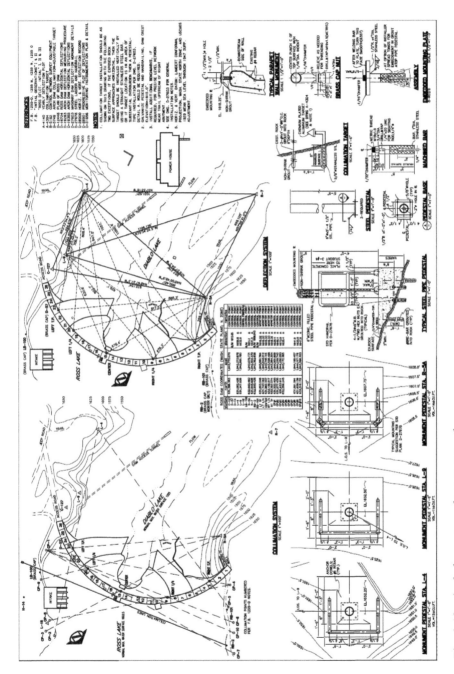

Fig. 6-12. Arch dam survey layout example

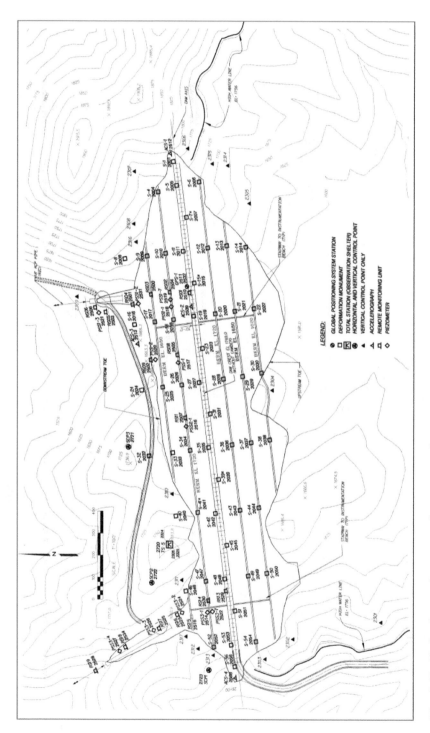

Fig. 6-13. Embankment dam survey layout example

6.8.3 Monitoring Frequency

Frequency may depend on design specification, stage of construction, regulatory requirement, expected movements (a failing slope, high wall monitoring, first filling), and time of year (seasonal movement of concrete structures). Time of year affects the deformation readings owing to thermoelastic effects on a structure in different seasons. It is important for the geodetic engineer/land surveyor to understand the duration of the monitoring for the project and the frequency of measurements that are expected. These two aspects are important in how the plan will be designed and implemented as they affect project labor and material costs and the type of geodetic monitoring to be used.

The surveyor works with the owner to understand the limitations of program costs (if any) and engineer to establish monitoring schedules.

The frequency of measurements (or observation schedule) varies depending on the stage of the life of the dam:

- Preconstruction stage—geodetic positions are established and measured to establish baseline positions of reference points for later comparison;
- During construction—geodetic positions are measured to identify problems or to verify that structural behavior is compatible with expectations;
- During first filling—monitoring frequency is greatest during this stage. Measurements at this stage provide a means to evaluate design assumptions, construction adequacy, and overall performance; and
- During service operation—geodetic positions are measured less frequently.

Whenever practicable, planning a monitoring system with digital data collection will provide greatest accuracy at the least cost. This eliminates errors related to handwritten notes and provides a backup system for data. Data-collection systems require review for flexibility of methods of collection and output formats.

Site conditions need to be considered when designing a survey system. The ability to automate a remote site or a site that is difficult or unsafe for personnel may determine whether it is better to use a conventional total station or a robotic system. The site conditions may also lend themselves to using GPS to monitor horizontal positions instead of a total station. This is especially true in areas where extensive heat waves make it nearly impossible to get good total station measurements. Steep valleys may interfere with signal reception from satellites. Otherwise, areas with line-of-sight problems may be better suited for GPS.

6.9 MONUMENTATION

All movement-monitoring plans require stable points that can be measured in three dimensions. Monuments of various types achieve this function. Monuments are described in this section for different monitoring surveys and for different structure types. A discussion on active versus passive monumentation is provided, along with monument diagrams. Factors to consider for monitoring plans are discussed along with suggestions for mapping and tracking of monuments for the duration of a project. Suggestions are included for proper measuring, acquisition, analysis, and reporting movement measurements.

Typical monuments are permanent devices such as concrete pedestals and embedded metal plaques. Once surveyed and marked, these monuments are observed repeatedly to measure movement. Each should have a unique identification name stamped on the monument itself. Computer-aided drafting (CAD) or geographic information system (GIS) mapping can be used to show the relative locations of both the monitoring points and the control monuments in relationship to the structure being monitored. A comprehensive map that accurately represents monument locations is useful for the duration of the project for all involved with the monitoring plan.

The durability of surface monuments is important. These devices may be located on spillways, embankment dams, concrete dams, reservoir slopes, abutments, supporting fills, dikes, and appurtenant structures such as penstocks. Failure to locate in both stable and accessible locations will result in loss of monitoring accuracy, as well as a lack of confidence in the results. Monuments used on a structure must also be located for durability, taking into consideration ground conditions.

6.9.1 Selection Criteria

Care is needed in considering

- Spacing interval of monitoring points,
- Efficiency and economy,
- Accessibility (crew safety),
- Durability,
- Type of monument,
- Terrain (stable ground, line of sight),
- Network design,
- Geometry of network, and
- Type of structure.

6.9.2 Monument Types

The type of monument to be used at a given site is a function of the soil or rock type encountered. Setting monuments that have the possibility of losing their foundation support will result in loss of data. Reference monuments are located in natural ground or in cut areas where they are used as benchmarks or primary survey control points. Reference monuments are stable and durable away from areas of interference.

As well as being marked or stamped with a unique identification name, a monument will have a measuring point near the center that is easily discernible and will not be lost or misinterpreted over time. This mark, known as the "punch mark," allows for repeat measurement because a rod or prism tip can be placed in the same location every time a measurement is taken. The measuring point is best marked in brass or aluminum cap-type monuments whenever possible. Brass caps facilitate leveling operations. The attachment of the brass cap to the concrete foundation should be made at the time of the concrete pour whenever possible. Installation of brass caps into existing structures should be made with strong epoxy. The brass cap should be set flush or slightly recessed to avoid traffic damage and deter vandalism.

Monuments in large existing concrete structures or set in extensive constructed concrete foundations are the most commonly used. The larger the structure, the more likely the monitoring monument will not be affected by local ground disturbances such as vehicle traffic, rain runoff, or soil expansion.

The monuments shown below are examples of robust design. Variations on these basic designs also are used.

These monuments can also serve as observation points.

Monument Type A-1, shown in Fig. 6-14, is best suited for embankment dams and as a reference monument when there is no bedrock for its foundation. It is the most economical of the monuments listed here.

Monument Type D-1, shown in Fig. 6-15, is used on an embankment dam crest where protection is required from motor vehicles. This monument can also be used to protect reference monuments that are located in natural ground that need protection from off-road vehicles.

Monument Type D-2, shown in Fig. 6-16, is used for reference monuments that must be located in areas with substantial permafrost activity or where the soil consists primarily of sand or other granular soil.

Monument Types D-4 and D-5, both shown in Fig. 6-17, are installed in embankment and concrete dams, respectively, where semi-automatic or robotic systems are utilized. These monuments can also be used for references at any site designed for semi-automatic or robotic instrumentation.

Another type of monument, shown in Fig. 6-18, consisting of a steel stud with a threaded end that is drilled and placed in concrete and fixed

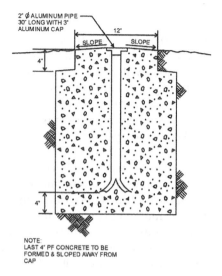

Fig. 6-14. Monument type A-1

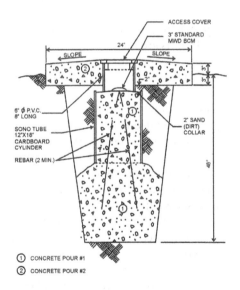

Fig. 6-15. Monument type D-1

with epoxy, can be used on the crest of concrete dams in lieu of those previously listed. A retro-reflector can then be attached and left in place if measurement frequency justifies the investment. Devices left in place are vulnerable to severe weather and vandalism, and they may cause tripping hazards.

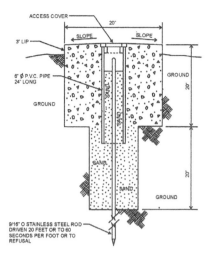

ACCESS COVER
20"
3' LIP
SLOPE   SLOPE
6" Ø P.V.C. PIPE
24" LONG
GROUND
GROUND
20'
SAND
SAND
SAND
20'
GROUND
SAND
GROUND
9/16" O STAINLESS STEEL ROD
DRIVEN 20 FEET OR TO 60
SECONDS PER FOOT OR TO
REFUSAL

Fig. 6-16. Monument type D-2

The superior monument is one that really may not appear to be a monument, at least in the classical sense. CORS monuments, shown in Fig. 6-19, eliminate the effect of slight surface movement and weathering of the top soil. Collaboration among scientific agencies and organizations suggest a best-practice monument, which is five-legged, stainless-steel piping filled with pumped grout and driven at least 30 ft into the ground and welded at the apex where the GPS antenna resides. CORS monuments are designed to last more than 50 years with little or no movement owing to any factors other than earthquakes.

6.10 DATA COLLECTION

Data-collection methods are an important consideration when planning a monitoring system. The use of equipment that digitally collects and stores data, as shown in Fig. 6-20, improves the overall accuracy and provides a record of the fieldwork and measurements made. Labor efficiency is also improved with data-collection systems because they can replace handwritten note-taking typically done in the field. One potential disadvantage is that digital collection excludes the opportunity to look for visible clues to explain the data.

Total stations typically collect data with a handheld data collector or onboard data-collection software. The data is then usually stored on a digital memory card or similar device that is inserted into a PC or small peripheral for downloading.

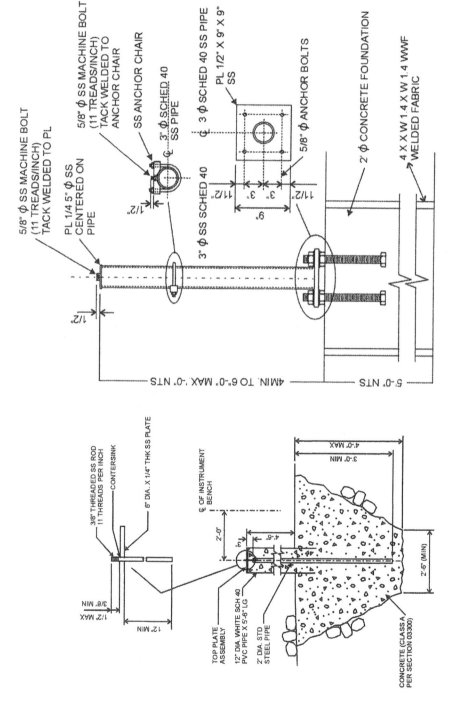

Fig. 6-17. Monuments type D-4 and D-5

Fig. 6-18. Stationary reflector

Fig. 6-19. CORS monument

The accompanying software allows the data to be downloaded and formatted for further processing or evaluation. Most equipment manufacturers sell proprietary software programs that are designed specifically for their equipment. It is important to consider the functionality of the software for data collection when considering the purchase of a specific surveying instrument as not all software programs have the routines that are best suited for movement survey methods, such as set collection. Electronic levels collect and store data on an onboard recording module. This module is then downloaded through a small peripheral utilizing the proprietary software that formats the data for further analysis. GPS receivers collect

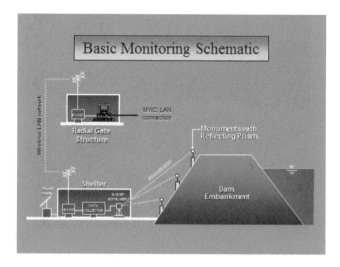

Fig. 6-20. Basic monitoring schematic

and store the data onboard, which may be downloaded through a serial connection to a PC. Each receiver manufacturer has proprietary software for downloading and processing of the raw data. Further analysis is then done using a least-squares adjustment program.

All three types of equipment need additional data collected by the field personnel. In all cases, measurements of the instrument and object heights need to be recorded. As a general rule, it is best to have the field person measure the height twice, once in meters and once in feet, to provide a check. Incorrectly measured or written-down height measurements are one of the most frequent avoidable sources of errors in survey field work. Recently, several equipment manufacturers have started selling fixed-height tripods. These tripods are designed to reduce errors in height by only allowing for a fixed height of the object (prism, GPS antenna, and target) and are usually in heights of 1.5 or 2 m.

Other data to be recorded include the point name or number. Leveling equipment and GPS equipment require a four-digit name or number of a monument. It is best to assign a four-digit number (versus a name with letters in it) to identify each monitoring point because not all equipment accepts alphanumeric input, and by assigning a number it gives greater flexibility for future methodology. Other data to collect manually include the usual survey information such as crew names, weather conditions, and visual observations.

ADASs are a cost-effective method to collect survey data on large projects. The Eastside Reservoir in Hemet, California, a very large project encompassing three dams, has an ADAS for all the geotechnical instrumentation and an automated data-collection system for the survey data. The data are collected by remotely operated robotic total stations directed by a PC-based monitoring application that directs the functions of the total stations. This system is automated to collect the data and then transfer the data to a remote location for analysis (Duffy 1998).

6.11 DATA ANALYSIS

Even the most precise monitoring surveys will not fully serve their purpose if they are not properly evaluated and utilized in a global integrated analysis as a cooperative interdisciplinary effort. This interdisciplinary effort includes visual inspections as well as geotechnical and surveying instrumentation.

The analysis of survey data has become much more advanced with the advent of computer software programs capable of least-squares adjustments. These types of analysis tools have allowed practitioners to use methods found in the field of mathematics to communicate statistical confidence in the position of monitoring points. With the advent of GPS technology, this type of tool has become essential. Positional tolerances can now be expressed in terms of the length and azimuth in the major and minor axes of an error ellipse figure. This can be calculated with different statistical probability criteria such as 39% probability when the standard error ellipse is used in defining the positional accuracy or 95% probability when the axes of the standard error ellipse are multiplied by 2.447. An engineer can express confidence, or lack thereof, in the absolute movement of each monitored point with some mathematical certainty. The output of many of these adjustment programs have a graphical display that can visually show the magnitude and directional movement trends of all the monitoring points together and individually.

It is important to understand that although geodetic networks and least-squares adjustment procedures are excellent analysis tools for understanding the integrity of control monuments, they are deficient, in some ways, to understanding the performance indicator movement associated with actual monitoring points in a geodetic scheme. The purest method of analyzing monitoring points for true movement comes about by simplifying the measurement vectors associated with each point. This can be restated by saying the best way to measure monitoring points is with an accurate independent measurement to each point accomplished directly from a stable reference monument. Any kind of adjustment performed at the monitoring point itself dilutes the ability to distinguish movement associated with adjustment and that associated with true movement.

Eventually, geodetic coordinates and elevations collected at monitoring points need to be analyzed together with other data such as seepage, pressure, and stress. This can be accomplished by comparing both types of data and trying to find a geographic connection or pattern to the data. Also, one may design a monitoring system so that most or all the instrumentation has a geodetic element to it, thus incorporating absolute capabilities to an otherwise relative measurement system. The monitoring system can also be organized or programmed to get readings from both types of instruments at approximately the same time intervals, thus allowing for better correlation of the data.

The technical term for this kind of evaluation process is "integrated movement analysis" (Chrzanowski et al. 1992). The expression "integrated analysis" means a determination of the movement by combining all types of measurements, geodetic and geotechnical, even if scattered in time and space, in the simultaneous geometrical analysis of the movement, comparing it with the prediction models, and enhancing the prediction models, which in turn, may be used in enhancing the monitoring scheme (Chrzanowski et al. 1992).

Several software packages for geometrical analysis have been developed, such as DEFNAN, PANDA, and LOCAL. Some of these packages, such as DEFNAN, are applicable not only to the identification of unexpected movement but also to the integrated analysis of any type of movement, whereas others are limited to the analysis of reference geodetic networks only.

6.12 REPORTING

The owner or dam safety engineer who will use a movement report will also determine what type of plan to use. The surveyor may need to choose between automation systems versus conventional survey systems and determine the best data-collection method to use. Either way, the product must be usable for the dam safety engineer and it must be produced in an efficient and timely manner; it will be more critical during an emergency or under special operating conditions. This may be especially true in cases of problems associated with structures during filling operations and in earthquake-prone areas.

An efficient and well-thought-out monitoring plan loses value if the results are delayed because of lack of personnel, equipment, or other factors that may occur in an emergency. An early detection or early warning is an important factor in planning. Software applications delivering real-time alarms when a position tolerance is exceeded on a structure can deliver a movement report. The format for a movement depends on how it will be used. Reports that are simple, well thought-out, and flexible enough to allow the surveyor to relay information quickly are the best (Fig. 6-21).

Methods of reporting movement vary among the surveyors. Written reports describe the method used at the time of a survey and are supplemented with

Horizontal and Vertical Displacement Monitoring

NAD83 1991.35 Epoch, CCS Zone 5, Grid Coordinates & NAVD88 Elevation Comparisons

*North, East & Up offsets are positive and South, West & Down offsets are negative

PCID NO.	INITIAL VALUES June 1998 NORTHING (MTRS)	EASTING (MTRS)	ELEV. (MTRS)	June 2009 Survey NORTHING (MTRS)	EASTING (MTRS)	ELEV. (MTRS)	June 2010 Survey NORTHING (MTRS)	EASTING (MTRS)	ELEV. (MTRS)	CURRENT MINUS INITIAL YEAR DELTA N	DELTA E	DELTA ELEV.	CURRENT MINUS PREVIOUS YEAR DELTA N	DELTA E	DELTA ELEV.
1001	584881.217	1948115.506	324.322												
1002	584797.712	1947987.474	310.238			310.232			310.234			-0.003			0.002
1006	584819.800	1948040.275	326.857			326.849			326.849			-0.008			0.001
1007	584831.964	1948025.588	326.815			326.806			326.806			-0.008			0.000
1008	584841.761	1948014.122	326.747			326.735			326.737			-0.011			0.001
1009	584851.490	1948002.395	326.662			326.654			326.655			-0.006			0.001
1010	584861.221	1947990.674	326.699			326.695			326.696			-0.003			0.001
1012	584817.724	1948019.483	318.086			318.083			318.082			-0.004			-0.001
1013	584810.021	1948007.161	312.095			312.094			312.095			0.000			0.001

*North, East & Up offsets are positive and South, West & Down offsets are negative

PCID NO.	INITIAL VALUES June 1998 NORTHING (MTRS)	EASTING (MTRS)	ELEV. (MTRS)	December 2009 Survey NORTHING (MTRS)	EASTING (MTRS)	ELEV. (MTRS)	June 2010 Survey NORTHING (MTRS)	EASTING (MTRS)	ELEV. (MTRS)	CURRENT MINUS INITIAL YEAR DELTA N	DELTA E	DELTA ELEV.	CURRENT MINUS PREVIOUS YEAR DELTA N	DELTA E	DELTA ELEV.
1001	584881.217	1948115.506	324.322												
1002	584797.712	1947987.474	310.238			310.234			310.234			-0.003			0.000
1006	584819.800	1948040.275	326.857			326.849			326.849			-0.008			0.000
1007	584831.964	1948025.588	326.815			326.806			326.806			-0.008			0.001
1008	584841.761	1948014.122	326.747			326.735			326.737			-0.011			0.002
1009	584851.490	1948002.395	326.662			326.654			326.655			-0.006			0.001
1010	584861.221	1947990.674	326.699			326.696			326.696			-0.003			0.000
1012	584817.724	1948019.483	318.086			318.084			318.082			-0.004			-0.002
1013	584810.021	1948007.161	312.095			312.095			312.095			0.000			0.000

Page 1 of 1

Fig. 6-21. Horizontal and vertical displacement monitoring

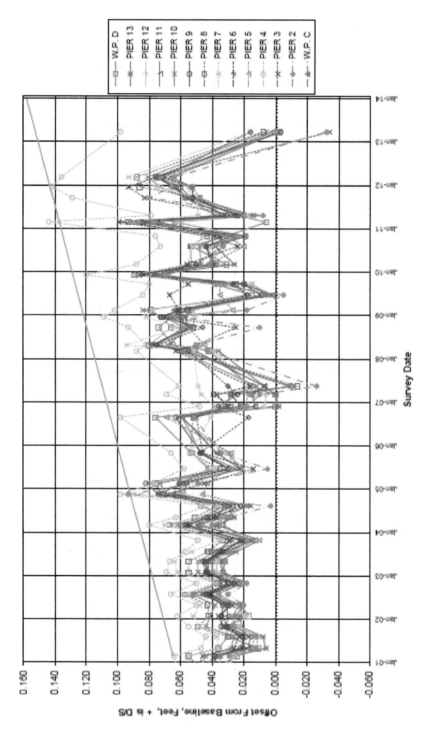

Fig. 6-22. Typical time versus reading crest movement graph
Source: Courtesy of GCPUD, reproduced with permission.

maps of the project area showing where measurements were taken. Survey reports are archived regularly. Maintaining the historical data on previous surveys is important when analyzing trends and problems in monitoring of dams or structures and for preserving the information when considering improving a monitoring plan. Typical report entries include

- General condition of the structure;
- Data formats and file locations;
- Equipment and instruments used;
- Horizontal measurement method;
- Vertical measurement method;
- Project notes such as changes to monuments;
- Reference drawings and monument maps;
- Point coordinates, number, and name;
- Date monitored;
- Time versus reading movement measurement (horizontal, vertical, or both); and
- Station name and monument description.

Field notes—A typical report includes original field notes such as leveling notes and traverse and GPS observation sheets, along with remarks made by personnel in the field. Although in raw form, field notes are included so that the survey used is documented.

Maps—Point maps that show monument locations in relation to the dam or structure are useful to add to a written report and aid in discussing particular areas of concern. More modern technologies in 3D surveying can provide not only a comprehensive map of a movement survey, but also color hue maps can be generated to show movements on structures by using color intensity.

Results—The purpose of the movement-monitoring exercise is to measure and report the measured movements in comparison to (1) expected movement and (2) Threshold and Action Levels that suggest a movement be explained. Time versus movement graphs (Fig. 6-22) can illustrate movement data for easy interpretation.

A valuable method of reporting is the use of an electronic real-time alerting system that has the ability to communicate between the dam safety engineer and surveyor in charge of the monitoring for a particular structure. Movement reports and alarms of exceeded threshold tolerances can be helpful to understand a problem with a structure or its system.

CHAPTER 7
DATA ACQUISITION

You can have data without information, but you cannot have information without data.

Daniel Keys Moran

Data are measurements of performance indicators. Data are acquired manually or with data-logging methods. Data logging may employ handheld devices or data loggers, or they may be fully automated in a system termed an Automated Data-Acquisition System (ADAS).

7.1 MANUAL ACQUISITION

Manual data acquisition involves using a variety of different tools such as water-level indicators for piezometers and observation wells, staff gauges for reservoir or water-level monitoring, surveying equipment, and specialized probes such as an inclinometer probe to measure inclination. Measurements are manually recorded on paper in the field or stored in the memory of a handheld readout and later transferred into a project data management system.

Checklists are valuable to manual data collection for preserving data integrity. Readily available checklists containing several previous readings can help the instrument reader discern an unexpected reading. Whenever an unexpected reading is noted, the measurement is repeated. For example, small-diameter standpipe piezometers may cause a false reading when the water-level probe reacts to humidity along the riser pipe rather than at the actual water level. Unexpected data may not be in error but require repeating a measurement. All field technicians are trained on the expected behavior of an instrument so that the reader can make an immediate judgment if additional readings are needed. Errors may occur when an instrument malfunctions or during reading and recording the data.

Metadata such as the name of the instrument reader, instrument serial number, and the date and time of the reading are recorded and stored in the database. Where an instrument reading is suspect, metadata can assist in troubleshooting. Instruments that develop a calibration issue or an intermittent fault may be problematic. An inclinometer probe cable that has been fatigued, for instance, can give some incorrect readings. Where the data problems are subtle, it may take several readings before the problem is noticed.

Manual methods have lower initial costs, require less maintenance, reduce the level of expertise required to operate and maintain, are less susceptible to damage by weather and vandalism, and allow for concurrent visual surveillance to observe signs of distress at a dam.

You can observe a lot by just watching.

Yogi Berra

Cameras (digital or video) are useful tools for electronically recording visual observations that can later be transferred to data management applications for use as supplemental data to instrumentation data. For specific features of concern, cameras can be mounted at various vantage points and set to record continuously or at regular intervals. Coupled with the use of historical images, cameras can be a useful tool to identify and evaluate new safety concerns. Cameras can be included in a monitoring system to allow remote visual surveillance of the site for confirmation of conditions indicated by automatically collected data. Infrared or regular cameras supplemented with lighting for night use are used on many projects.

7.2 AUTOMATED ACQUISITION

Visual surveillance is the backbone of dam-performance monitoring. ADASs augment but do not replace visual inspection of a dam.

ADASs can include cameras to allow for rapid, preliminary visual surveillance from remote locations. If instruments report measurements exceeding alarm values, cameras can identify visual signs of distress to aid in deciding whether action is needed.

7.2.1 System Configurations

System-level models represent the most common current practice for monitoring dams. Choosing the appropriate model requires consideration of several factors that underlie the choice of a model:

- Data may be acquired from many points of geotechnical and structural instrumentation or just one or two points of hydrological data;
- Instrumentation may be deployed over a large structure and benefit from a distributed system of multiple field measurement units, with collection units close to the instruments providing data; and
- The system not only records the measurements but also provides a mechanism to transport data to a computer where it can be recorded and viewed by an operator for evaluation.

Best practice is to scale a system to the monitoring needs of the site.

One model is a simple data logger, for example, one that reads only vibrating-wire sensors (Fig. 7-1), which are the most common sensor used in dam monitoring. One step above is a data logger that can measure multiple electrical signals (voltage, resistance, current) over extended periods

Fig. 7-1. Simple 8-channel data logger for vibrating-wire sensors only (left) and simple 1-channel data logger for MEMS/electro-level discrete or serial sensors only (slope indicator, right)
Source: Courtesy of DG-Slope Indicator, reproduced with permission.

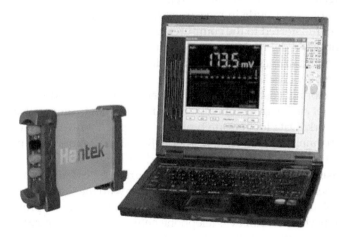

Fig. 7-2. Multisensor-type data logger and memory storage
Source: Quingdao Hantek Electronic Co.

of time (Fig. 7-2). Measured data are converted into a digital format and stored, along with the date and time of measurement, on a local or internal mass storage device. The key aspect of a data logger system is that data are stored within the logger's memory.

The simplest method of data transfer is to either connect a PC to the logger using a serial or USB cable or use a combination of flash memory drive and data logger, also shown in Fig. 7-2. Data are retrieved from the data logger using the flash drive, which is the removable storage media. The recorded media is retrieved from the data logger and then taken to a computer to load the data into a database. The database is accessible to one or more software programs for display, analysis, reporting, and archiving.

The term "SneakerNet" is a concise way of referring to a network of one or more personnel who collect and bring data from remote data loggers to upload for evaluation. SneakerNet has some advantages compared with strategies requiring communication connections for automatic transport of data to computers. There is a potential capital cost advantage because the additional equipment or communications services associated with automatic data transport are not required. Another potential advantage is that there is no restriction on the location of the data logger imposed by the requirement for accessibility to a communication link or associated power demands of telemetry equipment. The person collecting the data can also make visual observations while visiting the logger locations.

There are also some disadvantages to SneakerNet. First, the requirement for routine physical accessibility may require installing the data logger in a vulnerable location requiring longer connections to instruments or more exposure to potential vandalism. Secondly, timeliness of data observations may be affected by inaccessibility to remote sites during bad weather.

Changes or new trends in measurements will not be recognized until the data are transferred to the computer and analyzed. Another disadvantage of protracted delays between human observations is that equipment or instrument malfunctions may not be detected for extended periods of time, which may conflict with the monitoring objectives. There is also an ongoing personnel cost associated with data retrieval. The disadvantages of SneakerNet can be overcome by accessing the data logger over the telephone network. Data loggers are typically equipped with a serial communications channel for connection to a serial RS232 or USB port on a personal computer. This allows programming the data logger with configuration software usually provided by the manufacturer and extracting data from the logger memory or data storage device. Along with the serial communication channel, data loggers usually provide a communications protocol for modem connection control over telephone networks and a terminal protocol for transferring data from the onboard data storage to the computer. This allows the same configuration and data transfer functions between the computer and the data logger over the telephone network that can be done with a local serial cable connection. Connections may also be made with radio frequency, cellular telephones, and communication satellites. A second model is a host-driven network, where the host computer controls the time and frequency of measurements and the flow of data from remote terminal units (RTUs) to the host. Fig. 7-3 shows an illustration of an ADAS network. Note that the number of RTUs or number of instruments is tailored to the dam and the associated monitoring needs. Because of its centralized role as arbitrator for all system communications and I/O operations, the host is often referred to as a central computer.

The sequence of interrogating data from RTUs is referred to as polling the RTUs. This network architecture is referred to as master–slave, where the host computer is the master and the RTUs are slaves. Supervisory Control and Data Acquisition (SCADA) systems, for example, almost always employ host-driven networks because the primary purpose of a SCADA system is to provide data to a human operator in the process of operator-supervised command and control of remote I/O. In other words, SCADA is designed primarily to be operator-interactive with a fast response time over remote communication links, and these objectives are usually best served with centralized system architecture.

Host-driven networks are simpler, at least in concept, than node-driven networks described in the next section. A host-driven network distributes essential functions required for logging data at the user-prescribed intervals across the network. Therefore, if a communication link fails or an unattended central computer fails or becomes unstable (for example, it must be rebooted, as often happens with PCs), data records will be missed during these periods of equipment outage. Note that the host computer itself is the data logger in a host-driven ADAS. All data-logging functions, logic, and data storage reside in the host computer.

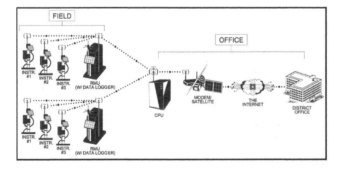

Fig. 7-3. ADAS host-driven network
Source: U.S. Army Corps of Engineers.

Host-driven networks are expensive, requiring investment in the central computer and the communication links to achieve across-network reliability.

A local field-bus network of intelligent instruments polled by a data logger or motor control unit (MCU) is a host-driven network. A field-bus network with one or more sub-networks is not burdened by the expenses of remote SCADA host-driven networks. Field-bus networks often benefit from industry-standard communication protocols and are quite reliable. Unlike PCs, the data logger or MCU host controller is designed to be a hardened unit, that is, to run unattended in remote areas with high reliability. The communication links between the master controller and slave intelligent instruments are local area networks secured within the confines of the project with high-reliability connections.

Node-driven networks are commonly referred to as peer-to-peer networks. As implied by the peer-to-peer terminology, nodes share communications capabilities equally without a master controller, and each node can initiate access to a shared communications medium.

Node-driven networks are employed in instrumentation systems to allow critical functions at the node to operate autonomously. In data-acquisition systems, intelligent nodes can often perform their automatic data-acquisition functions more reliably without requiring communications with a distant host computer. In normal operations, the remote nodes initiate the communication with the operator interface computer rather than the other way around. This way, the measurement node retains the data in a temporary queue until a reliable communication link is established with the remote computer. In addition, measurement nodes can report to the computer not only on a scheduled periodic basis but also upon detecting an alarm condition. Furthermore, node-driven remote equipment consumes far less power because communication resources do not need to be continually active, as they would need to be to respond to commands from a host computer.

To summarize, the beneficial characteristics of MCUs operating as peers in node-driven networks are

- The remote units will perform data acquisition even if the communication link with the operator interface computer is unavailable;
- The power consumption of the remote equipment can be managed effectively when nodes control access to communication links, which are often the highest power-consuming components of the remote equipment; and
- Nodes can be designed to bridge between different communication media, such as wire line, radio, telephone, and satellite, thus increasing flexibility and redundancy and reducing overall system deployment cost.

ADAS telemetry is defined as any system for monitoring the performance of a dam that includes components for data collection that are permanently installed and programmed to operate without human intervention. The definition is not limited to systems that include all potential aspects of ADASs (i.e., automated collection of data, transmission of the data to a remote site, processing data, and plotting of data).

For example, a simple system that included a data logger or computer located at the dam and programmed to automatically record measurement data would be included under this definition of ADAS, even if the data were subsequently retrieved and processed by personnel visiting the site and uploading the data manually. Use of ADAS tools has significantly improved what was once a labor-intensive and time-consuming exercise. Methods that rely on a data logger system are capable of performing continuous unattended data collection, but they require manual data transfer. Methods that rely on ADASs are capable of continuous unattended data collection, remote automated data transfer, and real-time display.

The advantages of investing in an automated system are

- Frequent collection of data, capturing performance data under varying loading conditions;
- Collection of data when manual collection would be difficult or impossible, for example, when the dam is inaccessible in the winter because of snow or when site access is difficult because of heavy rain and swollen rivers. During floods, staff can focus on data analysis and visual monitoring without spending time on data collection;
- Reduction of reader-error in data collection and transcription errors in data entry; and
- Rapid processing and plotting of data, allowing easier and timelier analysis.

ADASs have been designed to include early warning features. False alarms are likely to occur in any ADAS system. An alarm suggesting a developing dangerous situation requires visual verification before declaring an emergency.

There are typical aspects of ADAS installations, but most have unique aspects that are tailored to the monitoring needs for the dam and monitoring environment. Some typical components are shown in Fig. 7-4.

Typical components include a power source, a base station consisting of a desktop computer and monitor or just a stand-alone server, a main data logger, and means of remote communication using a software package. The base station may be installed within a project office or at a dam.

The number and type of instruments can be expanded to add more instrument connections at a later date.

Radios may be utilized to transmit data between data loggers and base stations. Spread-spectrum radios spread the normally narrow-band

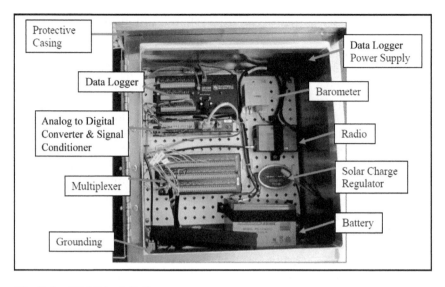

Fig. 7-4. ADAS installation
Source: U.S. Army Corps of Engineers.

information signal over a relatively wide band of frequencies. This allows the communications to be less susceptible to noise and interference from RF sources such as pagers, cellular phones, and other radios. Spread-spectrum radios do not require an individual license or frequency assignment through a regulatory agency.

7.2.2 ADAS Performance

Early systems had mixed results ranging from substantial success to outright failure. Some of the significant difficulties included contractors hampered by inadequate expertise, equipment that performed poorly or required excessive maintenance, and ADAS manufacturers that either went out of business or dropped out of the dam-monitoring business because of inadequate market potential. Sometimes installations were corrected by installing new ADAS equipment using instruments and sensors that were more reliable for automated measurements. With rapid advances in technology, ADAS equipment is continually becoming more reliable and more affordable, and there is an observable trend for increased use of ADASs for dam-performance monitoring.

Instrumentation only collects data from point measurements of performance indicators and cannot be relied upon to provide a complete picture of structural behavior. The instrumentation system can only collect the information that the instruments were designed to collect. For example,

ADASs would not identify the occurrence of seepage at a new location or the occurrence of settlement or horizontal movement at locations that were not instrumented. ADASs provide more opportunities for the extent and timeliness of analysis without curtailing opportunities for effective visual surveillance.

ADAS installations are capable of collecting more data than can be practically collected without automation. However, the collection of more data in and of itself does not provide value unless the data are regularly processed and evaluated. So, an important aspect of a successful ADAS is the allocation of resources to evaluate the collected data. There are applications that help with real-time collection, archiving, processing, and reporting of data.

It is also important that data collected from the ADAS are periodically verified through manual interrogation of the instruments. Hardware and software malfunctions can and do occur, so periodic manual verification is included in any ADAS program to ensure that judgments concerning performance of the dam are based on accurate information.

Finally, an ADAS will require periodic maintenance, calibration, and repair, and appropriate resources must be allocated for these functions. Sensors, computer hardware, and communication equipment will malfunction and require maintenance, repairs, and sometimes replacement. In addition, ADASs are subject to external influences such as construction activity, lightning strikes, vandalism, extreme weather conditions, fire, and water damage. Maintenance, repair, or replacement may be required.

Most remote data-acquisition equipment used for dam monitoring falls into one of four categories—single data logger, host-driven network, field-bus network, or node-driven network. Equipment often is battery operated and designed for minimal power consumption. This allows for the possibility of installing the equipment near instrument locations where there is often no availability of AC power. The equipment is operated from a battery, and an online battery charger is used to maintain the battery in a fully charged condition. The normal battery operation has the added benefit of allowing the equipment to continue operation (non-interruptible power) even in the event of intermittent interruptions of the battery charging power source.

The battery chargers for remote ADAS equipment are typically designed to use either AC power or solar power for inputs to the charger system.

If reliable AC power is conveniently available at the equipment location, this source is used for online charging. However, in many locations on and around dams, AC power may not be available; in such cases, battery charging is typically accomplished with a solar panel (Fig. 7-5) because of its low cost and good reliability.

Small 1-, 4-, and 8-channel loggers (Fig. 7-1) use nonrechargeable lithium batteries that last 3 to 6 years, depending on the amount of data recorded. Special extra-capacity batteries can last up to 10 years.

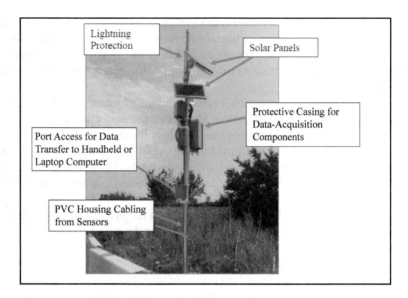

Fig. 7-5. ADAS setup using solar power
Source: U.S. Army Corps of Engineers.

When using rechargeable batteries, reliability problems may occur when the battery, the solar panel, or both are sized inadequately or marginally for the installation. Because power consumption of the remote equipment can be highly variable, depending on installed options, communication links, and programming, a degradation of power resources can occur over time if equipment configurations or programming are changed. When changes do occur from the original plan, more is usually demanded of the equipment rather than less by way of additional instruments requiring additional power-consuming signal conditioning and the increased duty-cycle of the radio transmitter.

ADAS designs require care when sizing batteries and solar panels for remote installations in order to evaluate the worst-case anticipated load, the equivalent sunshine days (ESD) for the installation site, and the lowest ambient temperature that a battery will endure. Battery technology and charging characteristics, particularly as they relate to temperature, must also be considered. Applications engineering data and assistance are available from most equipment manufacturers and systems installation contractors and integrators.

There are beneficial aspects to keeping the system as simple as possible. Whether data are manually or remotely entered, it is desirable that the number of steps and people involved between collection and entry is minimized to reduce opportunities for error.

When maintenance occurs or other suspect sources of erroneous/biasing instrumentation data, the cause(s) is documented on manual data gathering sheets and on generated plots. Example error sources include personnel unfamiliar with instrument reading, new equipment, excessive moisture in the readout device, and faulty batteries.

The reliability of the electronic sensors selected for inclusion in an ADAS is an important practical consideration. The reliability of the sensors must match the expected lifetime of the required data collection, and the ADAS system design and installation must include consideration of the potential for sensor failure. Means used to address the potential for sensor failure include redundancy and replacement. Redundancy requires designing the sensor array with redundant locations, so that some number of sensors can fail without compromising the data-collection objectives.

Designs are made to allow failed sensors to be replaced. For example, when installing vibrating-wire piezometers, the piezometers could be installed in PVC casings (standpipes) extending into the ground surface instead of in permanently backfilled boreholes. With the casing installation, a failed vibrating-wire instrument can be removed and replaced. With this installation, the response time for the piezometer readings will be that of the PVC casing (standpipe) and not that of the vibrating-wire piezometer. The piezometer readings will change only as water flows into and out of the casing.

Electronic equipment will not operate properly and will be seriously damaged from water or moisture intrusion due to condensation. Therefore, particular attention must be given to outdoor installations of ADAS equipment. Bringing multiple instrument cables into a weatherproof enclosure that also provides easy access to the equipment can present difficulties if the field conditions are not fully anticipated when the equipment is purchased.

The variations in quantity and sizes of instrument cables, appropriate space and mounting considerations for transient protection and communication options, and climatic conditions must be considered in selecting appropriate field enclosures. Resistance to vandalism must also be considered for many locations. The variations in these requirements from site to site can present some challenges for equipment manufacturers whose objective is to offer standard weatherproof enclosures that are suitable for most situations. There is usually less expense and less risk involved in purchasing properly engineered enclosures from the ADAS equipment manufacturer. However, this is true only if the manufacturer has a standard offering that meets all the site-specific requirements. This is not always the case, and custom solutions are sometimes required.

There are basically two approaches to environmental protection for outdoor installations of electronic equipment. One approach is to try to keep the dry air in and the moist air out with an appropriately sealed enclosure. The other approach is to design the enclosure to naturally aspirate, or breathe, to eliminate a significant temperature differential between the inside air and the outside air. Natural aspiration (convection) is assumed because forced aspiration, such as used in traffic signal control boxes, is not practical for solar-powered sites.

It is generally easy to keep the rain out of a weatherproof enclosure conforming to NEMA 4 (U.S.) or IP67 (international) standards. However, problems arise when large temperature differences occur between the captive air and the ambient air.

Most ADAS equipment for dams is installed in enclosures that meet the NEMA 4 or IP67 standards. Holes are punched in the enclosures for cable feed through connectors, which often violates the enclosure rating but which in and of itself does not necessarily cause a water intrusion problem.

The following scenario illustrates the main problem with low-differential pressure-sealed enclosures: The enclosure is secured with a desiccant package that slowly absorbs internal moisture up to the point that the package becomes saturated, with no more moisture-absorbing capacity. The desiccant is required to lower the dew point of the captive air. This prevents condensation on the equipment electronics as the air temperature lowers during the night or as a result of an advancing cold front. Now assume that following sunshine and hot weather, during which the enclosed air becomes quite hot, a cold front advances with heavy rain. The air is rapidly cooled inside the enclosure, resulting in a substantial drop in internal pressure. The pressure differential sucks moist air into the enclosure from the outside; the low-pressure seals required to meet NEMA 4 or IP67 do not prevent the leaks, not to mention all the cable entry seals. If water happens to pool around the weakest seals where the leaking occurs, the enclosure will suck in water directly, not just moist air. In the best case, the desiccant eventually becomes saturated with moisture and ineffective. In the worst case, vulnerable parts of the equipment become submerged in water, and failure occurs. This reality encourages the practice of making sealed enclosures as compact as possible, to minimize and better control locations that might leak, and to minimize the volume of air that must be kept dry.

There are several best practices that flow from the above scenario:

- Orient enclosures and gland panels so that water cannot pool around seals and cable entry glands and connectors;
- Use rain shields along top door seals, and use radiation shields to minimize internal air heating from sunlight; and
- Replace desiccant packages as needed. The replacement interval can vary dramatically depending on climatic conditions. Color changes of silica-gel desiccant make it easy to determine when replacement is needed. Only replace the desiccant during dry weather, because the cabinet will be closed up with the prevailing moisture content of the ambient air.

Avoid locating enclosures in underground vaults unless the vault will never fill with water during a rainstorm or water-main break. Otherwise, the enclosure must be NEMA 6-rated and -pressurized to prevent the type of leaking described above.

Aspirated enclosures are designed to keep out rainfall while preventing the entrapment of moist air within the enclosure. The aspiration keeps air moving through the enclosure to prevent condensation on internal equipment. This works most successfully when there is AC power available to run a motor and fan to force the air circulation. With power available, the air can also be heated slightly at the intake, preventing any possibility of condensation. Air intake and exhaust ports must be filtered to keep out insects and dust. Convective aspiration is not as effective because air flow is impeded by the requirement for filtering. However, successful installations with natural aspiration are possible using larger enclosures to develop adequate convective pressure differentials to turn over the air.

Successful environmental installations for dam monitoring have been implemented by installing a NEMA 4-rated enclosure inside of a larger convective enclosure with natural aspiration and minimal filtering. This configuration prevents the possibility of water pooling around enclosure seals and buffers temperature changes between the captive air and the ambient air.

Surge voltages or electrical transients are induced from many types of sources into the wires and metallic cables connected to electronic equipment. Common sources of these surges are the switching of large inductive electrical loads, such as motors, switching disturbances on the electric power grid, electrostatic discharges from human contact with connection terminals, and lightning discharges. These transients pose a threat to all ADAS equipment used for dam monitoring.

Electronic instrumentation is particularly susceptible because extended signal cables are often connected between sensing devices and the measuring equipment. The earth ground potential difference between spatially separated points is often driven to momentary extremes. In addition, nearby atmospheric discharges or the more violent ground strikes induce high-voltage transients into the cable itself, acting as an antenna. Without effective transient protection, a high-voltage pulse may be generated between the equipment terminals and the chassis ground to which the electronic components in the equipment are connected. Unprotected electronic components are likely to be damaged by high-voltage pulses.

For transient protection networks to perform, the ADAS equipment must be located where an effective earth ground connection can be made. This may require special attention for embankment dams as effective ground connections may be difficult.

The circuits in electronic measuring equipment are especially susceptible to damage from transients because they employ highly integrated components with a low dielectric breakdown voltage. In addition, a low-impedance direct electrical connection from the instrument must usually be made for the measurement circuits to work properly. The most troublesome aspect of measurement circuits is that most approaches to transient protection tend to interfere with the measurement itself conducted between the sensing device in the instrument and the measuring unit. Therefore, designers must ensure that transient protection devices are used that have the appropriate characteristics. Because measurement circuits are unique to the type of instrument being measured, the instrument and ADAS equipment manufacturers can provide information for transient protection devices that do not interfere with the intended measurements.

CHAPTER 8

DATA MANAGEMENT AND PRESENTATION

It is a capital mistake to theorize before one has data.

Arthur Conan Doyle

Up to this point in the manual, failure modes were evaluated to identify questions about a dam's behavior, monitoring programs and instrumentation to collect data to answer those questions were described, and data-acquisition techniques were discussed. This chapter describes how to manage those data and present them in an organized fashion to allow the exercise of engineering judgment about a dam's performance.

8.1 DATA ARE MEASUREMENTS. MEASUREMENTS ARE DATA.

Data management includes the procedures used after measurements are made to record the measurements for subsequent processing or evaluation, organize the readings so that the data can be easily located, process the raw data into meaningful engineering units, validate the measurements to eliminate faulty data or correct obvious errors, and archive the data to maintain a historical record. Data presentation refers to the display or reporting of the instrument data.

8.1.1 Data Organization

For each instrument or survey measurement, two types of information must be managed properly to permit interpretation of the data: instrument attributes and their measurements. Instrument attributes are the characteristics of a fixed instrument. For example, the attributes of an open hydraulic

261

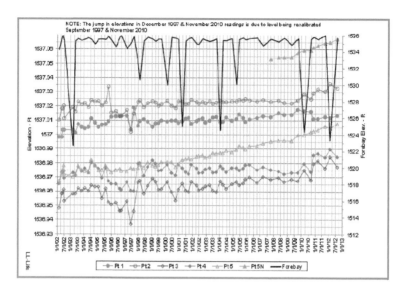

Fig. 8-1. Understanding data

piezometer installed within a sand filter would include location (*x–y* coordinates indicating its horizontal location), surveyed elevation, reference elevation, ground elevation, elevations of the top and bottom of the piezometer tip, and the installation date.

Attributes are often recorded when an instrument is installed, usually in a field logbook documenting the details of the instrument installation. The attributes may also be recorded in summary tables in paper or electronic files, but best practice is to retain the original logbooks as the data source. Most instrument attributes do not change over the life of the instrument. Some do change. Attributes may change as the result of instrument maintenance, repair, or recalibration. A new reference point may be established requiring a data adjustment. Whenever an instrument attribute changes and affects the processing of the raw data, the date of the change is recorded so that the field readings can be processed properly. For example, a damaged piezometer riser pipe may be replaced, requiring a new surveyed elevation. Historic data would remain the same, but future data reduction would utilize the new elevation. When this is not done, data become difficult to interpret, as illustrated in the unadjusted data shown in Fig. 8-1. The discrete jumps in Pt3 and Pt4 are at least partly caused by level recalibrations in December 1997 and November 2011, not by actual response.

One instrument attribute of particular importance is the instrument status code, which reflects the condition of the instrument itself; this is distinct from a reading status code, which conveys information on a specific reading. Typical instrument status codes might include **A**ctive for instruments operating normally, **M**aintenance for instruments that need maintenance but still produce accurate instrument readings, **R**epair for instruments that do not produce accurate instrument readings and require repair, and **X** for instruments that have been destroyed or abandoned. Instrument readings are periodic measurements. An open hydraulic piezometer measurement is the depth from the reference elevation to the water surface in the riser tube. The instrument reading should be recorded with the date and time of the reading.

Instrument readings may be recorded manually in field logbooks or on standardized forms or may be downloaded directly as electronic data. The original paper or electronic record is often retained as the original data source in case there are any questions about the data.

8.1.2 Data Recording

Each instrument type yields a raw measurement that is the primary piece of information that is being recorded and preserved. Raw data are the direct output from the instrument and may or may not be in meaningful engineering units. Examples of raw data include the measured depth to water in an open hydraulic piezometer, the frequency from a vibrating-wire piezometer, the voltage from the accelerometer within an inclinometer probe, the height of water in a flow measurement weir, the vertical offset of a settlement point from an established datum, or the spacing between reference points of a crack monitor. All may have different units.

For many modern electronic instruments, the value displayed on the readout unit may have already been converted into meaningful engineering units. The displayed data value is typically converted from the raw data using a formula that may include user-defined calibration factors. If readings are recorded manually, the displayed data value is recorded. If the readings are downloaded from the readout unit, the downloaded data may include both the raw data value and the displayed value. In both cases, it is important that the calibration factors are recorded with the data or that they can be calculated from the raw data and displayed data pairs.

The familiarity of the field technician with the normal and expected ranges of the instrument readings is very valuable because it can help identify suspect readings while still in the field, allowing an instrument to be reread to confirm or correct the original reading. Field reading forms for manual data collection can be printed and carried into the field to show the previous reading or the range of recent readings, facilitating detection of any significant variation.

Standard reading status codes can be developed to indicate the conditions associated with an instrument reading. For example, for one dam project, the reading status codes for open hydraulic piezometers included

Normal, Dry, Obstructed, Contaminated, and Questionable. Note that reading status codes convey information on a specific reading and are distinct from instrument status codes, which reflect the condition of the instrument itself. In all cases, the depth measured by the sensor—even if a piezometer was dry or obstructed—was recorded, together with the status code. The Normal status code was applied when the water level measured was at the expected depth. The Dry status code was applied when the sensor reached the bottom of the instrument without detecting water. The Obstructed status code was applied when the sensor did not detect water and could not be lowered farther in the standpipe, but comparison with the installation or maintenance records indicated that the sensor had not reached the elevation of the bottom of the instrument. The Contaminated status code indicated that oil or other hydrocarbons were detected on the sensor probe and wire. The Contaminated status code not only indicated potential leakage from the reservoir but also alerted the data user that the field reading did not accurately reflect the total head. The Questionable status code was used when the readings that appeared to indicate the presence of water in the standpipe were repeatable but were outside the normal expected range. The Questionable status code alerted the data user that the reading merited extra scrutiny before being accepted as accurate.

Whether written into a field book, entered into a field data collection form or electronic device, or acquired electronically, field data is placed in a permanent, consistent, retrievable, and traceable manner. Recording of data in field books, field reading sheets, or electronic forms allows for comparison of current readings with previous readings at the time the readings were collected. That comparison may be the linchpin that resolves a question about a performance indicator (e.g., has total seepage exceeded a Threshold Level)? This early comparison of the data aids in assuring the correctness of the readings and also allows for early detection of problems with instruments. The instrumentation monitoring plan specifies the actions required by the data collector if the field readings are outside of the expected range, such as obtain a second reading, inspect the instrument and readout device for damage, observe the area for signs of disturbance or distress, or alert the dam safety engineer to assess whether any action is required to ensure safety.

Project-specific field reading forms are developed to record both visual observations and instrumented measurements. Examples are provided in Chapter 13. Good practice includes columns for the current readings, as well as for the high and low limits of expected readings (threshold values).

Many owners find it beneficial to use dedicated software for the purpose of automating some of the processes around data management and presentation. Such data visualization, management, reporting, and alarming software is now available from several providers. The software is installed in an on-site or remote-site computer that is interfaced by telemetry or direct interface to the data loggers installed on the dam that are collecting readings automatically from instruments, as described in Chapter 7. Among its main features, the software typically allows the display of readings from all instruments in real time, graphs of readings in relation to time, and a comparison of readings to predetermined alarm thresholds. When alarms are detected, the software can be configured to generate actions, such as sending a text message or email, or to initiate a visual alarm or siren. Combined graphs such as piezometric levels in relation to reservoir level or seepage flows can be displayed or on another graph that enables better understanding of the behaviors and their evolution. Reporting functions are usually available to facilitate periodic reporting of instrument readings in tabular and graphical form.

The screens shown in Fig. 8-2 are typical outputs from typical data management and visualization software. The first two screens show real-time readings of concrete pressure cells (CPCs), embedment strain gauges (SG), uplift piezometer pressures, and temperatures in a concrete gravity dam. As can be seen at the bottom left of the figures, similar screens are available for pendulums, extensometers, seepage weirs, joint meters, open standpipes, and weather stations. The third screen shows historical plots of radial and tangential CPCs in the concrete lining of a powerhouse.

8.1.3 Data Reduction

Data reduction is the conversion of the raw field readings into meaningful measurements such as total head or displacement. Data reduction usually involves the application of an algebraic formula, either manually or through programming, to the raw field readings to convert them to meaningful engineering values.

8.1.4 Data Validation

Data validation is the process of reviewing the instrument readings and checking them for reasonableness. Data validation may be performed either on the raw readings or on the readings after they are converted to meaningful engineering units, but it is done as soon as practicable. Checking both the instrument and the measurement are required to validate data.

8.1.5 Instrument Validation

To ensure that the readings obtained from dam instrumentation are reliable, the instrumentation itself must be evaluated periodically. Readings

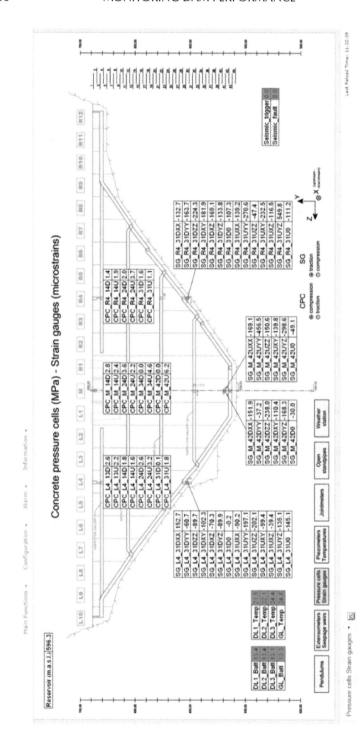

Fig. 8-2. Real-time screens and historical plots

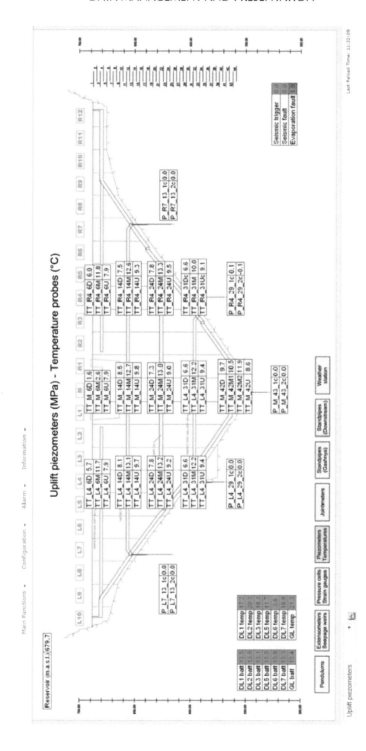

Fig. 8-2. (Continued) Real-time screens and historical plots

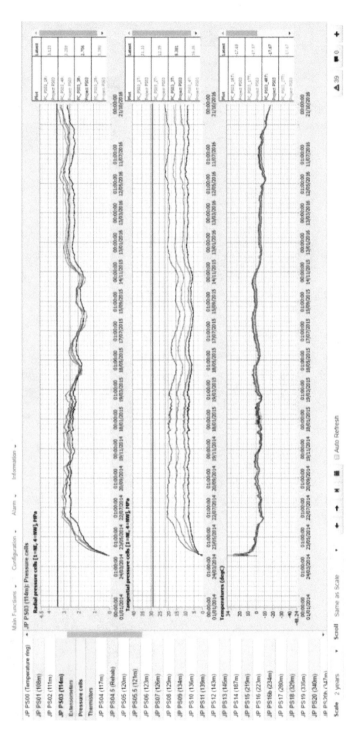

Fig. 8-2. (Continued) Real-time screens and historical plots

may be acquired, reduced, displayed, and reported correctly yet be of limited value because of instrument malfunction, rendering evaluation of dam behavior impossible.

Instruments are calibrated regularly to avoid bad measurements. Instruments are evaluated by checking measurable properties and by careful assessment of the instrument readings. Functional parameters of the instrument are checked on a scheduled basis and whenever instrument readings suggest the possibility of questionable reliability.

For example, in an open piezometer, the reference elevation from which the water depth is measured is periodically checked against a reference elevation. The standpipe is checked for obstructions by sounding to the tip elevation and checking the result against the installation records. The tip may need to be flushed to dislodge sediment. The response of the piezometer may need to be checked by dewatering or adding water to the standpipe. In some instances, the influence zone seals may need to be checked by adding water to the annulus between the surface protective casing and the piezometer standpipe or to another existing piezometer in the same hole.

For electrical piezometer transducers, the input and output voltages or frequencies and the steadiness of the reading can be checked. If the transducer is removable, its response and calibration can be checked by submerging it in various depths of water or applying known pressures. If removed, it is imperative that its elevation is accurately recorded upon its replacement. Whenever possible, replacing an instrument at the same location is good practice to allow consistency of measurements.

In many cases, the manufacturer has established procedures that prescribe methods for testing its instruments. The manufacturer's operation manuals are retained as references on operation, calibration, and troubleshooting. Most manufacturers offer technical support, and they may be needed to help with testing more sophisticated electronic instruments.

8.1.6 Reading Validation

The second element in assessing instrument reliability is a careful review of the instrument readings. The easiest cases to deal with are those in which the readings yield values that are impossible or as expected. Be aware that anomalous readings can be deceiving if they fall within the expected range. A reading that is within the realm of possibility but indicates unexpected behavior must be carefully investigated. The first step is to repeat the reading to determine if the instrument is faulty or if the dam performance is outside of the expected range. Any unusual instrument readings require explanation. If data collection and processing have been ruled out as causes of the anomaly, the instrument function must be investigated.

An investigation of instrument function will usually reveal why a reading does not seem plausible. If needed, the instrument can be abandoned, repaired, or replaced. An investigation may yield inconclusive results, and the decision to rely on the instrument data will require judgment. Dam behavior can be enigmatic. Restraint is needed when classifying data as erroneous. The confidence placed in the data may make the difference between timely notification of an emergency, late notification of an emergency, or the issuance of a false alarm.

Computer or data logger entry permits rapid checking of a reading against expected or historic values. Data obtained automatically by data-acquisition systems can be immediately validated as readings are recorded. In actual practice where computers do the graphing, proofreading the data displays or data plots is often the most efficient means of checking the data.

Validating data requires a technical background and experience to understand how the dam was designed and behaves, its performance indicators, how various measurements are made, and access to information on the external geometry, internal geometry, surrounding geology, and material properties of the dam.

In some cases, questionable readings can be corrected if the cause of the error is clear. For example, during manual data collection, it may be obvious that an instrument reading was entered in a form field for the wrong instrument or that digits were transposed. Any corrections are documented with the original data, preferably in the field logbook or on the field reading form containing the original data. Questionable data either are not used or flagged to document the uncertainty of the readings.

8.1.7 Data Archiving

Archives may be either paper or electronic records that include both the raw and the reduced data, as appropriate. Projects with many instruments or long-term monitoring programs can benefit greatly by archiving the instrument data in an electronic database, but for smaller projects with few instruments, paper records may be an acceptable choice.

The length of time to store archived records will depend on the nature of the project and the type of data, but many dam projects have a full dataset that covers the entire life of the project. This can prove useful in the event of performance assessment against new criteria, structure modifications, litigation, or dam failure. Important instrument data may be backed up and stored in multiple locations to protect against loss from accidental misplacement or physical damage. Periodic review is needed to ensure compatibility with current software and hardware.

The value of archiving historic instrument data can greatly outweigh the storage costs. The following quotation from the U.S. Army Corps of Engineer's Engineering Manual is worth noting:

Valid instrumentation data can be valuable for potential litigation relative to construction claims. It can also be valuable for evaluation of later claims relative to changed groundwater conditions downstream of a dam or landward of a levee project. In many cases, damage claims arising from adverse events can be of such great monetary value that the cost of providing instrumentation can be justified on this basis alone. Instrumentation data can be utilized as an aid in determining causes or extent of adverse events so that various legal claims can be evaluated. (USACE 1995)

8.2 DATA PRESENTATION

Data presentation encompasses the portrayal of the reduced data in tabular or graphical form. Graphs of time versus readings are generally recognized as one of the most powerful visual means of communicating information (Williams 1985). When necessary, graphs can show the data set, give a summary of behavior, and illuminate detail all at once.

Instrumentation monitoring is worth little if the results are not evaluated and communicated to the appropriate decision makers in an understandable and timely manner. The monitoring presentation or report easily allows an engineer, even one not intimately familiar with the dam, not only to understand the findings but also to develop conclusions without having to refer to a great deal of other documentation for key information.

8.2.1 Defining the Objective

The form of data presentation depends upon the intended purpose. Data may be presented in one form for engineering evaluation but in a different form for a public presentation or summary report. Each format may include different data subsets, instrument attribute information to aid interpretation, and different scales. When the choice of format is suitable for the intended purpose, the objective of the data presentation is clearly defined.

Some typical objectives include confirming design estimates, monitoring historical or long-term trends, comparing actual values to predicted values, detecting changes in the data values, or consolidating data from multiple instruments to improve the understanding of the dam behavior.

8.2.2 Example Formats

One of the most common forms of graphical presentation is the historic trend or time series plot. The values of the data collected are seldom as important as trends. In tabular form, the presentation could simply be a two-column list of dates and instrument readings. In graphical form, the most common format plots time on the horizontal axis and the engineering parameter on the vertical axis. Historical trend plots often include multiple instruments on a single page to show how changes in one instrument compare to changes in some other instrument.

8.2.3 Synoptic Presentations

Another common form of data presentation shows the data from multiple instruments acquired at approximately the same time and presented to show the variation in the measured parameter across an area or along a path. Examples would include total head contour plots or settlement along the crest of a dam. Graphical forms are generally better for this type of presentation because both the values and the locations can be easily portrayed, but tabular forms can also present useful information such as the range, minimum, maximum, and average of the values.

8.2.4 Comparison to Predicted Values

Perhaps the most informative format for presenting the data involves comparing the instrument readings to the predicted values for the measured parameters. This allows the engineer to evaluate whether the dam is performing as expected, whether any of the engineering assumptions may be questionable, whether physical phenomena not considered in the design may be affecting the dam's behavior, or whether the instrument is providing erroneous data.

In tabular form, instrument readings that fall outside of the predicted range can be easily highlighted to focus the reviewer's attention on those values. Including minimum and maximum predicted values in the table provides more context with which to assess the deviation from the predicted values. In graphical form, the predicted range can be visually represented. For example, on a historical trend plot, the minimum and maximum predicted values can be shown on the plot as a shaded band, minimum and maximum lines, or low and high bands on individual data points.

8.2.5 Changes from Past Values

A similar type of presentation compares current instrument readings to historical instrument readings. An increase in seepage flow, turbidity, or crest settlement can indicate that the physical condition of the dam has changed. It is important for the engineer to understand these changes in context to properly evaluate the performance and safety of the dam. As with the comparisons to predicted values, historical comparisons can be presented in either tabular or graphical form.

8.2.6 Hysteresis Plots

Hysteresis plots are often an informative way to plot the correlation between two variables and to illustrate how the correlation changes with time or with the direction of change. Hysteresis plots are useful, for example, in displaying the response of piezometers to changes in reservoir level or the patterns of movement of a concrete dam with changes in temperature.

8.3 REPORTING INSTRUMENTATION DATA

The objectives of the instrumentation data report are to clearly present the current field readings, highlight any deviations from the normal or expected readings, document historical performance, and preserve performance history. It provides information to the dam safety engineer to evaluate the performance of a dam along with other information such as geological data, design data, construction records, and visual inspections.

8.3.1 Reporting Instrument Reading Results

To ensure that data are evaluated in a timely manner, a routine process to reduce, validate, archive, and report the instrument reading results is required (Terzaghi and Peck 1968). The process need not be elaborate but should include the following elements:

- The data management process is a written document. It explicitly defines the steps in the process, the person responsible for each step, the inputs and outputs from each step, and the schedule for performance;
- Field readings are required within an appropriate time frame to alert operators and owners of any problems;
- A review is made of the field readings to document that the readings have been received, to indicate whether the data are complete, and to note any omitted, unexpected, questionable, or out-of-bounds readings. If warranted, the review may suggest that unusual readings be retaken; and
- A report of the instrument reading results is prepared for the engineer who will evaluate the dam performance. This report would include the reduced readings and may include supporting information such as the raw data or historical trend plots of past readings. The report may present the data in tabular or graphical form. Having the capability to view the data in both formats can provide a clearer interpretation of the data. Access to an online database that can present the data in several formats provides the greatest flexibility.

Dam size and complexity and organizational preferences produce different approaches to preparing data for evaluation. Personnel who collect the data and produce the data reports and the engineers who evaluate the dam performance may be different. At other projects, these tasks may be performed by a single individual. The details of the data management and presentation plan reflect the staff organization and organizational preferences. Having a periodic independent review of the instrumentation system and data can help protect against complacency or oversight, even in small organizations. If the tasks are distributed among several parties, it is important that a single individual has a lead role to assess whether data collection and evaluation are occurring according to schedule and to acceptable levels of quality.

8.3.2 Report Content

Performance indicators are likely to include

- Seepage quantity and quality,
- Reservoir levels and piezometric pore pressures, and
- Movement.

The introduction of the data report explains which measurements are being reported. For example, small dams may have only reservoir level and crest monuments. For each performance indicator, the information may be divided into three parts:

- Instrumentation description,
- Instrument readings, and
- Data validation notes.

Each subsection contains a description of the instrumentation involved. For example, a discussion of seepage performance would begin with a description of the weirs, flumes, samplers, or other devices used to obtain seepage data.

Instrument types and locations are included so that the report becomes the primary documentation of the instrumentation. The presentation of reading data that follows the instrumentation description is organized according to the performance indicator measured.

Complete reports contain drawings and cross sections of instruments. It may not be necessary to go to any document other than the report on instrumented performance for information on the layout, location, type, and operating condition of the instrumentation. Previously measured but abandoned instruments are identified.

The report includes performance indicators and their measurements in tabular or graphical form. Both current data and historical results are reported, preferably graphically. Time-versus-measurement graphs are the easiest to evaluate. In some cases, it may be neither desirable nor practical to display all data for a given installation (e.g., an observation well with a 20-year record of weekly readings). In some cases, all data are presented.

Data validation notes discuss the characteristics and quality of the data. Gaps in the data record, questionable readings, instrument reliability, and other facts that affect the data quality are described. A summary of the most significant aspects of the recorded behavior is required.

8.3.3 Reporting Instrumentation Data

This section presents a sample format for a periodic instrumentation data report. Many other formats are equally valid, and, of course, the sections would be customized to the specific project. Consistency in the report format improves efficiency, facilitates locating relevant data, and facilitates comparisons with different periods.

Identification of the dam site

Example: "This Instrumentation Data Report covers the XYZ Dam site. The site instrumentation covers the dam itself, the dam abutments, the upstream water level, and downstream area."

Identification of the monitoring period

Example: "The report includes new instrument readings from 1 March 2015 through 31 March 2015, as well as historical readings required for trend analysis."

Identification of the installed instruments

Example: "Instrumentation at the XYZ Dam site includes headwater and tailwater gauges, weirs, piezometers, inclinometers, and surface settlement points."

Data presentation (repeated for each performance indicator)

Performance Indicator (Seepage and flow quantity)
Measurements
Data plots or tables

- Headwater and tailwater levels,
- Observations and recommendations,
- Significant performance aspects,

- Adequacy of instrumentation, and
- Adequacy of reading and reporting schedule.

If measured, the seepage flow rate is a very important performance indicator. Most dams are constructed to retain water, and the seepage rate is a primary indicator of the ability of a dam to fulfill this primary function. More important, changes in the seepage rate can provide an early indication of physical changes in the structure of a dam.

Seepage quantity and quality must be correlated with headwater level and precipitation. Other influencing variables may include ambient and concrete temperature, reservoir water quality, tailwater level, groundwater level, and groundwater quality. The following sections provide examples of how data may be reduced and presented.

8.3.4 Reduction and Display of Seepage Quality

Water-quality data related to a performance indicator is generally limited to turbidity. Turbidity may be reported in simple terms such as clear, cloudy, muddy, or carrying particles.

If measured, water quality can be reported as turbidity or as total suspended solids (TSS). Turbidity can be measured in the field with direct reading instruments or by visual comparison to prepared standard samples prepared in a laboratory.

Water-quality data on the dissolved salts or TDS may come from direct reading meters or from laboratory analyses. In either case, the data are most likely to be received in meaningful engineering units, not as raw data. However, the reviewer must be involved with determining and understanding where, when, and how the samples are obtained and analyzed. The measurement of most interest is that of the mass of dissolved material per unit volume of seepage water. Measuring the dissolved concentrations of particular compounds may also help identify the dissolution of foundation bedrock or grout curtain.

Water-quality data are displayed in a variety of ways. Because the reviewer is most often interested in the rate of mass loss in the foundation or embankment, a simple plot of turbidity, TSS, or dissolved concentrations versus time may be useful. Such plots may also indicate useful correlations.

For chemical analysis, a quick comparison of concentration and constituents can be made by comparing the size and shape of graphical representations such as Stiff, Piper, or Schoeller diagrams. These are different methods of graphically displaying water-quality data, and details may be found in any groundwater or hydrogeology text.

A Stiff diagram, shown in Fig. 8-3, portrays the equivalent concentrations of the major ions in a water sample along a series of parallel horizon-

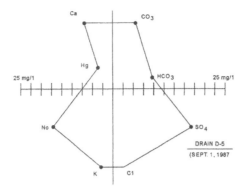

Fig. 8-3. Stiff diagram
Source: ASCE Task Committee on Instrumentation and Monitoring Dam Performance (2000).

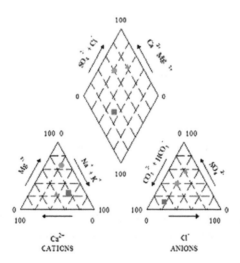

Fig. 8-4. Piper diagram

tal axes. Traditionally, cations (positively charged ions) are plotted on the left and anions (negatively charged ions) are plotted on the right.

In a Piper diagram, also known as a trilinear diagram, shown in Fig. 8-4, the relative ion concentrations in milliequivalents per liter (meq/L) are plotted in two triangles, one for cations and one for anions, with each point indicating the relative percentages of the total cation or anion concentrations. Projecting the cation and anion relative percentages onto a gridded quadrilateral allows a visual comparison of the geochemical properties among different water samples.

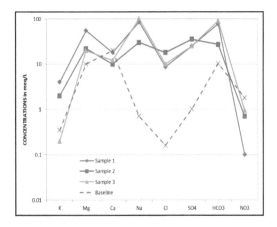

Fig. 8-5. Schoeller diagram
Source: ASCE Task Committee on Instrumentation and Monitoring Dam Performance (2000).

A Schoeller diagram, shown in Fig. 8-5, presents the individual ions spaced evenly along the horizontal axis and plots the ion concentrations in milliequivalents per liter against a logarithmic vertical axis. It is particularly useful in identifying water samples with similar patterns of concentrations.

8.3.5 Reduction and Display of Hydrostatic and Piezometric Data

Hydrostatic data refer to the reservoir headwater and tailwater levels. Piezometric data refers to pore water pressures within a dam, its foundation, and its abutments.

The piezometric level or total head is the elevation to which water from a given pressurized location would tend to rise in a tube vented to atmospheric pressure. Generally, this form of pressure expression is used because of the ease of relating it to the reservoir surface elevation. It can be measured directly by actually allowing water to rise in a tube, as is done in open piezometers, or indirectly, as is done in closed piezometers or by pressure transducers by measuring the force exerted on a calibrated diaphragm or a closed tube. The latter method determines the pressure head that can be converted into an equivalent height of water, which in turn can be used to calculate total head by adding the piezometer tip elevation (also known as the elevation head). For groundwater flow, the velocity head is generally ignored.

Total head = Pressure head + Elevation head

Measurement of total head may come in raw form as depth to water. Thus, the elevation from which the depth is measured must be known.

Usually, the depth is subtracted from the top of pipe elevation. If the pressure head is needed, the tip elevation of the piezometer can be subtracted from the total head to yield a height of water that can be converted to an equivalent pressure. In a properly installed and functioning piezometer, the pressure head corresponds to the pore pressure at the location of the piezometer tip. When stability analysis is the concern, it is most appropriate to express pressure in the same units that the material strengths are expressed in, usually force per unit area.

Raw data from closed piezometers may come in various forms. In some cases, such as with pneumatic and hydraulic sensors, the pressure is measured directly, and the readings can be used as they are or converted to total heads. In others, such as circuit strain or vibrating-wire sensors, the readout is electrical resistance or period of vibration, respectively, and requires conversion according to the data reduction formula and instrument calibration factors.

Caution is required to be certain that data are reduced with the correct equations and associated with the correct elevations. For example, it may not become evident that readings of depth to water are being associated with the wrong standpipe until a depth measurement indicates the impossible situation of a hydrostatic level below the tip elevation.

Because of the flow characteristics of open piezometer systems, there are trade-offs between the length of the collection zone and response time. The total heads measured by a piezometer are taken as the midpoint elevation of the collection zone.

Common methods of displaying hydrostatic-level data consist of hydrographs (water level versus time), hysteresis plots (water level versus reservoir head), pressure diagrams (maps of the pressure distribution), and total head contours.

Graphs of water level or pressure versus time, shown in Fig. 8-6, are plotted at a scale that allows deviations of significance to be easily identified. Reservoir and tailwater levels are graphed on the same page and, if practical, to the same scale. This allows a visual estimation of the piezometer's response to fluctuations. Several piezometers can be shown on one page, and it is often convenient to do so if several piezometers have a common feature.

Examples of common features would be all piezometers at a given elevation in a given zone, perhaps all piezometers in a zone if there are only a few, or all left abutment piezometers. Tailwater level versus time is also graphed because the driving variable is really "reservoir head," which is the difference between reservoir level and tailwater level. In some cases, graphing the reservoir head may be more informative than graphing the reservoir level.

To show the relationship between piezometric heads and the reservoir level, a hysteresis plot may be helpful. An example hysteresis plot is shown in Fig. 8-7.

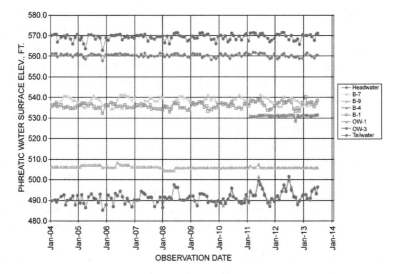

Fig. 8-6. Graph of reservoir and piezometer levels versus time
Source: ASCE Task Committee on Instrumentation and Monitoring Dam Performance (2000).

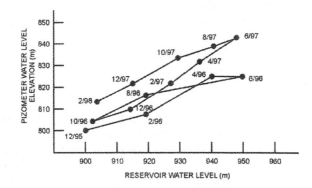

Fig. 8-7. Hysteresis plot
Source: ASCE Task Committee on Instrumentation and Monitoring Dam Performance (2000).

If the sensor location is such that tailwater level has little effect, then a plot of piezometer reading (dependent variable) versus reservoir level (independent variable) will suffice. With such a plot, it can easily be determined if there are factors other than reservoir level that influence the piezometer and, more important, if their relationship is changing with time. To get a clear picture of what is happening with time when several cycles are graphed, it may be necessary to differentiate the data in different time periods by color, symbol, or annotation. It is possible that because of the location of the instrument that tailwater, groundwater, or river stage is the prime influence, and those independent variables are plotted against the instrument reading. Hysteresis plots can also clearly demonstrate any critical levels (levels at which the dependent variable response changes) for the independent variable.

Fig. 8-8 illustrates pressure distribution within an embankment. Lines are drawn to connect points of equal pressure. Using elevation information, the pressure data may be used to also produce lines of equal potential energy (pressure head plus elevation).

A plan view of the pressure distribution is a horizontal slice through the distribution at a given elevation. To do this, the data used to draw pressure contours or equipotential lines must be from the same or nearly the same elevation within the dam or foundation.

Pressure contours are perhaps most valuable when embankment stability analysis is the concern. Pressure values for any point on the diagram can be determined by interpolation between the pressure contours. The equal potential diagram on the other hand is best used to understand the direction of seepage flow along with gradient in the direction of flow.

For concrete dams, the hydraulic grade line, shown in Fig. 8-9, indicates the uplift pressure beneath the foundation contact or at other surfaces.

8.3.6 Reduction and Display of Movement Data

Movement data encompasses surface, crack and joint, internal, and appurtenant structures. Movement may be horizontal, vertical, or a combination of both.

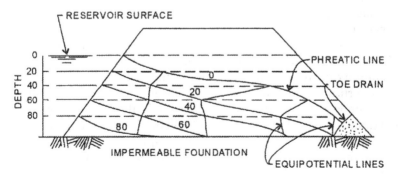

Fig. 8-8. Embankment cross-section pressure diagram
Source: ASCE Task Committee on Instrumentation and Monitoring Dam Performance (2000).

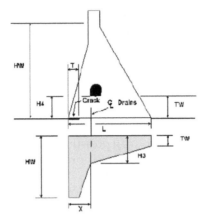

Fig. 8-9. Hydrostatic grade line
Source: ASCE Task Committee on Instrumentation and Monitoring Dam Performance (2000).

Raw data for displacement measurements can be of several forms: direct measurements, angular measurement, measurements relative to some reference point, pressure, stress, voltage, resistance, or vibration period.

All data except the direct measurement must be reduced to show magnitude and direction. Chapter 6 described the method of surveying measurement points.

Data from each type of instrument are reduced to coordinates to provide a means of visualizing the nature and significance of movement. For some instruments, the measurement is referenced to a line such as a centerline showing changes in horizontal and vertical position related to the line.

Display, however, can sometimes be more effective if actual measurements are not shown but rather a calculated value such as change expressed as a percent of the original value. An example of this could be internal vertical settlement. The reviewer would have to be aware of actual settlement and where the maximum occurred, but it may well be that it is desirable to communicate the embankment settlement to others as percent compression or vertical strain. For movement plots, it is sometimes helpful to use a logarithmic scale for time because the slope of the line after initial settlement (immediately after construction) will be constant. For example, normal compression, consolidation, and lateral spreading will yield a straight line on semi-log paper and rate changes are magnified (Sherard et al. 1983).

One important note about presenting movement data is that the sign convention used must be clearly identified (i.e., the data plots and tables must indicate whether positive data values indicate compression or expansion, upstream or downstream).

8.3.6.1 Horizontal Movement. Fig. 8-10 illustrates how horizontal movement may be presented. With such displays, the reviewer can evaluate total displacement for excessive values and look for significant differentials between adjacent points that may signify embankment instability, cracking, or internal erosion. When there are indications of a problem, net head (headwater minus tailwater) and temperature may be included along with displacement versus time on a graph to determine if the rate of displacement is sensitive to reservoir level or temperature. Data from additional lines of measurement points would allow for areal mapping of the movement and identification of patterns of movement.

In Fig. 8-11, a graph of four sets of settlement profile data for seven measurement points is shown. On such a graph, the reviewer looks at the total settlement to determine if the freeboard is adequate and if the pattern of settlement is as expected. Again, differentials between points are examined to determine if cracking or internal erosion may be a problem. Under normal conditions, the rate of settlement will generally diminish with time, although reservoir operation may affect the rate. With a regular reading schedule, the interval between lines tends to decrease with time, as shown. Determination of the elevations on additional rows of measurement points will aid in delineating any potential slope stability problems.

8.3.6.2 Internal Movement. Various instruments such as settlement plates, cross-arm devices, extensometers, settlement sensors, and slip-joint inclinometer casings are used to determine internal settlement within dams and foundations. With these instruments, compression and vertical

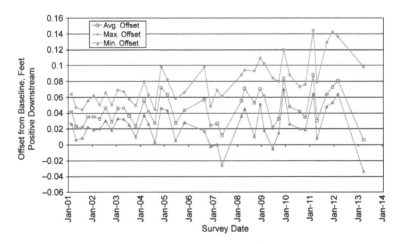

Fig. 8-10. Horizontal deflections
Source: ASCE Task Committee on Instrumentation and Monitoring Dam Performance (2000).

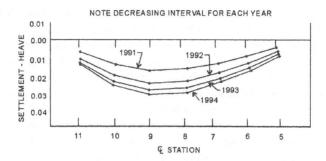

Fig. 8-11. Settlement versus time
Source: ASCE Task Committee on Instrumentation and Monitoring Dam Performance (2000).

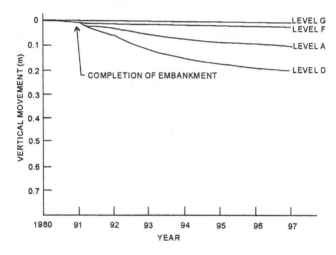

Fig. 8-12. Internal settlement versus time
Source: ASCE Task Committee on Instrumentation and Monitoring Dam Performance (2000).

displacements of foundations and embankments can be determined at any depth. Data records can begin during construction.

A typical plot of internal settlement or compression is shown in Fig. 8-12, and it is similar to the vertical settlement plots discussed earlier. Settlement of each level was anticipated by the designers, and compression data are compared to the anticipated values. Again, note how rate decreases with time.

Internal movements, especially with extensometers, may be measured in directions other than vertical or horizontal. The principles for effective

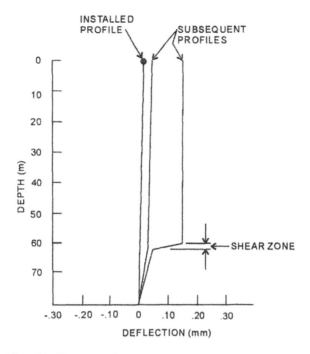

Fig. 8-13. Plot of inclinometer data
Source: ASCE Task Committee on Instrumentation and Monitoring Dam Performance (2000).

display remain the same, but the plots must clearly indicate the location of the instrument and the direction of movement.

The principal purpose of an inclinometer is measuring horizontal deflection of the casing along a vertical profile. Deflection can be measured in two planes, one parallel to the line of expected movement and the other perpendicular to it. If the casing moves horizontally, the inclinometer will measure the magnitude of movement. Readings are typically plotted as shown in Fig. 8-13. The series of readings on the same plot illustrate that progressive movement is occurring. Shear-zone elevations can be shown on a cross section to identify a failure surface. Usually the bottom of the inclinometer casing is considered as a fixed reference point. The reviewer must be familiar with the foundation to know if the bottom is fixed. If it is not, the top of the casing must be surveyed each time data are taken. All displacements from a set of readings are then referenced to the verified top of casing elevation.

Embankment construction is a special case for inclinometers because nearly every time the casing is read, there is a new, longer casing length as sections are added as the embankment height increases. Each profile

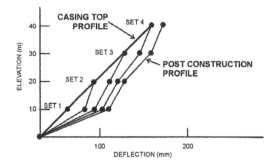

Fig. 8-14. Construction-stage inclinometer data
Source: ASCE Task Committee on Instrumentation and Monitoring Dam Performance (2000).

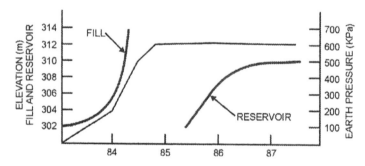

Fig. 8-15. Earth pressures response to fill and reservoir level
Source: ASCE Task Committee on Instrumentation and Monitoring Dam Performance (2000).

represents a new initial point. A good way to get an overall picture of what is happening during construction is to make a profile consisting of initial top of casing locations, as shown in Fig. 8-14. Profiles can be compared to the initial top of casing profiles as if it was an initial profile of casing.

8.3.7 Reduction and Display of Earth Pressure Data

Readings from earth pressure cells are reduced to units of stress. The data are usually plotted versus time but may also be plotted versus reservoir level or construction fill height to produce a hysteresis plot (Fig. 8-15). If the information is being used in an effective stress analysis, then the pore water pressure, measured by a nearby piezometer, is plotted along with the cell readings or subtracted from the cell readings before plotting.

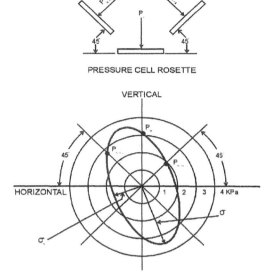

Fig. 8-16. Data from a stress rosette
Source: ASCE Task Committee on Instrumentation and Monitoring Dam Performance (2000).

If the data come from a rosette, then a stress ellipse, shown in Fig. 8-16, can be constructed to show the direction and magnitude of principal stresses. If principal stresses are calculated for each set of readings, then it is advantageous to plot principal stress, major, minor, or both, versus time, reservoir level, or fill height.

Chapter 13 contains sample data forms and plots that may be helpful in preparing a performance-monitoring report. Data management and presentation prepare the dam safety engineer to evaluate and make decisions about dam performance—the subject of Chapter 9.

CHAPTER 9

EVALUATION, DECISION, AND ACTION

Monitoring programs have failed, because the data generated
were never used. If there is a clear sense of purpose for a monitoring program, the method of data interpretation will be guided
by that sense of purpose. Without a purpose there can be no
interpretation.

John Dunnicliff

Up to this point in the manual, ways and means were presented to
address a single question: "Is this dam performing as expected?"
Failure modes and their associated performance indicators were identified. Instruments and a monitoring scheme to answer those questions were
integrated into a dam-performance monitoring plan. The plan was implemented. Both visual observations and instrumented measurements produced
data. Data were validated and their quality assessed for each performance
indicator. The instrumented data were compared with the visual observations. Reports were written and data were presented.

The next step is to evaluate the data in the context of answering a
single question about a dam's safe performance: "Is this dam behaving
as expected?" This chapter is intended to provide a foundation for evaluating, deciding, and acting on the products of the dam-performance monitoring program.

9.1 EVALUATION

Data gathered for the identified performance indicators are meant to
provide a basis to judge whether or not a dam is performing as expected.

Some examples of measurements of key performance indicators may be similar to these examples:

- Does the record of seepage volume and quality support the conclusion that an internal erosion failure mode is not developing?
- Is the measured rising trend in joint water pressure in the left abutment indicative of the development of a privileged path of leakage that might lead to loss of foundation support, potentially destabilizing the arch dam?
- Are measurements along the embankment crest indicative of foundation settlement or piping or simple consolidation?
- Do plunge pool survey measurements suggest that the toe is likely to be undercut, causing a loss of support and potentially allowing the gravity dam to slide?

As discussed in Chapter 2, experience has shown that there are vulnerabilities common to all dams, but each dam has its unique set of vulnerabilities that could provide a path to initiate a sequence of events leading to a loss of reservoir control. Evaluation of performance indicator data from a monitoring program in the context of the common failure modes and any additional failure modes unique to a particular dam requires skill and judgment. Analyses may be required to reach a conclusion about a dam's safety.

9.2 DECISION

We cannot solve our problems with the same thinking we used when we created them.

Albert Einstein

Two questions must be answered before deciding what, if any, action to take:

1. Are there sufficient reliable performance indicator data to judge whether this dam is performing as expected?
2. Have I thought of everything?

The decision process leading to appropriate action will benefit from setting limits for a measured performance indicator beyond which additional measurements must be taken or analyses performed to judge whether a dam's performance is acceptable.

Threshold and Action Levels are part of FERC's *Engineering Guidelines* (FERC 2005) for monitoring dam performance. Other organizations and agencies may use different terms, but the intent is to provide a basis to support safety decisions. As defined by FERC,

- A Threshold Level is a value used in the analysis or design or a value established from the historic record. Measurements that exceed a Threshold Level require further inquiry.
- An Action Level is a value that requires increased surveillance or the need for intervention.

To the degree practical, Threshold and Action Levels are established for each instrument, taking into account the expected performance of the dam, location of each instrument, and measurement characteristics of the readings. All data are compared with design and analysis estimates. For example, measured pore pressures and uplift pressures are compared against those used in design stability analyses and prior measurements.

Numerical limit values may be based on theoretical or analytical studies (e.g., uplift pressure readings above which acceptance criteria are no longer met). In other cases, limits may need to be developed based on historically observed behavior (e.g., seepage from an embankment dam). The assigned limit values are those that are judged to require further evaluation to determine whether a potential failure is developing or whether there is a problem with the data. Limits for both magnitude and rate of change may need to be established. In some cases, a value may be a lower limit (e.g., a decreasing trend of piezometric level may indicate the opening of a flow path to the downstream side of the dam). The limit levels reflect a careful consideration of failure modes, and more than two limit levels may be appropriate in some cases.

The established values are periodically reevaluated to determine if the values need to be changed to address daily, seasonal, or other cyclic relationships, new information, or changes to the dam. Values may be set by the designer based on his or her understanding of how the dam will perform. During actual operation, the dam may perform differently, with some of the instruments showing readings exceeding the originally set levels. Additional conscientious evaluation of the construction experience and overall performance of the dam may indicate that these readings do not indicate unacceptable performance. The values could then be revised based on the more complete picture of how the dam is performing. The reverse is also true, where undesirable performance occurs at levels below the limit levels, and the values are reduced.

The limit levels are evaluated periodically during operation to verify that they remain consistent with dam performance. Considerable care and thought must be given before revising Action Levels. It is important to verify that proposed revisions are compatible with safe performance.

Periodically, the appropriateness and adequacy of the instrumentation and monitoring program are evaluated. The evaluation includes answers to the following four questions:

- Is the instrumentation and monitoring program appropriate for the potential failure modes of concern at the dam?
- Are the type, number, and location of instruments appropriate for the behavior being monitored?
- Is the frequency of readings appropriate?
- Are the data being collected, processed, and evaluated in a timely and correct manner?

If the performance indicators for failure modes are not being appropriately monitored, they should be. Additional instrumentation or monitoring may be required. If there is a discrepancy between the measured and expected behavior of the dam, it may indicate that data do not adequately represent the behavior of the dam or that conditions exist that were not accounted for in the expected behavior. In either case, after detailed evaluation and interpretation, it may be necessary to make new analyses, perform field investigations, or install additional instrumentation.

Best practice limits measurements to potential failure modes. If there is no question related to a performance indicator identifying a developing sequence that could progress to a failure, then no instrumented measurement is required.

If trends or interrelationships between data are not clear, it may be appropriate to take more frequent measurements or collect additional, complementary data.

9.3 ACTION

I never worry about action, but only inaction.
Winston Churchill

A convenient way to decide if action is required is to think of limits as a traffic signal, shown in Fig. 9-1. Measurements less than a Threshold Level would be GREEN and signal GO. Measurements exceeding the Threshold Level but less than the Action Level would be YELLOW and signal CAUTION. Measurements exceeding an Action Level would be RED and signal STOP.

Although measurements usually fall within the expected ranges (the GREEN or GO state), there are actions that can be considered, including

- Continuing data collection unchanged,
- Reducing reading for instruments with a long unchanging record, and
- Abandoning or deactivating instruments that no longer measure a performance indicator related to a failure mode.

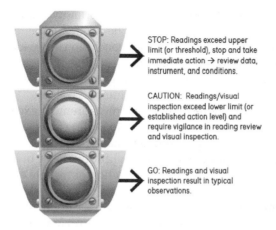

STOP: Readings exceed upper limit (or threshold), stop and take immediate action → review data, instrument, and conditions.

CAUTION: Readings/visual inspection exceed lower limit (or established action level) and require vigilance in reading review and visual inspection.

GO: Readings and visual inspection result in typical observations.

Fig. 9-1. Stop–Caution–Go
Source: ASCE Task Committee on Instrumentation and Monitoring Dam Performance (2000).

Measurements may fall within the range of threshold values (YELLOW) or CAUTION zone. Measurements within threshold boundaries are unlikely to jeopardize reservoir control. The signal has not turned RED. However, evaluation is required.

For example, a high reading of one piezometer in an embankment may exceed the piezometric level estimated in the stability analysis for that location at the same time other piezometers along the same cross section could indicate the overall piezometric line is lower than estimated in the design stability calculations. The same logic holds for piezometers used to measure uplift pressure beneath a concrete gravity dam. For these cases, a single instrument indicating values above design level does not necessarily indicate a process headed toward instability.

Measurements outside expected ranges are supplemented as quickly as possible by a visual check of conditions and the instrument of concern in an attempt to determine the likely cause of the unexpected measurement. If needed, an additional instrument may be strategically located to provide independent data for evaluation of an unusual condition. For example, one might be placed along a cross section between existing instruments to interpolate the behavior along the cross section. Or a more appropriate instrument that better responds to varying rates of change may be added.

Data outside expected ranges may lead to additional engineering investigations or analyses. Analyses may be undertaken to evaluate the margin of safety associated with a particular failure mode based on measurements. Alternatively, analyses may be undertaken to try to determine why a measured performance indicator is outside of the expected range. Depending on circumstances and the need to resolve uncertainty about a dam's behavior, additional field investigations may be needed to better define conditions. A thorough review may lead to the conclusion that widening threshold limits is justified because the margin of safety is greater than previously estimated. Again, sound engineering judgment is required.

When the signal turns RED, it is time to STOP and take immediate action to either quickly resolve why a measurement exceeds the upper bound of the threshold limit or prepare to initiate emergency procedures. It would be unusual to implement the emergency action plan (EAP) solely on the basis of monitoring data. There is little evidence of loss of reservoir control without some physical manifestation of trouble evident to visual observation.

The instrument returning the worrisome measurement may have failed; if so, its failure must be discovered quickly. A review may reveal that the action limits are too low. In any event, the review of monitoring data is intended to trigger a thorough visual inspection to judge whether a rapidly developing situation exists that requires initiation of the EAP. For remote unmanned projects, it may not be possible to physically observe the situation quickly enough to delay the initiation of emergency procedures. The exercise of good judgment is essential in deciding when to implement an EAP.

Lowering the reservoir or placing stabilizing fill is among the actions that may be appropriate, depending on the situation.

CHAPTER 10
EMBANKMENT DAMS

True stability results when presumed order and presumed disorder are balanced. A truly stable system expects the unexpected, is prepared to be disrupted, waits to be transformed.

Tom Robbins

Chapter 3 introduced monitoring considerations for embankment dams from the perspective of their potential failure modes. This chapter discusses embankment dam vulnerabilities and behavior and provides guidance about means and methods to measure the performance indicators that answer the question, "Is the dam behaving as expected?"

10.1 DESIGN

Embankment dam designs vary widely in terms of size, purpose, and type of construction materials. Common varieties are zoned earthfill dams, zoned rockfill dams, homogeneous dams, and membrane dams. Each relies on its mass to resist the imposed loads. Proper design demands that an embankment dam remain stable and retain the reservoir under all anticipated loading conditions.

Embankment dams are constructed with soil and rock and are supported by foundations and abutments consisting of soil and rock. Engineers characterize soil, rockfill, and bedrock properties by some or all the following material properties: strength, stiffness, consolidation, compressibility, density, plasticity, void ratio, permeability, and moisture content. The properties of embankment materials govern their response to the loads of the dam itself, water, and earthquakes. Materials are selected to provide sufficient stable mass to resist those loads.

An embankment dam must maintain stable upstream and downstream slopes and be compatible with the materials and geometries of its abutments. The design must meet acceptance criteria requiring margins of safety for normal, flood, and earthquake loads.

Abutment geometry, foundation stratigraphy, and character influence abutment shaping, requiring an understanding of

- Rock properties such as permeability and presence of discontinuities with marginal stability,
- Soil properties such as strength, seepage resistance, and susceptibility to liquefaction,
- Weak zones,
- Stress history, and
- Groundwater conditions.

Chapter 3 introduced three principal modes of embankment dam failure—uncontrolled seepage, overtopping, and slope instability. This chapter describes effective monitoring measures to reduce the risk of the loss of reservoir control and to answer the question, "Is this dam behaving as expected?"

10.2 PERFORMANCE

Performance depends on an embankment dam's ability to respond to physical processes experienced during its service life.

10.2.1 Settlement

Settlement results in general or local decreases in the elevation of the crest and slopes of a dam. The decreases in dam surface elevations resulting from settlement can vary gradually across the dam or occur as distinct differential movement, which can result in surface cracking. Settlement of the general embankment results from a reduction in the void volume (consolidation) within the soil or rockfill of the embankment or foundation. An undesirable consequence of general settlement of the embankment is that freeboard decreases, increasing the risk of overtopping and possible breach. Settlement can also occur owing to exterior loading, such as the weight of an appurtenant structure on a compressible soil foundation.

Transverse cracking can occur where there are abrupt changes in abutment contact geometry, such as at abrupt changes in the height of the dam over near-vertical faces in the foundation or abutment rock or over concrete structures such as outlet works conduits. Abrupt slope changes produce a sudden change in stress that can cause differential settlement resulting in cracking and damage to cores and filters and the subsequent loss of resistance to internal erosion. Differential settlement along conduits through embankments can result in joint separation or cracks in the conduits. Settlement, longitudinal cracking, and slope displacement can also be a sign of slope movement or earthquake-induced deformations. Localized settlement such as sinkholes or depressions along retaining walls can be signs of internal erosion. Dam contacts with foundation soil, bedrock, and concrete structures are particularly susceptible to seepage and potentially to internal erosion, as are contacts between the embankment and concrete structures.

10.2.2 Slope Stability

Embankment dam materials consist of discrete soil and rock particles rather than a continuous solid such as concrete. As such, these materials are dependent on friction and cohesion (shear strength properties) for stability. They must be contained by abutments and have adequately proportioned slopes to remain stable. Embankment slope stability depends on the slope angle, the strength of the embankment and foundation materials, and the phreatic surface through the embankment. A defect in one or more of those factors may trigger slope failure resulting in rotation or lateral translation of the slope that could cause a loss of freeboard or cause seepage failure through resulting cracks in the embankment. Slope instability also can occur if stresses from ground shaking during an earthquake exceed the strength of the soil.

Instability of the upstream slope can be caused by rapid drawdown of the reservoir. If the reservoir is drawn down faster than the seepage within the embankment can drain, the slope experiences rapid, undrained loading. If the undrained strengths of the soil materials are not sufficient to resist the driving forces during undrained loading, slope failure will result. Upstream filters and drains can reduce the potential for upstream slope failure, and they have become common in current practice but are seldom found at older dams.

Instability of the downstream slope may initiate during intense precipitation, causing unusually high and sustained reservoir levels. As the reservoir level rises, pore pressure in the downstream slope may rise, reducing effective stress, shear strength, and stability. Slumping may result, as shown in Fig. 10-1.

Another potential initiator is a blocked toe drainage system.

10.2.3 Freeboard

Freeboard is the difference between the embankment crest elevation and the elevation of the reservoir water surface, and thus maintenance of

Fig. 10-1. Scarp in downstream slope indicative of instability

adequate freeboard is essential to embankment stability. Embankment materials are susceptible to erosion from flowing water. As the depth and duration of water over the crest increases, the probability of loss of freeboard increases. The crest elevation of the core is an important design consideration because once the core is overtopped, breaching of the dam owing to progressive internal erosion and erosion of the embankment is likely with loss of reservoir control. Failure by overtopping is the second most common cause of embankment dam failures, uncontrolled seepage being the most common. Prudent designs provide for adequate freeboard for all design conditions, considering settlement, reservoir level rise during floods, predicted wave amplitudes, and wave run-up on the upstream slope of the dam.

10.2.4 Seepage and Leakage

Seepage is the movement of water through the voids of soil and rock of an embankment, abutment, and foundation. Water seeps through, around, and under embankment dams, controlled by the permeability of the embankment, abutments, and foundations. The volume and location of seepage depends on the seepage gradient, design, and construction. Embankment dams are often zoned, using materials of varying permeability in different parts of the cross section to keep the phreatic surface as low as possible in the downstream slope.

All embankment dams exhibit some seepage at or under the toe. The potential for higher rates of seepage exists along the contacts of embankment materials with rock foundations, concrete structures, or pipes. Leakage is water that moves through preexisting openings such as an open joint in the foundation, a crack in the concrete facing of a concrete-face rockfill dam, a deteriorated grout curtain, a tear or bad seam in an impermeable liner, or through a crack in a slurry cutoff wall.

Seepage and leakage cause problems if flow velocity rises to the point of initiating internal erosion by carrying away soil grains at the toe or at the contact between the embankment and foundations or structures. This concern is addressed by providing properly designed filters and a drain at all potential seepage flow locations to reduce the probability that uncontrolled seepage or leakage will occur.

Leakage that does not threaten the safety of the dam may still be unacceptable because of the economic loss of water.

10.2.5 Stress

Changes in stress can be caused by differential settlement between core and shell material in a zoned dam. If the core is very compressible and consolidates more than the shell, the core can "hang up" on the shell, reducing the overburden pressure and, consequently, the lateral pressure, which is lower within the core. Steep abutments may exacerbate differential settlement. The sharp change in stress at an abrupt change in slope can crack the core and initiate internal erosion.

Stresses imposed by the embankment overburden on buried conduits and structures or retaining walls may be a concern. If stresses exceed the conduit strength, cracking could occur and provide a source of water under pressure to initiate erosion of embankment material into a conduit. Conversely, a leaky conduit through an embankment may be a source of water under pressure that initiates internal erosion along a conduit.

10.2.6 Earthquake Response

As the seismic waves from an earthquake pass through the foundation and abutments of an embankment, the embankment shakes in response. Its response is dynamic oscillation and the initiation of shear stresses and deformation in the embankment and foundation materials. The shear stresses may also reduce strength in the embankment and foundation materials and lead to post-earthquake instability. A dramatic example of such strength loss was the liquefaction of cohesionless soils that resulted in failure of the upstream slope of the Lower San Fernando Dam during the 1971 San Fernando, California, earthquake (Fig. 10-2).

Fig. 10-2. Lower San Fernando post-earthquake
Source: Courtesy of EERI, reproduced with permission.

10.2.7 Construction

An understanding of how an embankment was constructed is essential to judging performance.

The embankment and its foundation require the greatest attention to detail during construction. Other features such as the provisions for seepage control, filters, and drainage vary widely with the particular design; however, performance depends on how well the foundation is prepared, how well the embankment is placed and compacted, and the care taken to provide for seepage control, filters, and drainage.

Rock foundations require preparation to receive fill. Removal of weathered and loose rock is essential. Seepage past a cofferdam or springs is troublesome because it makes compaction difficult owing to difficulty with moisture control in the fill. Care must be exercised to keep flowing water away from active fill zones, especially the core contact area. Special treatment such as slush grouting and dental concrete work may be required for fractures and weak zones to inhibit seepage and to avoid stress concentration. Proper shaping of the foundation and abutments is essential because abrupt changes in the rock surface may make proper compaction difficult and may concentrate stress that could lead to cracking of the overlying fill.

Soil foundations also require preparation to receive fill. Weak or compressible soils must be removed. Pockets of coarse soils that might concentrate seepage along the contact with the overlying fill must be removed or cut off. Control of groundwater is important, as noted previously, because of the potential for loss of fines and for poor conditions for compaction. Scarifying, discing, and compacting the subgrade are common foundation-preparation practices.

Provisions for seepage control in the foundation vary. Some embankment dams have no cutoff. Others have a simple cutoff trench filled with compacted impervious fill. High-embankment dams customarily have a grout curtain or a cutoff wall backfilled with concrete or impervious soil material. How well the cutoff is constructed will affect its performance and the potential for undesirable seepage. The decisions made and actions taken during foundation preparation and cutoff construction affect long-term performance.

Good-quality embankment construction requires careful control of material properties, moisture, and compaction. Weather is an important factor. The right materials that meet the designer's specifications must be placed in the right zones. Moisture control is important, both at the borrow area and on the fill to achieve the design density. Fill placed too dry may settle excessively upon wetting. Fill placed too wet may crack or may cause excess construction pore pressures. Frozen subgrade or fill is not allowed. Sufficient effort must be applied to compact soil to the specified density. Clean separation without contamination between zones is important to the proper function of the zones. It is important that filters and drains not become contaminated with excessive fines. Where coarse soils are used as fill, segregation of coarse particles must be avoided in areas where rock pockets may form and provide a path for undesirable concentrated seepage.

Other construction practices may affect performance:

- Overworking of fill is avoided by carefully establishing construction traffic lanes.
- Contamination between zones is avoided by carefully establishing protected crossing areas.
- Compaction around penetrations and against structures is performed with care.
- Completed elements of the embankment are protected against the degrading effects of weather.

In summary, construction practices strongly affect long-term embankment performance.

Opportunities to monitor performance differ between prospective and existing embankment dams. The differences are illustrated in the figures in this chapter; the major difference is that some instruments such as liquid-level settlement devices, fiber optic cables for distributed-temperature sensing, total pressure cells, soil extensometers, and many instruments in the foundation cannot be installed in an existing embankment but are routinely installed during construction of a new embankment.

10.3 INSTRUMENTED MONITORING

After the embankment is constructed, its performance indicators from settlement (change of shape), water pressures (reservoir and adjacent groundwater), and seepage (amount, clarity, and distribution) are measured and compared to expected performance. If responses equal or exceed expected responses, a dam is judged to meet performance criteria because it meets the required margin of safety. If a measured performance indicator suggests that a failure mode may be developing, then evaluation is required.

The level of effort involved to evaluate depends upon the significance of the difference between observed and predicted behavior. Several performance indicators of behavior may require measurement:

- Movement to assess change of shape,
- Water pressure,
- Stress, and
- Seepage and leakage to assess changes in flow quantity and quality.

Chapters 5 and 6 provided a palette of instrument and survey means and methods suitable for measuring performance indicators. Table 10-1 summarizes the possibilities.

Not all the properties listed in the table are, or necessarily should be, monitored for all embankment dams. For a concrete-face rockfill dam, the horizontal and vertical movements of the face, the settlement of the crest, and the quantity of seepage through the face may be more important than other performance indicators.

Detailed internal distribution of piezometric pressures may be very important in a homogenous embankment because the entire cross section may develop pore pressures that could cause slope instability.

Some of the performance characteristics that are considered important for the most common types of embankment dams are shown in Table 10-1. As discussed previously, the monitoring program for a particular embankment dam requires careful design to address the specific performance indicators of that dam.

Existing and prospective dams and typical locations of instruments to monitor their performance are shown in Figs. 10-4 to 10-9, with the key to the symbols representing the various instruments shown in Fig. 10-3.

Care must be exercised when making penetrations for installing instruments in an existing dam. Proper filter criteria and compaction around an instrument are needed to avoid opening a seepage path. Borings to install instrumentation, especially in a core or at the toe, can create a shortcut through filters that may fracture the core. Installing instruments in a core

Table 10-1. Instrument and Survey Methods.

Performance indicator	Measurement location	Instruments - methods
External movement	Crest or other surface location of interest	Visual observations, geodetic survey
Internal movement	At points of interest within embankment	Inclinometer, settlement gauge, extensometer
	At points of interest within foundation	Inclinometer, extensometer, settlement gauge
Water pressure	Within embankment	Piezometer, observation well
	Within foundation	Piezometer, observation well
Stress	Within embankment	Total pressure cell
Seepage quantity	Where measurable	Calibrated container, weir, flume, flow meter
Seepage quality	Where measurable	Turbidity meter, water chemistry
Flow	At toe	Weir, flume, fiber optic cable for distributed temperature measurement
Earthquake response	Crest and adjacent bedrock	Seismograph

requires great care and is generally avoided unless there is a question that cannot be answered otherwise.

In new dams, current practice is to avoid installing instruments that extend upward through subsequent embankment lifts. Where extensions such as a pipe cannot be avoided, special care is required to fully compact the soil around the extension to reduce the potential of creating a seepage path. An extension at a lift surface is an obstruction for construction traffic. Marking is required to avoid damage.

The scope and procedures of installing an instrument penetrating an embankment requires special care. Design and peer review by experienced geotechnical engineers is required. Installation requires qualified personnel experienced with installation of similar instruments in embankments.

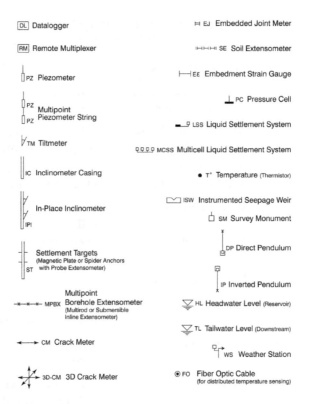

☐DL Datalogger	⊢ EJ Embedded Joint Meter
☐RM Remote Multiplexer	⊩⊣⊣⊣⊩ SE Soil Extensometer
☐PZ Piezometer	⊢──┤EE Embedment Strain Gauge
☐PZ Multipoint ☐PZ Piezometer String	⊥ PC Pressure Cell
⊬TM Tiltmeter	▬─9 LSS Liquid Settlement System 9999 MCSS Multicell Liquid Settlement System
‖IC Inclinometer Casing	● T° Temperature (Thermistor)
In-Place Inclinometer IPI	⊏▽⊐ ISW Instrumented Seepage Weir ⊔ SM Survey Monument
Settlement Targets (Magnetic Plate or Spider Anchors ST with Probe Extensometer)	DP Direct Pendulum
Multipoint ─×─×─×─ MPBX Borehole Extensometer (Multirod or Submersible Inline Extensometer)	IP Inverted Pendulum ▽ HL Headwater Level (Reservoir)
◄───► CM Crack Meter	▽ TL Tailwater Level (Downstream)
3D-CM 3D Crack Meter	⌐→ WS Weather Station ◉ FO Fiber Optic Cable (for distributed temperature sensing)

Fig. 10-3. Key to symbols

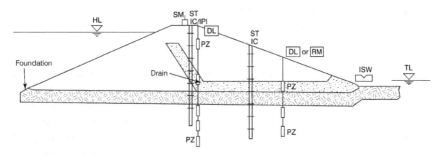

Fig. 10-4. Existing homogeneous embankment dam

10.3.1 Settlement

During the design of embankment dams, the engineer typically estimates the magnitude of foundation and embankment compression owing to the weight of the embankment. Compression of the embankment materials owing to self-weight consolidation is generally limited to about 1% of the

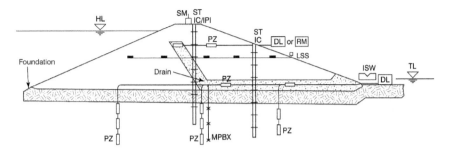

Fig. 10-5. Prospective homogeneous embankment dam

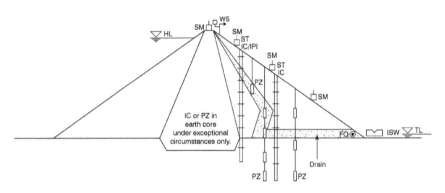

Fig. 10-6. Existing zoned embankment dam

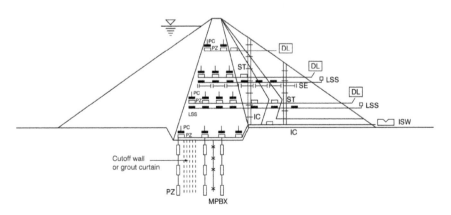

Fig. 10-7. Prospective zoned embankment dam

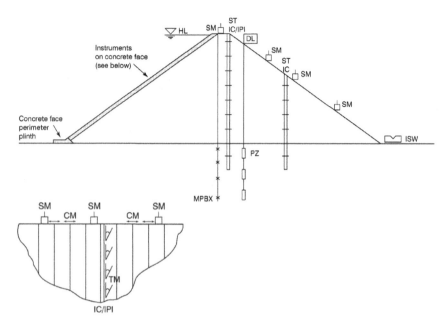

Fig. 10-8. Existing membrane (CFRD) embankment dam (section and plan of face)

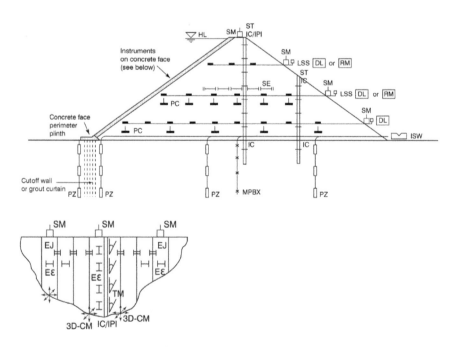

Fig. 10-9. Prospective membrane (CFRD) embankment dam (section and plan of face)

embankment height in modern compacted embankment dams. Compression of relatively softer foundations can result in increased amounts of settlement. During an earthquake, shaking may cause embankment settlement or liquefaction of loose sands in the foundation. When establishing a design crest elevation for the dam, the estimated crest settlement resulting from the compression must be taken into account and the crest appropriately "cambered." This involves building the crest to an elevation higher in the middle tapering to the abutments to accommodate anticipated settlement. The desired result is that the settled crest elevation does not dip below the design elevation. This is extremely important with respect to the ability of the dam to maintain freeboard and withstand a flood without overtopping.

During construction, settlement platforms can be used to monitor settlement as embankment fill is placed. Settlement platforms are generally located at the maximum embankment section, the location of the maximum thickness of compressible foundations, and at other locations of interest to the designer where settlement magnitude has been estimated. Settlement platforms are particularly useful because the vertical riser pipe can be extended upward as fill is placed, allowing measurements during and after construction. Settlement is indicated by changes in the surveyed elevation of the top of the riser pipe with time.

Borros points and borehole extensometers are also used during construction. During construction, they can be used to isolate the settlement that has occurred in individual compressible foundation layers (below the construction subgrade). They are installed in cased holes made by rotary drilling methods.

After construction, settlement of an earthfill embankment dam is typically monitored along the crest and/or on benches of the downstream slope utilizing monuments that are periodically surveyed for changes in elevation. Monuments are typically concrete piers, usually flush with the ground surface, with a punch mark or "X" on a brass cap (as shown in Fig. 10-10) to indicate the exact survey point. These monuments may follow one of the designs presented in Chapter 6. Spacing of the settlement platforms and survey monuments depends on the size and length of the dam, as well as on the composition of the embankment materials, foundation and abutment conditions, and estimated magnitude of settlement.

If the foundation of the dam rises steeply at a stiff abutment contact (such as a bedrock abutment) and the main foundation consists of relatively compressible soil, then differential settlement caused by this abrupt change in shape and stiffness can cause transverse cracking perpendicular to the dam axis. This type of cracking can cause seepage problems internally within the dam and possibly damage filters that guard against internal erosion. Therefore, when there are steep abutment contacts, it is important to have a continuous settlement profile across the zone of abrupt shape change. Such a profile can be monitored using a differential settlement gauge (liquid-level

Fig. 10-10. Embankment crest monument
Source: Courtesy of DG-Slope Indicator, reproduced with permission.

or overflow settlement cell), horizontal inclinometer, or hydraulic settlement profiler, which is best installed during construction of a dam.

Horizontal inclinometers can in some cases be installed after construction; however, this involves drilling a horizontal hole that requires more
specialized drilling equipment. Access to achieve the proper orientation
may be difficult. When installing during construction, care must be taken
in the compaction of the lifts directly above the instrument so as not to damage it due to overstressing. Also, strain on the instruments, for example
shear or tension due to stretching, which can occur owing to differential
settlement, must be considered to avoid broken tubing, casing, cabling, or
conduit. Differential settlement gauges can also be installed during construction and require the same care in installation and protection.

Installation of cabling and tubing in embankment dams, shown in Figs.
10-11A and 10-11B, presents vulnerabilities that must be considered in
design and construction.

Cabling or tubing is installed in trenches to provide sufficient burial to
avoid damage due to subsequent lift placement. Cable trenches passing
entirely through the core of an earth or rockfill dam can lead to cracking
and internal erosion and therefore, must be avoided. It is also usually desirable to minimize the cable runs and daylight the cable at the ground surface

Fig. 10-11A. Horizontal cabling and overflow settlement cell installation in an embankment dam
Source: Courtesy of DG-Slope Indicator, reproduced with permission.

Fig. 10-11B. Vertical cabling, magnetic settlement target, and inclinometer casing installation in a CFRD dam
Source: Courtesy of DG-Slope Indicator, reproduced with permission.

in as short a distance as practicable. Horizontal routing of the cabling should avoid areas where differential settlement could induce damaging tension or shear. Vertical routing also has vulnerabilities in areas of significant settlement; the vertical tubing runs of instruments installed during construction on the Pyramid Dam in Southern California failed due to compression of a shear zone that consolidated excessively. Adequate compaction around vertically routed tubing, conduit, cables, or casing is an important consideration. Selection of flexible and ductile materials and installation procedures can have a large impact on the resistance of the instrument installation to overstressing due to compression, settlement, or compaction. The design life of the tubing is also very important. Tubing that becomes brittle can become a conduit of unintended water flow and pore pressure that can threaten core and filter zone material. Installation and protection of cabling/tubing during construction are discussed in Chapter 5. They are also discussed by Dunnicliff (1988), the U.S. Army Corps of Engineers (1995), and the U.S. Bureau of Reclamation (1987).

Permanent settlement of the embankment dam surface can also occur owing to deformation resulting from seismic response to an earthquake. Survey monuments on the downstream slope and crest of the embankment dam are typically used to monitor permanent settlement at the surface of the dam from seismic activity. In seismically active areas, it may also be advisable to place internal vertical movement devices, such as Borros point anchors or borehole extensometers, at strategic locations within the embankment to monitor continuity of the filters. These locations include the interface of the embankment fill with the foundation or at chimney or blanket drain interfaces. At these locations, permanent earthquake-induced settlement could damage drains and filters, possibly causing internal erosion or pore pressure build-up and associated slope instability owing to strength loss. Borros point anchors and borehole extensometers are discussed in Chapter 5.

Settlement observed or measured along abutment or structural contacts or above pipes and conduits passing through the embankment dam can be a sign of internal erosion of fines from the dam. Such observations require closer monitoring and appropriate investigation. Also, small, localized sinkholes on the crest or the upstream or downstream slope of an embankment dam can be a sign of internal erosion. It usually is not practical to specifically place survey monuments or internal instruments at all potential internal erosion locations. However, the design of the settlement monitoring system needs to consider the location of conduits.

Zoned earthfill dams are similar to homogeneous earthfill embankment dams except that embankment materials are placed in zones to meet stability and seepage control requirements. Zoned dams generally have a fine-(or finer)-grained core section flanked by more granular soil materials upstream and downstream of the core. Appropriately graded filters designed to prevent potential erosion of core fines and possible development of internal erosion separate these major zones. Of particular concern for zoned earthfill dams is the difference in the compressibility of the different zones. This can result in a heightened potential for differential settlement across filter or drainage zones and the potential for longitudinal cracking along the embankment. To assess this potential for longitudinal cracking, it may be desirable to supplement traditional surface survey monuments, which monitor only surface settlement, with internal vertical movement devices such as Borros anchors, soil extensometers, or borehole extensometers. Such devices can be used to monitor settlement of deeper layers. Full-profile settlement gauges such as horizontal inclinometers, full-profile hydraulic settlement gauges, settlement plates (Fig. 10-12), or differential settlement gauges may also be useful. The use of these more sophisticated instruments is governed by the size and hazard potential of the dam and the degree to which differential settlement is perceived to be a potential problem.

Hydraulic fill dams are constructed by sluicing embankment materials into place with water. Although not generally used anymore for water-retaining structures, hydraulic fills are occasionally constructed to retain mine tailings. Hydraulic fill dams are more susceptible to liquefaction and/ or slope movements during earthquakes. Regular settlement measurements and extra measurements after earthquakes are important.

In a zoned embankment, core materials are generally fine-grained soils, and the upstream and downstream shells consist of various gradations of rockfill. The core and the outer shells are typically separated by graded filters to allow development of a phreatic surface without internal erosion of core fines into the coarser downstream shell materials. A chimney and blanket drain downstream of the core draws down the phreatic surface and

Fig. 10-12. Settlement plate
Source: ASCE Task Committee on Instrumentation and Monitoring Dam Performance (2000).

keeps the downstream shell materials relatively dry. Upstream filters guard against slope failure during drawdown.

Zoned rockfill dams are similar to zoned earthfill dams except that the difference in grain-size distribution between the zones of materials is much greater. These dams are generally constructed in mountainous regions where overburden soils are at a premium and the majority of available dam construction material is excavated and crushed bedrock. Differential settlement increases the potential for longitudinal cracking parallel to the crest. Settlement may result from relatively softer core material "hanging up" on the shell as it settles, which can result in a reduction of vertical stress in the lower portions of the core material, thus increasing the potential for hydraulic fracture. An abrupt change in abutment slope is a worthy candidate for measuring differential settlement because the effective stress in the embankment changes quickly over a short distance, leading to the potential for core cracking (Fig. 10-13). This mechanism was postulated as a contributing factor in the failure of Teton Dam in 1976.

Monitoring of zoned rockfill settlement can be performed in a similar fashion to earthfill embankment dams and zoned earthfill dams.

Embankments with upstream membranes are different because the water-retention capabilities are provided by a low-permeability barrier on the upstream slope. The seal may be constructed of concrete, asphalt, steel, or a geomembrane that is connected to the foundation and abutments at the heel with a concrete plinth and a grout curtain into bedrock. Modern designs commonly utilize a concrete upstream face and are referred to as concrete-faced rockfill dams (CFRD). The composition of dams with upstream membranes may include two transition zones between the upstream membrane and the general rockfill.

CFRD dams are not immune to developing problems. They are vulnerable to failure by internal erosion at the foundation level if they do not have adequately filtered internal drainage layers. Leakage can also occur owing to deterioration or cracking of the face caused by aging or differential settlement of the structure. Settlement and downstream movement of the embankment in the maximum section area, especially from any rock particle breakdown under load, bends the membrane, creating tension near the abutments and causing leaks. A concrete upstream face is sensitive to panel damage above the reservoir level during an earthquake that may crack concrete or tear waterstops, requiring repair before the reservoir can be filled.

At membrane dams, it may be important to supplement periodic surveying of surface monuments with appropriate measurements to verify satisfactory performance of the membrane itself (Teixeira da Cruz et al. 2010). Instruments may consist of internal movement devices installed in the membrane, such as embedment strain gauges and joint meters and at the perimeter of the membrane, with submersible 3D crack meters to

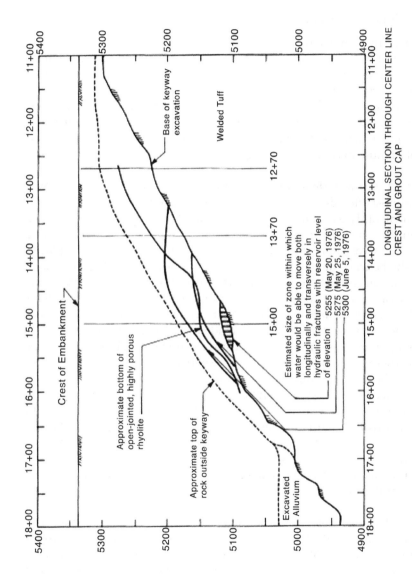

Fig. 10-13. Example of core cracking
Source: U.S. Bureau of Reclamation.

measure the relative movement between the membrane and the perimeter plinth, and on the membrane, such as inclinometers aligned to the upstream slope angle or a series of single-point submersible tiltmeters.

10.3.2 Slope Movement

Measuring movement of embankment dams can be performed at the surface and in the subsurface. Instruments to measure slope movement include survey monuments on the crest or benches, manual or in-place inclinometers, and tiltmeters.

Well-designed and -constructed embankment dams are expected to settle. However, slope movement may not be the result of ordinary settlement. If slope movement measurements exceed expected movement, then a slope stability failure may be developing and require investigation of the cause.

10.3.2.1 Surface Measurements. Surface monitoring of slope movement is most often done by surveying the same monuments used to monitor settlement. These monuments are embedded in the embankment surface or mounted on structural components such as a parapet wall or concrete facing, in the case of a CFRD dam. Survey monuments can also be placed along the benches of an embankment dam or at other strategic locations where movement could be critical to the overall performance of the dam.

Successive measurements taken over time and plotted against reservoir elevation will show movement trends, if any. Data can be geo-referenced. Where possible, a survey traverse is made to distribute closure errors.

A temporary, quick, rough form of surface monitoring is by use of slope stakes. These are stakes for temporary use that can be put in a line across the slope both on and outside of a known area of movement. Continued movement can be assessed by sighting (or running a string line) across the stakes and noting any relative movement from previous observations. In addition, existing guard rails, fences, and utility poles along the crest of the dam make useful "qualitative" lateral movement monitors. Note that as with all instruments, bumps by construction or other traffic can break the continuity of the monitoring data. As such, instruments must be protected from disturbance.

10.3.2.2 Subsurface Measurements. Surface monitoring measures slope movement only at the surface of a slope. In contrast, subsurface monitoring instruments are designed to measure movement at one or more depths within a dam. Inclinometers and tiltmeters are described in Chapter 5. Inclinometers are located to measure movement at any depth of interest. If movement in the foundation is expected, then measuring at the toe of the slope will be the best location. Where the embankment slope is expected to fully contain a failure surface, measuring at the crest and at depth intervals down the slope is best to intercept any potential slip surface.

Typical subsurface measurement locations for inclinometer casings and borehole tiltmeters are shown in Fig. 10-14.

Inclinometers are difficult or impossible to install in or through rockfill. Instead, visual observations detect slope changes but are not able to judge subsurface movement. Judgment is required to decide if a failure mode is developing.

Parapet walls and CFRD faces may crack in response to movement in underlying fill. Cracks can be measured. The distance between simple marks or rebar set on either side of an identified crack can be measured. Gridded plastic devices that fasten across a crack and electronic crack meters are also available.

10.3.3 Seepage and Leakage

Seepage flows through, around, and under all embankment dams. Acceptable seepage is benign seepage, being adequately controlled by the dam's drainage system. Observed seepage can be measured by several devices. Seepage that emerges at the toe may be acceptable if it runs clear without evidence of material transport or erosion of the embankment slope and does not increase (independent of reservoir level) with time. Seepage emerging on a downstream slope must be interpreted carefully because an elevated phreatic surface may be the prelude to slope instability. Seepage emerging along embedded conduits or concrete walls often is worrisome because internal erosion is more likely to occur along these interfaces than at other

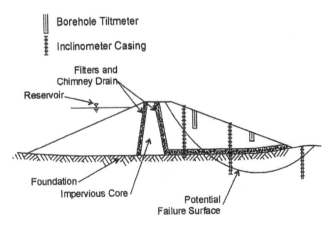

Fig. 10-14. Potential inclinometer and tiltmeter locations
Source: ASCE Task Committee on Instrumentation and Monitoring Dam Performance (2000).

V-Notch Weir   Parshall Flume

Fig. 10-15. Weir and flume

locations of a dam. Decreasing flow may indicate declining effectiveness of a chimney, blanket, or toe drain. Rainfall and reservoir level affect seepage. Plotting flow, rainfall, and net head versus time aids in effective interpretation. Seepage measurements can be compared with design assumptions and with prior measurements to decide if the measured seepage is evidence of a developing failure mode. Flow rate and pressure can be measured.

Many dams have a drain system that discharges at the toe. The toe drain system is designed to collect seepage flowing through both the embankment and foundation. Seepage is measured at any location of interest where seepage can be channeled. Increasing flow, particularly flow carrying sediment, may be a sign of internal erosion. Any measurement suggesting that internal erosion is occurring is worrisome and requires evaluation by a geotechnical engineer experienced in internal erosion. Typical locations to measure flow are at the toe and along groins.

Common types of seepage flow measurement instruments are weirs, flumes, and calibrated containers. A pair of weirs is shown in Fig. 10-15. Downstream topography and expected flow range are important considerations when selecting the seepage flow instruments. Weirs are appropriate for modest flow where a minimum of 0.2 ft (0.061 m) of head upstream of the weir is likely. In flatter areas where there is less head loss or at locations where higher flow exceeds weir capacity, flumes are used. Typical arrangements of weirs and flumes are described in Chapter 5. Care is required to isolate a weir from rainfall and surface runoff. Low flows can also be measured by the time it takes for flow from a discharge point, such as from an outlet pipe, to fill a container of a known volume.

10.3.4 Pressure

Hydraulic pressure often is measured in an embankment and its foundation by piezometers. Their performance indicators are pressure and water surface elevation, and they serve to detect pore pressures, which may indicate a developing slope failure. They may also be used to determine flow conditions through the embankment. Pore pressures in excess of those anticipated may reduce slope stability to the point where a slope begins to move.

Piezometers are also used to measure induced pore pressures during construction, particularly in low-permeability soil. After construction, piezometers measure pore pressures for comparison with design. Rising pore pressures and reduced flow may indicate changes in internal drainage efficiencies, possibly blockage of drains. This unfortunate situation may be the consequence of improperly graded filters and drains—a situation with no simple remedy.

Piezometers are also used to monitor the performance of a cutoff, such as a slurry wall or grout curtain. In this case, piezometers are placed upstream and downstream of the cutoff as an indication of the hydraulic gradient across the cutoff. In an embankment, pressure may be measured at any point of interest. For cases where the foundation is of concern (for example, foundations on pervious sand and gravel or confined artesian zones), measurements are made in isolated foundation zones. Both embankment and foundation piezometers can be located to measure hydrostatic pressure upstream and downstream of chimney, blanket, and/or toe drains to assess effectiveness.

Measurement data travels from a piezometer via flexible cables, tubes, or conduits installed during embankment construction or after construction in drilled boreholes. A sand filter is often backfilled around a piezometer's transducer, porous filter, or slotted screen to provide the hydraulic connection to pore water and provide a filter to protect the piezometer from plugging. An unfortunate problem with poorly designed standpipes occurs where the specified slot size is too wide, allowing infiltration of finer soil, eventually plugging the tip and requiring replacement.

In borehole installation, a layer of swelling bentonite pellets or chips is placed above the collection zone to isolate the instrument within the zone of interest. Bentonite pellets are generally preferred over bentonite chips or powder that are more likely to "bridge" in the borehole during placement. When bentonite pellets are used, they are added a few at a time to avoid bridging.

The response time of a piezometer is an important selection consideration. Standpipes or piezometers with sand filters are not desirable in lower-permeability soils because they are slow to respond to changes in pore pressure. If an accurate, responsive reading of changing pore pressures is desired, particularly in less permeable soils, a small displacement transducer is used. Then, wide changes in pressure require very little volume change in the transducer to accurately measure pressure. If a transducer with a larger displacement or a standpipe (which has significant storage in the riser pipe) is used, it will take more time for sufficient water to pass through the filter zone to accurately reflect the pore pressure at the point of interest. Standpipes and large displacement transducers are only used in pervious stratum or in situations with slow changes in reservoir elevation.

When piezometers are installed during construction, care must be given to protect the transducer and cabling and tubing from damage during subsequent lift placement. Also, because cabling and tubing usually involve long horizontal runs to extend to the readout station, the effects of differential settlement must be considered (Fig. 4-1 in Chapter 4). Selection of flexible or ductile materials and how the instruments are installed can have a large impact on the resistance of the instrument installation to settlement-induced overstressing.

10.3.5 Stress

There are occasions when stress monitoring is desirable. Stress monitoring can be done using total pressure cells typically installed during construction or, alternatively, using a self-boring pressure meter after construction. Total pressure cells can be installed to verify overburden and lateral stresses. Total pressure cells cannot be installed after construction without disturbing the stress field of interest. Although somewhat exotic, the self-boring pressure meter provides an alternative that can be used postconstruction. Total pressure cells are located at any point of interest where the designer is concerned with soil pressures against a particular structure or conduit. Because the cells measure total pressure, they can be used in combination with a piezometer to interpret effective stress. Total pressure cells require care during installation to provide accurate measurements. Installation of a total pressure cell is depicted in Fig. 10-16.

Fig. 10-16. Total pressure cell
Source: Courtesy of DG-Slope Indicator, reproduced with permission.

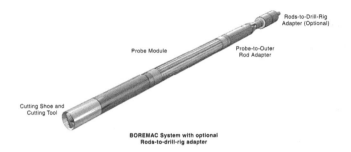

Fig. 10-17. Self-boring pressure meter
Source: Courtesy of Roctest, reproduced with permission.

A self-boring pressure meter (SBP) requires special equipment and expertise for installation. Equipment and trained operators are available at a few universities with advanced geotechnical research programs or through a few highly specialized practitioners. Although these devices have been used primarily for university-level research, their ability to accurately measure in situ stress is proven. The instrument essentially tunnels itself into the soil without significant disturbance or stress relief, allowing an accurate stress measurement. The device measures total stress; however, effective stress estimates can be made if testing includes accurate determination of in situ pore pressure. The device is ideally suited for use in sand and clay without stone content, and as such it is limited to stress measurement in finer filters or core material. Fig. 10-17 is a photograph of an SBP meter without its protective shield.

In situations where the stress imposed by the embankment overburden on buried conduits and structures or retaining structures is of concern, a total pressure cell can be recessed into the face of the structure to measure the pressure. The core block in the Oroville Dam, a 770 ft (235 m) high embankment in Northern California, was an example of where stress cells were used to monitor the stress imposed on an internal structure owing to embankment placement.

An example where stress measurement might prove to be a performance indicator would be an embankment with a steep-sided soft core and a relatively stiffer shell. With differential settlement between the core and shell materials, it is possible that the core could "hang up" on the shell. If this occurs, it could tend to reduce the effective stress in the lower portions of the core. If the pore pressures are sufficiently high, hydraulic fracture of the core could occur, opening the potential for internal erosion of the core. This would be a special case where lateral stress in the lower portions of the core could be measured using an SBP meter drilled down from the crest. An SBP meter can also be used to measure lateral stress beside conduits and retaining walls.

10.3.6 Earthquake Response

Only large embankments in areas of moderate-to-high seismic activity are monitored for earthquake response. Agencies such as the USGS and the California Division of Mines and Geology contract for installation and maintenance of strong-motion instruments at dams. The advantages of this arrangement to dam owners are avoiding highly specialized installation, operation, and maintenance and processing of the acceleration records obtained. Similar offerings from government agencies are also available in a number of countries with moderate or high seismic activity, although some large dam owners choose to develop their own expertise. Measuring earthquake response serves two purposes: (1) design verification of expected versus actual response during a seismic event and (2) adding to the state of knowledge about how dams perform in earthquakes. For example, did the dam deform as expected, or did it deform excessively?

Ground motions are measured by accelerographs (strong-motion machines). Accelerographs are often located on bedrock at the downstream toe or abutment and on the crest of the embankment. Earthquake energy is carried away from the source of the earthquake through bedrock, transferring the energy to the dam foundation. By monitoring the bedrock in the dam foundation, the energy being transferred to the dam can be measured. Monitoring the dynamic response of the dam is typically done by means of one or more accelerographs at strategic points on the embankment. During an earthquake, the oscillatory motion of the earthquake causes a sympathetic oscillatory response in the embankment. The magnitude of the response is dependent on the resonant frequency of the embankment dam and the frequency spectrum of the earthquake.

Unless there is significant strength loss (liquefaction), embankment dams tend to amplify the peak acceleration of an earthquake. This is exemplified by the fact that peak acceleration on the crest tends to be higher than the peak acceleration in the foundation bedrock. This magnification effect is typically greater for earthquakes with smaller peak accelerations, and the amplification fortunately tends to be less for larger earthquakes.

Permanent earthquake deformation falls into two general categories—liquefaction and deformation. If an earthquake causes pore pressure to rise in the foundation or embankment to the level of the intergranular stress, the affected zone of soil will lose most of its strength. Poorly graded granular soils of low relative density in the embankment or foundation may liquefy. Embankment dams of modern compacted lift construction are resistant to liquefaction; however, if left untreated, loose foundation soils may be vulnerable even with modern construction methods. Measuring earthquake deformation is performed by geodetic survey using the same techniques that are used for measuring settlement.

10.4 PERFORMANCE INDICATORS

Measurements and visual observations of performance indicators are tailored to answer the question of whether or not an embankment dam is performing as expected.

Acceptable measurements of performance indicators depend on the size of an embankment, the properties of its materials, the care exercised during construction, age, and the dam's hazard classification. For example, one foot of settlement at the center of the crest of a 15 ft (4.75 m) high embankment during its first year of operation would be cause for concern, whereas the same amount of settlement of a 450 ft (137 m) high earth and rockfill dam may have been predicted.

Settlement of the crest is expected immediately after first filling. The amount of settlement will reflect the height of the dam, the materials used, and the care taken during construction. Typical designs provide camber to accommodate settlement at the highest section of an embankment. An acceptable value of settlement upon first filling is estimated during design. Measuring settlement provides a basis for comparing actual performance versus estimated design performance. Vertical settlement decreases over time for consistent loads. Measured settlement exceeding 1% of embankment height requires explanation, the character of the embankment materials notwithstanding.

Geodetic surveys or inclinometers may measure movement of an embankment or its foundation under various loading conditions at a variety of locations. The actual magnitudes of movement will vary with the size of the dam and the loading conditions in the context of the dam design and construction. There is no generally accepted amount of allowable movement for an embankment. Again, it is a case-specific measurement. Accelerated rates of movement at zones of interest on and within the embankment may be indicative of a developing failure surface.

Pore pressure within the core of a zoned embankment dam may be higher immediately after construction because of compaction and consolidation from the weight of the overlying embankment. However, excess pressures should dissipate over time and should shift to a steady-state condition. The phreatic level estimated to maintain stability can be compared with the design level to decide if the performance indicator points to stability or instability.

The designer of an embankment dam estimates the magnitude of seepage expected from the dam and designs a drainage system to collect and discharge seepage safely from the dam to minimize the potential for erosion or saturation of the downstream slope. The performance indicator to measure seepage requires monitoring weirs or flumes to measure the quantity of the seepage and to examine the clarity of the discharge. Measuring seepage quantity over time for each weir enables a review of any trends that might indicate a developing failure mode. There is no standard value or range of values of seepage quantity that can be considered as acceptable. Seepage and leakage are evaluated on a case-by-case basis with respect to reservoir level and precipitation. If the dam designer predicts a seepage quantity from the dam under full reservoir head and that quantity is exceeded prior to the reservoir reaching that elevation, or if seepage becomes cloudy, immediate attention and careful evaluation are required.

In the design of a monitoring program, design criteria are reviewed to set threshold limits for measured performance indicators. Threshold limits are based on engineering analyses and predicted magnitudes of behavior made during the design of the dam or upon long histories of performance of existing dams.

Threshold limits are established to focus attention on measurements outside the limits of the predictions and readings that require evaluation to explain the behavior. Both overall magnitude and rate of change thresholds may be developed to raise awareness of when evaluation is required.

Evaluation could lead to one or more of the following actions:

- Alerting the responsible engineer;
- Confirming the reading (measurement) by checking it again and, if possible, confirming the proper operation and calibration of the monitoring instrument;
- Inspecting the dam for the cause of the measurement;
- Increasing the frequency of measurements to provide data for further evaluation;
- Revising the threshold limit;
- Installing additional instruments to verify a measurement that exceeds a threshold;
- Implementing remedial measures such as cleaning foundation drains, repairing damage, or modifying the dam in some fashion; and
- Determining whether implementing emergency measures is required.

The performance indicators and threshold limits for embankment behavior are established by careful review of the design, construction, and historic performance records. Experienced geotechnical engineering is required because one approach does not fit all embankments. Engineering judgment is required to properly interpret the results.

The next chapter is a review of performance indicators for concrete dams.

CHAPTER 11

CONCRETE DAMS

There exists everywhere a medium in things, determined by equilibrium.

Dmitri Mendeleev

Chapter 3 introduces monitoring considerations for concrete dams from the perspective of their potential failure modes. This chapter discusses concrete dam vulnerabilities and behavior, and it provides guidance about means and methods to measure the performance indicators that answer the question, "Is the dam behaving as expected?"

11.1 DESIGN

Concrete dams exhibit a wide variety of sizes and shapes, all of which are variations on three basic themes—gravity, arch, and buttress. A gravity dam transfers its reservoir load on the upstream face directly to its foundation, relying on its weight for stability. An arch dam transfers its reservoir load relying on arch action to transfer load into its abutments. An arch dam requires less mass for stability than a gravity dam of equivalent height.

A buttress dam transfers the reservoir load on its inclined upstream face through piers and into the foundation. Buttress dams have fallen from favor, and there are far fewer of them than gravity and arch dams. This chapter discusses instrumented monitoring of gravity and arch dams. Buttress dams are discussed in Chapter 12.

For many years, all concrete dams were constructed of conventional concrete. The advent of roller compacted concrete (RCC) introduced a new technique for constructing gravity and arch dams. It was developed to promote rapid construction and reduce cost, and as a result it has won global acceptance and is widely used.

Common factors affecting the performance of gravity, arch, and buttress dams are the structural properties of the concrete and the foundation stability. Important concrete properties are

- Loading conditions (normal, unusual, extreme),
- Concrete components (aggregates, cement, admixtures, water) and their proportions (mix design),
- Strength (compressive, tensile, and shear),
- Elastic properties (modulus and Poisson's ratio),
- Unit weight, and
- Seasonal temperatures.

For all concrete dams, abutment geometry and foundation character influence abutment shaping, requiring an understanding of

- Loading conditions (normal, unusual, extreme),
- Proper angle of incidence with the abutment,
- Rock properties such as permeability, weak zones, and presence of discontinuities with the potential to affect stability,
- Rock strength (compressive, tensile, and shear),
- Rock elastic properties (deformation modulus and Poisson's ratio),
- Stress history, and
- Groundwater or artesian conditions.

11.1.2 Gravity Dams

The design of gravity dams must consider the following:

- Shape must be adequate to transfer load to maintain moment and sliding equilibrium while maintaining concrete stresses within acceptable limits.
- The dam must be proportioned to meet acceptance criteria requiring margins of safety for normal, flood, and earthquake loads.
- The dam must have stable foundations and abutments. Particular care must be taken to ensure that blocks of rock when loaded by the mass of the dam and reservoir pressure are not displaced. Loss of foundation support is a mode known to cause gravity dam failure by sliding.
- Gravity dams are suitable for wide gently sloping valleys, as well as for steep valleys. Steep valleys may promote three-dimensional load sharing, a positive factor for stability.

11.1.3 Arch Dams

Arch dams are shaped to satisfy one basic requirement while maintaining concrete stresses within acceptable limits: safely transfer load into the abutments and foundation.

Implied in this requirement are foundation and abutment stability, with particular care to ensure that blocks of rock under load cannot move, robbing the arch of the ability to successfully redistribute stress. There have been few arch dam failures, and the only mode known to have caused failure was loss of foundation support.

Arch dams are best suited to valleys with width-to-dam height ratios of less than 6:1 to keep concrete proportions as low as possible yet achieve efficient load transfer to the abutments.

Chapter 3 describes the vulnerabilities of concrete dams and introduced sliding and overturning or combinations of both motions as potential failure modes.

11.2 PERFORMANCE

Concrete is the medium common to gravity and arch dams. How those dams perform depends upon the care taken at every step of the design and construction processes. Recognizing a dam's vulnerabilities and associated performance indicators related to failure modes provide the basis for monitoring performance.

11.2.1 Concrete

A blend of cement, water, fine and coarse aggregate, and admixtures, concrete derives its strength from the qualities of its various constituents. Concrete technology is quite advanced, and how the constituents are proportioned to produce strong concrete for a wide range of conditions is well understood, either for conventionally placed concrete or RCC. When suitably proportioned, batched, placed, consolidated, and cured, concrete performance seldom requires measurement as an indicator of failure mode in modern concrete dams except in cases where concrete is composed of aggregates reactive with the alkalis in cements.

Older (pre-1930s) dams may exhibit symptoms of freeze–thaw damage, a consequence of water freezing and thawing in the open concrete pores. In the 1930s, admixtures were developed that employed a liquid similar to detergent to create tiny bubbles in the concrete mix. The bubbles provided the concrete with the ability to absorb freezing strain without cracking or spalling. For the most part, freeze–thaw damage is cosmetic (Fig. 11-1), requiring occasional maintenance to restore lost concrete. Other than visual

Fig. 11-1. Freeze–thaw damage

inspection to determine depth of damage, measurements are not likely to be required.

Quartz is a common rock-forming mineral. Some species of quartz found in igneous and metamorphic rocks may react adversely with the alkalis in cement, forming a gel that swells and cracks the aggregate. Ultimately, it may crack the concrete, as shown in Fig. 3-17 in Chapter 3.

This process is variously referred to as an alkali–silica silicate reaction (ASSR), alkali–silica reaction (ASR), alkali [Mg]–carbonate reaction (ACR), or alkali–aggregate reaction (AAR). Many existing concrete dams suffer from these reactions, some with consequences that demand monitoring to indicate whether action is required to reduce the effects that cracking has on structural capacity. Appropriate measurements for "healthy" and ASR-affected dams may include

- Movement (x, y, z),
- Total stress,
- Crack aperture and depth of penetration,
- Temperature, and
- Water pressure within the dam body.

Fig. 11-2 shows plots of changes in crest elevation with time at Kariba Dam caused by swelling from ASR.

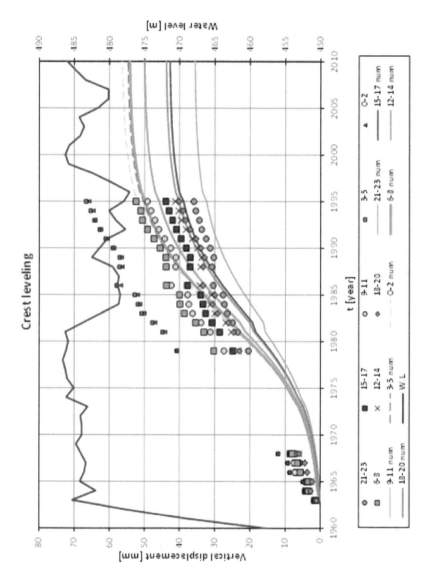

Fig. 11-2. Kariba Dam crest movement caused by ASR

11.3 INSTRUMENTED MONITORING

Chapters 5 and 6 describe instruments and techniques to measure these performance indicators. Geodetic surveys measure deflections along the crest.

Figs. 11-3 and 11-4 illustrate existing and prospective concrete gravity dams, respectively, and typical locations of instruments to monitor its performance. The figures show broad arrays of instruments that may be installed in concrete gravity dams. However, the actual type and number of instruments to be installed depends on the specifics of the project, including its design, size, and hazard potential.

Where the instruments are placed and what measurements are made varies somewhat among the three basic types of concrete dams, but the basic instrument types are the same for all three.

The key to the symbols representing the various instruments in the cross sections is the same as that in Chapter 10 (Fig. 10-3).

11.3.1 Temperature

Concrete temperature is a concern not only during construction but also over the life of the dam. During construction, temperature changes and shrinkage in the concrete introduce secondary stresses. Mass concrete mixes are carefully designed. Fly ash (pozzolan) is employed to reduce the heat of hydration. Heat may be controlled by cooling the aggregate and adding ice to the mixing water. Embedded cooling coils are used to control the

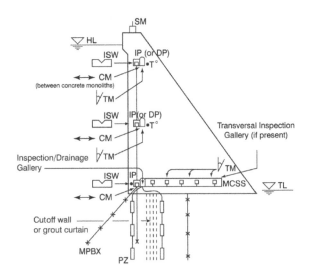

Fig. 11-3. Instrument locations for an existing concrete gravity dam

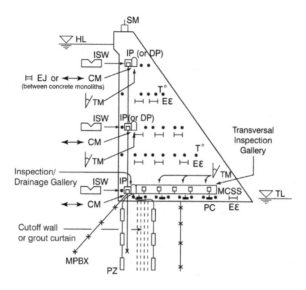

Fig. 11-4. Instrument locations for a prospective concrete gravity dam

secondary stresses by limiting the heat rise, thereby avoiding undesirable thermal cracking. Field adjustments to mixes and construction methods are made based on concrete temperature measurements.

Over the life of a dam, particularly an arch dam, temperature measurements may be important for a proper understanding of both dam behavior and the behavior of instruments. Arch dams and buttress dams are more responsive than gravity dams to movements caused by temperature changes. Temperature changes cause concrete volume changes that apply horizontal forces, resulting in load transfer across contraction joints and into the abutments. The abutments may thus experience increased load. Seasonal temperature changes cause a concrete dam to deflect downstream in winter and upstream in summer.

When it is desirable to measure temperature for a dam under construction, temperature-measuring devices can be placed strategically in fresh concrete. Thermometers and thermistors such as those shown in Chapter 5 are the most commonly used temperature-measuring devices. For existing dams, temperature-measuring devices can be surface mounted or installed in drilled holes.

Knowing the ambient air and reservoir water temperature is important to sorting out the dam's response to temperature changes, and both can be measured with thermometers or thermistors. Heating of concrete from solar radiation is usually considered qualitatively.

11.3.2 Movement

Movement refers to either translation or rotation within a frame of reference. For example, movement occurs at the crest of a concrete dam as a result of changes in the reservoir load and the temperature of the dam's concrete, air, and reservoir water. Movement is measured relative to a fixed point on a rigid body.

Deformation is movement that changes the shape of a solid body relative to its unloaded shape or its shape at a particular load or stress. Reservoir filling at a concrete gravity dam causes its axis to deform as a curve. Expansion and contraction due to concrete temperature changes also cause deformation. Measurements are related to a line or a plane.

The fundamental design requirement for gravity dams is to resist reservoir load by self-weight. Measurements are made to understand dam movement relative to the movements estimated in design. Periodic surveys of crest monuments provide the appropriate measurements. Because a gravity dam is designed to maintain both moment and sliding equilibrium, all three movement components (x, y, z) are surveyed relative to benchmarks (nonmoving monuments), as described in Chapter 6. The positions of the crest monuments require accuracy to at least 0.01 ft (2.5 mm). This accuracy is easily obtained. Movements for an existing dam are often measured monthly or quarterly. Problem dams may require more frequent measurement. Survey frequency during first filling is important and may be scheduled on the basis of reservoir filling increments rather than by regular time increments.

The fundamental design requirement for arch dams is efficient load transfer from the arch into the foundation and abutments while maintaining acceptable stresses in the concrete. Surveys of crest monuments using abutment benchmarks or primary survey control networks give a quantitative measure of movement that may be compared to expected values from design or from performance of similar dams.

Measurements of three-dimensional (x, y, z) movement within the dam body are less common than two-dimensional (x, y) movement and are tailored to the specific dam and foundation conditions. Two-dimensional (x, y) measurements may be made with a plumbline, inclinometer, or a series of tiltmeters installed in inspection galleries at different elevations. Horizontal movement within the foundation relative to the dam may be measured with an inverted pendulum installed into the foundation, anchored at depth in the foundation, with movement measured within the dam.

Concrete dams are constructed as discrete blocks separated by contraction joints. Upon completion of construction, the contraction joints are grouted to form a continuous structure. To verify the dam is functioning monolithically, the relative movement, if any, across the joints can be measured. Block movements are measured with points between contraction joints located in galleries or across contraction joints at the surface. Embedded joint meters, crack meters, and extensometers are designed to measure movement across joints.

The dominant gravity dam load causing block movements is reservoir stage. As the reservoir water surface changes, the relative movement between adjacent blocks is expected to be very small, less than a few hundredths of an inch.

For an arch dam, the normal operating loads are both reservoir stage and concrete temperature. For variations in either load, the relative movement between blocks is likely to be very small, with joints between adjacent blocks closing as the dam goes into compression and opening as the dam extends.

Swelling caused by AAR is measured with surface or embedded strain gauges, crack meters, extensometers, or geodetic techniques presented in Chapters 5 and 6.

With the advent of electronic distance measuring equipment and LiDAR, deformation can also be measured using surface monuments on the downstream face, total stations, and least-squares adjustment. Synthetic aperture radar allows targetless measurements of movement along the entire downstream dam face. Fiber optic arrays can detect and measure very small movements of exposed concrete surfaces.

Most mass concrete structures crack. The cracks are often cosmetic (nonstructural). Cracks are usually observed on the crest, the downstream face, and in the drainage gallery. When cracks may be related to structural integrity, movement across the crack is measured. These movements can be made at the surface using pins or crack meters or at depth using joint meters. Usually the crack width or relative slip is measured, depending on the location and orientation of the crack. Another method to monitor a crack is to mark and date both ends of the crack to track changes in its length. Generally, cracks form during and immediately after construction, during first filling, and as a result of strong ground shaking. Cracks in most concrete dams move only a few hundredths or thousandths of an inch with no net long-term movement. During design, cracking may be anticipated at certain locations owing to unique features of a particular dam and its site. If needed, instruments such as multipoint extensometers or joint meters are appropriate to measure crack aperture and extension.

11.3.3 Uplift

Uplift exerted by the reservoir opposes self-weight and reduces the effective normal force acting on the base or along any horizontal plane within the dam body. The reduction of effective normal force reduces both frictional and moment resistance. The simplified diagrams (Fig. 11-5) illustrate the

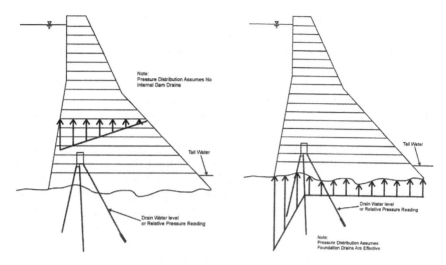

Fig. 11-5. Uplift within the dam body and along the base
Source: ASCE Task Committee on Instrumentation and Monitoring Dam Performance (2000).

typical pressure diagram associated with uplift on horizontal planes within the dam body and on the contact between the dam and foundation.

If a dam has no internal or foundation drains, the uplift pressure through a horizontal section of a dam varies linearly from reservoir pressure at the heel to tailwater pressure at the toe.

Uplift is managed with seepage reduction measures and drainage. Most, if not all, concrete dams have grout curtains designed to reduce seepage by lengthening the seepage path across the dam base. Downstream of the grout curtain, seepage at the foundation is drained by drilled or formed drains to reduce uplift. Within the dam body, seepage is intercepted in drilled or formed drains. Galleries above foundation level provide an ideal location to drill down for foundation drains and up for dam body drains.

The simplest flow-measuring device employs a calibrated container and a stopwatch (Fig. 11-6). The container is placed at the discharge of a drain. The operator then measures the amount of time it takes to fill the container. Generally, the longer it takes to fill the container, the more accurate the reading; selection of a container of sufficient volume to result in a minimum filling time of at least 30 s is recommended. The volume of the container divided by the time it takes to fill the container is a measure of flow rate. In areas where there is not a defined discharge pipe, a vertical drop can be made by installing a fabricated plate with a small discharge pipe through the plate to create the vertical drop necessary to fill the container. Small weirs such as those shown in Fig. 11-7 also may be useful in measuring flow.

Fig. 11-6. Calibrated container and stopwatch

Fig. 11-7. Small measuring weir

11.4 PERFORMANCE INDICATORS

Uplift correlates directly to the margin of safety and, consequently, to the potential for failure by sliding or overturning moment. Key performance indicators are measurements of drain pressure and volume of drain flow. Unexpected increases in drain pressure or flow may indicate a deterioration

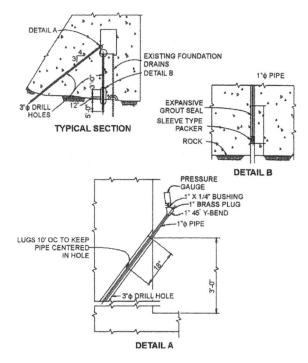

Fig. 11-8. Pressure check
Source: Courtesy of Alabama Power Company, reproduced with permission.

of the grout curtain. Instruments similar to those employed for an embankment are suitable—piezometers for pressure and weirs for flow. Fig. 11-8 illustrates a method to measure hydrostatic pressure mostly for upstream but also downstream of the existing grout and drainage curtains.

Where instruments are placed and measurements are made varies somewhat among the three basic types of concrete dams, but the basic instrument types are the same for all three. A typical arch dam monitoring arrangement is shown in Fig. 11-9, and typical measurements and instruments are shown in Table 11-1.

Decreasing drain flow may signify plugging. Foundation drains are susceptible to both chemical deposition and bacterial growth. Dissolved solids such as calcium carbonate may precipitate to coat the walls of drains in rock, retarding flow and requiring periodic cleaning. Similarly, bacterial growth caused by organisms that thrive on iron or sulfur content in water in an oxygenated environment may retard flow. This phenomenon is characterized by distinctive orange or brown slime at drain discharge. Careful measurements will disclose loss of drainage efficiency. Drains that exhibit plugging are candidates for cleaning, reaming, or redrilling in an alternate location.

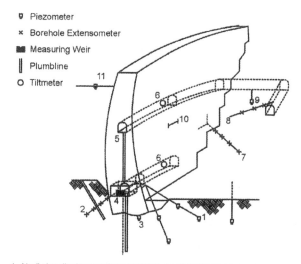

▼ Piezometer

✗ Borehole Extensometer

▰ Measuring Weir

‖ Plumbline

○ Tiltmeter

1 - Monitoring effectiveness of grout and drainage curtains by pore pressure measurement

2 - Monitoring movement of fissure openings, using borehole extensometers

3 - Pore pressure measurement at concrete-rock interface

4 - Seepage flow measurement

5 - Plumbline to measure movement

6 - Fixed tiltmeter

7 - Monitoring of foundation compressibility, using borehole extensometers

8 - Monitoring of specific geological features, using borehole extensometers

9 - Pore pressure measurement for monitoring effectiveness of grout curtain and abutment stability

10 - Measurement of strain to determine stress

11 - Measurement of reservoir level

Fig. 11-9. Typical arch dam monitoring arrangement
Source: ASCE Task Committee on Instrumentation and Monitoring Dam Performance (2000).

Uplift directly correlates to the margin of safety and, consequently, to the potential for failure by sliding or overturning moment. Key performance indicators are measurements of drain pressure and volume of drain flow. Unexpected increases in drain pressure or flow may indicate deterioration of the grout curtain. Instruments similar to those employed for an embankment are suitable—piezometers for pressure and weirs for flow.

Uplift is an important consideration in the design of gravity dams. Uplift is often neglected in arch dam design because base widths are small in comparison to dam height and the foundation and abutments are capable of resisting the imposed loads; however, uplift may adversely affect foundation stability. Notwithstanding the minimal effect that uplift may have on arch dam behavior, many arch dams have drains and periodic measurements are made.

Table 11-1. Monitoring Locations and Methods.

Performance indicator	Measurement location	Typical instruments or methods
Movement (upstream-downstream)	Crest or other surface location of interest	Geodetic survey
Movement (rotation)	Within concrete	Inclinometer, pendulum, tiltmeter
Differential movement	Across joints or cracks	Strain gauge, extenso-meter, joint meter
Differential movement	Within foundation	Extensometer
Water pressure	Within dam body	Piezometer
Water pressure	At foundation contact	Piezometer, pressure gauge
Water pressure	Within foundation	Piezometer
Stress and strain	Within foundation	Total pressure cell, load cell, strain meter, flat jack
Stress and strain	Within concrete	Total pressure cell, strain meter
Internal temperature	Within concrete	Thermocouple, RTD, thermistor
Seepage quantity	Where measurable	Calibrated container, weir, flume, flow meter
Seepage quality	Where measurable	Turbidity meter
Anchor load retention	Anchor head	Load cell, jack and pressure gauge
Earthquake response	Crest, free field, or other surface location of interest	Strong-motion accelerometer

11.4.1 Stress

Design criteria for both gravity and arch dams require the dam shape to be adequate to transfer load to maintain stability while maintaining concrete stresses within acceptable limits for all anticipated loading conditions.

The state of stress and resultant strains can be important in understanding the behavior of concrete dams and their foundations. Stress may be measured directly; however, strain measurements are more common because the instruments are easier and less expensive to install. Converting strain measurements to estimates of stress requires knowledge of the mechanical properties of concrete (modulus of elasticity and Poisson's ratio), together with the concrete's creep, shrinkage, and response to temperature changes. Occasionally, direct stress-measuring instruments are used to verify the results of strain-measuring instruments. Loads can be estimated from measured or inferred stresses and those loads compared with design estimates.

Where to measure stress or strain depends on the loads to be estimated. For any concrete dam, locations where tensile stresses may develop are candidates for measurement. Another reasonable location is the maximum section to attempt to confirm the maximum load. Arch dam instrumentation often includes strain meters or stress meters oriented to measure stress in the direction of arch thrust.

Foundation stresses and strains may be measured similarly to concrete stresses and strains; however, the instruments are not embedded in fresh concrete but rather, they are installed in drill holes or sawed slots.

Strains also may be measured precisely with geodetic techniques by measuring between fixed points on a structure.

Measuring stress or strain in an existing dam can be accomplished by drilling a hole to the point of interest and securing an appropriate instrument in position. Stress also may be measured by inserting a pressure cell or flat jack into a sawed slot.

There are many instruments and techniques for measuring stress and strain. Flat jacks, Carlson concrete stress meters, WES pressure gauges, Gloetzel stress cells, and vibrating-wire stress meters comprise part of the array of stress-measuring instruments. Electrical strain gauges, Carlson strain meters, mechanical strain-measuring devices, linear variable differential transformers, and vibrating-wire strain gauges complement the field of strain-measuring instruments. These instruments are described in Chapter 5.

Strain gauges are usually of the vibrating-wire type. The length of their measuring base must be longer than three to four times the coarse aggregate size. For this reason, gauge lengths of 25 to 30 cm (250 to 300 mm) are often specified. The use of a dummy gauge as described in Chapter 5 is also recommended.

Upon installation, the strain gauges are strongly tied to rods or brackets that will keep them in place during concrete placement. The rods or brackets are small enough to not interfere with the stress field surrounding the strain gauges. Concrete is not placed directly on the gauges but rather beside them and then vibrated. If necessary, the coarse aggregates can be removed from the concrete to comply with the requirement of gauge length equal to three or four times the aggregate size.

Signal cabling must possess adequate quality and strength for direct embedment in concrete, as it is not recommended to use embedded cable protective ducts that could later become a preferential path for water circulation. It is also important to consider the location where the electrical cables come out of the concrete. This is the location where electrical cables often get severed because they are exposed. Some strain relief is needed at this location, for instance, by the use of a short length of plastic duct that will protrude from the concrete or alternatively by a recess in the concrete where the cable can be placed.

11.4.2 Alkali–Silica Reaction (ASR)

Stress measurement is often used in concrete affected by ASR to assess the extent of the effects, monitor long-term effects, and verify the effectiveness of the remedial measures such as stress-relief dam sawing. Stress measurement has an important role in monitoring of dams experiencing ASR.

In the case where existing strain gauges are already embedded in the dam and are located at suitable positions with respect to the areas affected by ASR, they are obviously the best instruments to monitor the evolution of deterioration. Postconstruction, embedding strain- or stress-monitoring instruments in the concrete becomes necessary.

For this purpose, two approaches are generally preferred. The first is to use vibrating-wire stress inclusions, as described in Chapter 5. These inclusions are installed in small-diameter, usually 32 mm (1.5 in.), boreholes with a special installation tool that wedges the instrument in the hole at the proper depth, orienting it so that the direction of the change of stress measurement will be known. Up to three instruments can be installed in the same borehole with measurement directions rotated 60 degrees from each other. This is done to obtain the true state of stress in the plane perpendicular to the borehole or, more precisely, the change of stress, because the initial stress in the concrete is usually not known when the instrument is installed. Instruments have been installed successfully up to 20 m (65.6 ft) or more into a borehole. When the direction of stress to be measured is known, only one instrument will provide sufficient information.

A second type of inclusion that has been used to monitor changes of stress related to ASR is an instrumented cylinder as shown in Fig. 11-10. The cylinder consists of three or six vibrating-wire strain gauges oriented respectively in a two- or three-dimensional rosette configuration so as to obtain a plane state of strain or the complete three-dimensional state of strain. Short strain gauges approximately 100 mm (3.93 in.) length are generally used so that the rosette-configured assembly of gauges can be precast in a concrete cylinder 140 mm (5.51 in.) in diameter. The modulus of the concrete can be adjusted by use of suitable aggregates to approximate that of the AAR-

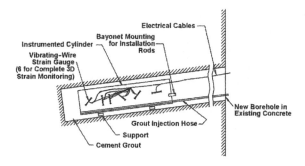

Fig. 11-10. Vibrating-wire strain gauge-instrumented cylinder

affected concrete in which it will be installed. The instrumented cylinder is an inclusion with modulus adjusted to be in the same range as the enclosing material. This procedure maintains the state of stress around the borehole in which it will be installed and grouted.

To install the cylinder, a 152 mm (6 in.) borehole is drilled in the concrete and the instrumented cylinder is pushed in the borehole by means of installation rods that allow orientation of the cylinder in the hole. The cylinder is grouted in place with a cement grout so that the cylinder becomes an integral part of the concrete. The hole can be grouted up to its collar to restore the integrity of the dam concrete.

11.4.3 Foundation Stress Measurements

Concrete stress cells or "Gloetzl" cells, from the name of the earliest manufacturer, are described in Chapter 5. In concrete dams, they are used at rock–concrete interfaces at the foundation or abutments to measure the stress applied perpendicularly to these interfaces. Either rectangular or circular cells can be used to measure foundation stress. The cells incorporate a repressurization tube to restore contact between the pressure pad and the enclosing concrete once concrete has cured and shrunk, especially in situations where cells are used on vertical surfaces. The area of rock or concrete over which the cell is to be placed is prepared clean and flat ±10 mm (0.39 in.). The area is precoated with a 15 mm (0.59 in.) thick layer of nonshrink cement mortar prior to placing the cell on the uncured mortar pad.

Many existing concrete dams have post-tensioned anchors or tendons to improve stability. Long-term monitoring may be important to measure load retention or anchorage creep.

There are two basic opportunities to measure anchor or tendon performance. The first and easiest opportunity is to include a load cell in the anchor or tendon head as a part of the design, allowing direct measurement of load retention. Robust, simple, mechanical, vibrating-wire load cells such as those described in Chapter 5 are preferred. Where the anchor or tendon head is accessible but no load cell is installed, the load can be verified with a lift-off test. Simply, the test involves lifting the head off the shims with a jack and verifying that the lock-off load is retained in the anchor or tendon. Where the tendon head is not accessible, load verification is not yet practical. However, several recent advancements in nondestructive techniques offer promise for confirming stress retention.

11.4.4 Earthquake Response

High-concrete dams in areas of moderate-to-high seismic activity are monitored for earthquake response. Agencies such as the U.S. Geological Survey and the California Division of Mines and Geology contract for the installation and maintenance of strong-motion instruments at dams. The advantages of this arrangement to dam owners are to avoid highly specialized installation, operation, and maintenance and to avoid processing the acceleration records obtained. Measuring earthquake response serves two purposes: (1) it provides design verification of expected versus actual performance during an earthquake and (2) it adds to the state of knowledge about how dams perform in earthquakes. Earthquake response is post facto and is not a performance indicator. What it does provide is a measurement of earthquake loading. Post-earthquake measuring gravity or arch dam response will reveal whether the dam is deformed.

Seismographs are placed in arrays or nets to provide coverage of large regions and measure the size of an earthquake, as well as its location and time of occurrence. The output from a seismograph is referred to as a seismogram.

Ground motions are measured by accelerographs (strong-motion machines). Accelerographs are located on bedrock at the downstream toe or abutment, within a gallery in a dam, and on the crest of the dam.

Earthquake energy is carried away from the source of the earthquake through bedrock, transferring the energy to the dam foundation. By monitoring the bedrock at the dam foundation, the energy being transferred to the dam can be measured. Monitoring of the dynamic response of the dam typically is done by means of one or more accelerographs and at other strategic points on the dam. During an earthquake, the oscillatory motion of the earthquake causes a sympathetic oscillatory response in the dam. Concrete dams tend to amplify the peak acceleration of an earthquake. Amplification is exemplified by the fact that peak acceleration on the crest tends to be higher than the peak acceleration in the foundation bedrock. This magnification effect typically is greater for earthquakes with smaller peak accelerations, and the amplification fortunately tends to be less for larger earthquakes.

Strong shaking is likely to cause contraction joints to open and close, usually without requiring repair. Permanent earthquake concrete deformation

Fig. 11-11. Accelerometer for arch dam base motion

may cause cracking or separation at the abutments. Otherwise, experience suggests that concrete dams are likely to survive strong shaking and retain the reservoir, provided that fault rupture displacement does not occur.

   Measuring post-earthquake deformation is done by geodetic survey using the same techniques as those described in Chapter 6.

   To measure the earthquake load on a dam, a three-component accelerometer can be placed on the ground surface near the dam, considered the free-field record. Another location might be in a drainage gallery that is near the base of the dam. An accelerometer may also be placed on an abutment near the crest if topographic or other amplification effects are of concern. Where an accelerometer is placed is important. What is desired is the response of the dam, not the response of the accelerometer housing. Accelerometers typically are bolted to a concrete slab and enclosed in lightweight fiberglass housing (Fig. 11-11).

   Strong-motion accelerographs details are described in Chapter 5. Basic requirements for accelerometers on concrete dams include reliability and sensitivity. Because of the wide range in severity of shaking, accelerometers need to be capable of recording the peak accelerations exceeding 1.0 g (0.03 oz). For dams with multiple accelerometers, timing is important to correlate the accelerations measured at different locations on or within the dam. This can be accomplished by encoding an external time reference on the record. A desirable requirement is a record of the initial motions of an earthquake. Modern, digital accelerometers have a preevent memory. They continuously record accelerations, which are stored in memory. The memory is continuously updated to store the most recent data (e.g., the last 4 s). Then, when triggered by an earthquake, motions for 4 s before the earthquake also are obtained, providing a record of the entire event. With older film or cassette tape accelerometers, the initial accelerations before the threshold are not recorded.

Accelerometers measure the acceleration at a specific location at small time intervals (e.g., every 0.01 s) and usually in three dimensions. The records are used to determine the earthquake load and the dam's response.

Records from an earthquake can be used to estimate whether the design response was exceeded and, using dynamic analysis (finite element, finite difference, wave equation, or simplified methods), to calculate stresses, deformations, and other parameters from which to evaluate performance and safety.

CHAPTER 12
OTHER DAMS AND APPURTENANT STRUCTURES

The difficulty lies not so much in developing new ideas as in escaping from old ones.

John Maynard Keynes

This chapter addresses instrumentation and monitoring needs for other dams and appurtenant dam structures such as spillways, intakes, outlet works and conveyances. "Other dams," cousins of conventional embankment and concrete dams, are slab-and-buttress, masonry, timber, inflatable, and concrete dams founded on piles or directly on soil.

Other dams have the same vulnerabilities of overtopping, uncontrolled seepage, and structural inadequacy as embankment and concrete dams. Appurtenant structures exhibit distinctive vulnerabilities because of their unique designs. Performance indicators are selected to answer the question, "Is this other dam or appurtenant structure behaving as expected?" Instruments capable of measuring those performance indicators are described in detail in Chapter 5.

12.1 OTHER DAMS

This section addresses recommendations related to surveillance and monitoring of buttress dams, masonry dams, timber crib dams, and inflatable dams.

12.1.1 Buttress Dams

Slab-and-buttress and multiple-arch dams (sometimes referred to as "hollow dams") consist of concrete buttresses or piers that support a sloping

343

Fig. 12-1. Downstream view of slab and buttress dam

upstream concrete face to impound water. A photograph of a slab-and-buttress dam with a flat sloping upstream face is shown in Fig. 12-1, and a cross-sectional view is shown in Fig. 12-2. Several slab-and-buttress dams in the United States were constructed by the Ambursen Hydraulic Construction Company and other companies, mostly in the first half of the twentieth century in the United States, although more recent examples also exist. Many were also constructed in Europe after World War II. The name Ambursen Dam, however, is often used to generically refer to a slab-and-buttress-type dam, whether or not it was constructed by the Ambursen Company.

A variation on the slab-and-buttress dam is a multiple-arch dam, shown in Fig. 12-3, which consists of sloping arch structures instead of a flat sloping upstream face. The downstream buttresses support the arches that serve as the upstream water barrier.

Slab-and-buttress and multiple-arch dams differ from concrete gravity dams in that they are engineered to conserve materials. The members are sized and reinforced to carry the anticipated design loading condition. In contrast to concrete gravity dams, whose cross section is solid concrete, slab-and-buttress dams have hollow areas between their buttresses. Slab-and-buttress dams use significantly less concrete than is used for a concrete gravity dam of similar size. Rather than simply using the weight of concrete to resist sliding and overturning, the slab-and-buttress and multiple-arch dams elegantly make use of the downward force of hydrostatic pressure on the sloped upstream face to mobilize frictional resistance to sliding on the concrete–bedrock interface at the base of the supporting buttresses. The upstream face of the dam is tied into the bedrock foundation by a "plinth" or, alternatively, by structurally tying the slab to a foundation structural mat if the dam is founded on soil, as shown in Fig. 12-4.

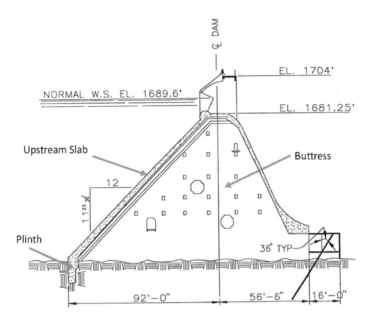

Fig. 12-2. Cross-sectional view of slab-and-buttress dam
Source: ASCE Task Committee on Instrumentation and Monitoring Dam Performance (2000).

Fig. 12-3. Multiple-arch dam
Source: Photographer Alex Stephens, U.S. Bureau of Reclamation.

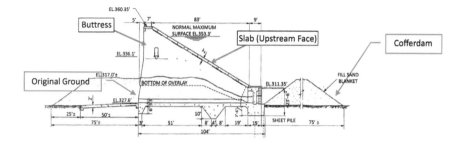

JUNIATA DAM

This is a series of six photographs taken during the construction of the dam for the Juniata Hydro-Electric Co. on the Juniata River, above Huntingdon, Pa. It is essentially the same dam as the Schuyler-ville Dam except as modified for soft foundations. The rollway of this dam is 375 feet long and 28 feet high. The total length of the dam including the power house and earth embankment is roughly 1200 feet.

PLATE 11 is an admirable photograph showing the nature of the foundations which are strictly gravel there being no ledge in sight. A stratum of hardpan underlies this gravel at a depth of 18 feet. In the foreground is the excavation and flooring of the wheel pit and beyond it is seen the gravel sup-porting the dam and through which are trenched the two cutoff walls which go down until they intersect the hardpan.

PLATE 11

Fig. 12-4. Slab-and-buttress dams founded on soil
Source: Ambursen Hydraulic Construction Company (1907).

Monitoring of buttress dams is similar to that of gravity dams; however, buttress dams have unique features that must remain stable for reservoir control.

The upstream face of a buttress dam (slab and plinth) serves as the water barrier for the dam. The buttress design allows foundation drainage into the hollow area of the dam between buttresses, and that drainage protects buttresses from uplift forces. Any foundation instability could result in pos-sible movement of a slab and plinth, initiating a sequence of events that may lead to failure. The segmented configuration of slabs supported by spaced buttresses is sensitive to differential settlement and misalignment. Buttress movement could result in (1) redistribution of stress exceeding the strength of a face panel, creating a "domino" failure or (2) opening a privi-leged path of seepage creating an internal erosion path that robs a buttress of support, again leading to a "domino" failure.

Key performance indicators are (1) unexpected seepage or leakage, (2) crack growth, and (3) buttress movement.

Freeze–thaw cycles, ASR, or general aging can initiate cracks in the face slab or plinth. If leakage through the slab and plinth concentrates on the ground, it is typically channeled into trenches to convey flow to a weir or other flow-measuring device to detect increasing flow indicative of increased cracking or deterioration.

A buttress dam supported on a concrete structural mat on soil foundation is vulnerable to a lack of drainage, causing underlying soil to settle away from the mat locally or to be carried downstream by internal erosion flow. The concrete mat can form a "roof" spanning a developing void in the soil foundation, which is one of the primary causes of internal erosion. Slab-and-buttress dams founded on mat foundations on soil often have upstream and intermediary cutoffs (Fig. 12-4). Standpipes or piezometers are used to monitor changes in pore pressure indicative of internal erosion deterioration of a cutoff. Measuring water levels that decrease with time and slowly approach the tailwater levels could be a potential indication of the loss of flow resistance through a void between the instrument and tailwater. Leakage is measured with any flow-measuring device.

Cracks can be painted along their length to allow visual detection of any lengthening of cracking into unpainted regions of the dam face. Cracks may grow from seasonal temperature changes, foundation instability, or reservoir level variations. Measuring crack growth is performed using the same methods described in Chapter 10.

Buttress movement can be measured by geodetic techniques, tiltmeters, and extensometers.

For older buttress dams, it is unlikely that earthquake loads were considered in the original design, and their design may not meet current standards. Understanding of the seismic setting of the dam site itself may have changed since design. In earthquake-prone regions, loading in the cross-channel direction (direction of the axis of the dam) may be important because that probably will be the weak axis of a buttress dam. Because of their less-massive, segmented construction in comparison with concrete gravity dams, buttress dams are more susceptible to earthquake damage, and consequently, these structures are closely inspected following a reported or felt earthquake. The means and methods for measuring earthquakes and their effects on a dam are explained in Chapter 10.

12.1.2 Masonry Dams

Dams were commonly constructed using stone masonry in the late eighteenth and early nineteenth centuries because stone was a readily available construction material and labor skilled in stone cutting was plentiful (Figs. 12-5 and 12-6). Design and construction of masonry dams varied widely. In some cases, the entire structure was constructed of hand-laid cut stone (with or without mortar). In other instances, the upstream and

Fig. 12-5. Masonry-faced dam

Fig. 12-6. Masonry arch dam

downstream faces of masonry dams were constructed using cut stones with mortared joints, and the internal zone of this containment was filled with rock rubble or cyclopean concrete that incorporated boulders to reduce the overall volume of concrete required.

Masonry dams may require little more than periodic movement surveys and visual surveillance. Where masonry is mortared, visual monitoring of the rate of deterioration may be sufficient to estimate when periodic maintenance, such as repointing, is necessary. If the interior interstitial spaces are not sealed with concrete or mortar or if the material has deteriorated over time, water pressure may rise between the upstream and downstream faces. Careful research of construction records and photographs may reveal a potential for water pressure rise.

Mortar used in masonry dams was proportioned to be durable at the time of construction, but it may now be weathered and susceptible to deterioration from acid rain or aggressive water chemistry. Visual surveillance is important to detect changes not amenable to instrumented monitoring. If monitoring of movement, instability, and seepage is required, measurements can be accomplished using techniques similar to those introduced in Chapters 6 and 10 for monitoring concrete dams.

A variation of the masonry dam theme is incorporation into the body of a taller concrete gravity dam, forming a composite dam with separate zones of masonry and concrete (Figs. 12-7–12-9). Typically, concrete forms the downstream portion of the composite section, and drains are required to collect the seepage along the joint between masonry and conventional concrete. Drains function to reduce the buildup of pore pressures within the dam. Monitoring seepage and pressure and leakage is important to understanding the effects on stability, if any. Appropriate instruments include various types of piezometers and strategically located seepage measuring weirs.

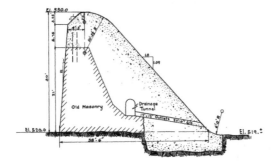

Fig. 12-7. Composite masonry and concrete dam
Source: Courtesy of Alabama Power Company, reproduced with permission.

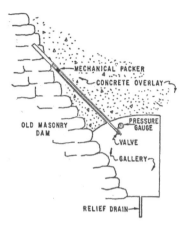

Fig. 12-8. Example of pressure monitoring in composite dam
Source: ASCE Task Committee on Instrumentation and Monitoring Dam Performance (2000).

Fig. 12-9. Ashlar masonry composite dam

12.1.3 Timber Dams

Dams constructed of timber (Fig. 12-10) were widely used in the eighteenth century, mainly to divert water for mills.

Timber crib dams were typically constructed of compartments or "cribs" formed of individual logs or timbers. The cribs were filled with ballast, usually boulders to add weight to mobilize sliding and overturning resistance. An example is shown in Fig. 12-11.

Fig. 12-10. Mad River Dam
Source: James Leffel & Co. (1874).

Fig. 12-11. Construction of log-crib (timber) dam at Postill Lake, east of Kelowna
(c. 1911)
Source: Kelowna Museum photo, No. 10, 347.

Although wood timbers are most common, large timbers may be replaced by precast-concrete "sleepers." The upstream face of timber crib dams is typically constructed of tightly jointed or overlapped timbers, whether wood or concrete. Upstream faces constructed of wood timbers are not designed to be watertight but are constructed to allow some leakage so that the internal wooden timbers remain sufficiently saturated to minimize the rate of deterioration.

Timber crib dams can have varying cross sections. Some have been treated with shotcrete as reinforcement, encased in concrete, or integrated into larger composite dams. Similar to the composite masonry structures, composite timber crib dams (Fig. 12-12) can present some challenging internal pressure and uplift conditions and may require monitoring by piezometers.

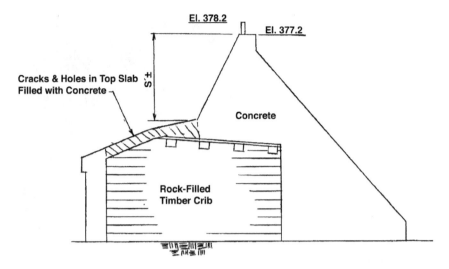

Fig. 12-12. Composite timber crib dam
Source: National Resources Committee (1938).

Analysis of the sliding stability of the cross section shown in Fig. 12-12 would benefit from a piezometer pressure measurement in the rock-filled timber crib to estimate the uplift profile on the base of the composite structure. Various composite cross sections exist, and a good understanding of their compositions is important to assess if and where instrumented monitoring could measure a performance indicator.

A few timber buttress dams (Fig. 12-13) are still in service, and their performance indicators are similar to those of concrete buttresses with added concern for timber strength.

Movement is a key performance indicator that is measured with the same geodetic techniques described for embankment or concrete dams. Survey measurements of crest and other monuments are used to detect vertical and lateral changes that may be indicative of structure movement caused by timber deterioration, settlement within the rockfill, or foundation settlement.

A potential initiator of a failure sequence is erosion of a soil foundation under the open rockfill contained within the timber cribs. Unfiltered, the foundation material may rise into the rockfill, resulting in settlement of the dam. Visual monitoring of foundation erosion or undercutting of timber crib dams may be accomplished with remotely operated vehicles or divers. Access is limited to the upstream and downstream edges of the dam unless voids under the cribs are accessible. Access may be difficult and potentially dangerous for divers owing to confinement in the case of undermined areas on the downstream side and the potential for high-velocity flow into and under the crib structure on the upstream side. For this reason, movement

Fig. 12-13. Timber buttress dam

often is limited to regular surveys of monitoring points with visual surveillance (robot or manned) following large floods.

Deterioration of wood timbers themselves may be a problem with timber crib dams. Although wood members that remain submerged can last for many decades, timbers in or above water-level fluctuation zones may need periodic replacement because of rot from exposure to sunlight, temperature fluctuations, and the atmosphere (Fig. 12-14).

The upstream face of tightly jointed or overlapped wood forms the water-retaining envelope. The upstream face is best designed to allow some leakage to keep the internal wood as moist as possible while still retaining the reservoir. Fungi, insects, or rot may weaken wood. Regular, repeatable visual monitoring of the timbers is an important surveillance aspect for timber crib dams. Where visual surveillance suggests weakness, random core samples or removal of complete cross sections to inspect timbers may be helpful. Experienced timber and wood material specialists can provide opinions on longevity and strength through visual examination and possibly testing of the recovered core samples.

12.1.4 Inflatable Dams

Inflatable dams are tubes or bladders formed of a rubber-like material that is often reinforced by fabric or metal mesh. Inflatable dams are mounted

Fig. 12-14. Wood gates showing vegetation and leakage

on concrete dam crests, in spillway bays, and on concrete sills placed directly across stream channels. They are water-control structures—storing water when inflated and releasing it when deflated. Air for inflation is typical for inflatable dams on the crest of concrete dams or in spillway bays. Water can be used for inflation if precise reservoir-level control is required. Freezing of water-filled tubes is avoided by adding antifreeze.

Loss of reservoir control is the only failure mode applicable to an inflatable dam. A sudden deflation could cause unanticipated reservoir discharge and, conversely, sudden inflation could cause unanticipated reservoir rise. Pressure maintenance is the performance indicator for inflatable dams, pressure being controlled by programmable logic.

Two types of inflatable dams are in common use. Both types are linked by piping to a compressor and accumulator that pump air or water into the bladder, causing the bladder to inflate. The first type is a rubber-like fiber or mesh-reinforced composite bladder, shown in Fig. 12-15. The dam is operated in the inflated condition to store water and deflated to release water.

The second type employs a similar bladder to raise and lower hinged flashboards, as shown in Fig. 12-16.

Fig. 12-15. Rubber dam

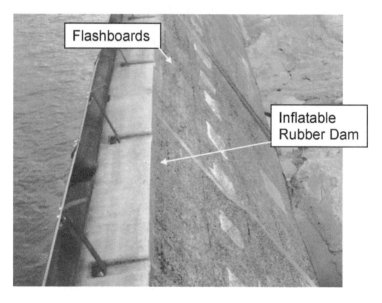

Fig. 12-16. Rubber dam with flashboards

Instrumentation and monitoring of inflatable dams is done with prepackaged proprietary systems that do not require owner intervention. Rather, the supplier's technician usually works with the contractor on installation and startup. Vendor-supplied pressure gauges and pressure transducers are used to measure pressure in the bladder, compressor air accumulator, and reservoir surface elevation. Level control is achieved by a programmed logic controller (PLC). Periodic checks of the reservoir level sensed by the reservoir-level transducer are necessary by manually cross-checking with a staff gauge to calibrate the reservoir-level transducer.

12.1.5 Concrete Dams on Soil Foundations

Concrete dams founded on soil (Fig. 12-17) present the same vulnerabilities discussed under buttress dams founded on soil and require similar surveillance and instrumentation to address potential vulnerabilities associated with soil foundations. These dams include gravity dams and buttress dams supported on a mat or apron founded on soil. How these structures are designed and constructed vary in terms of the care provided to control seepage. As discussed previously in the section on buttress dams, seepage is a performance indicator to evaluate internal erosion potential through the foundation. Foundation settlement combined with internal erosion is the potential initiator of ground loss. Loss of ground under the concrete may cause the dam to move, adversely affecting stress distribution. If the concrete cannot successfully redistribute stresses, it will crack. Continued movement could lead to a loss of structural capacity. Movement and seepage measurements are the important performance indicators for

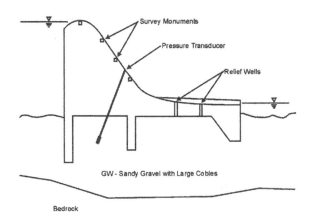

Fig. 12-17. Concrete dam on soil foundation
Source: ASCE Task Committee on Instrumentation and Monitoring Dam Performance (2000).

concrete dams on soil foundations. Measurements can be obtained with piezometers and pressure transducers at the soil–concrete interface and by quantitatively measuring flow from drains or relief wells. Movement is measured by geodetic methods or with a movement-measuring instrument.

12.1.6 Concrete Dams on Pile Foundations

Concrete dams (gravity or slab-and-buttress dams) founded on piles (Fig. 12-18) are a subset of concrete dams founded on soil, with some distinctive vulnerabilities that need to be considered in the development of an instrumentation and monitoring program. Though not common, dams supported on piles do exist, particularly in the midwestern United States and in areas where competent bedrock is often quite deep.

Concrete dams on piles have similar vulnerabilities and performance indicators as dams on soil foundations. One distinct difference in design and construction is that the base of the dam is effectively supported on piles that restrain slab movement. The soil supporting the slab may provide a seepage path. Care must be taken to ensure that seepage does not exploit a path under the slab and initiate internal erosion.

Sealed piezometers, open standpipe piezometers, or pressure transducers can be used to measure uplift along the foundation interface. Foundation drains (relief wells or "weep holes" with reverse filters) within the dam can help control uplift pressures and safely collect and convey seepage flow without soil transport. Periodic measurements of flow from drains or wells, together with pressure readings from the piezometers, will allow an understanding of changes in the structural response to changes in reservoir levels

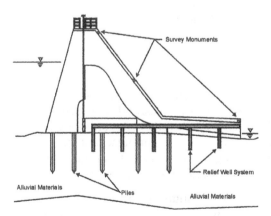

Fig. 12-18. Concrete dam on pile foundation
Source: ASCE Task Committee on Instrumentation and Monitoring Dam Performance (2000).

and seasonal effects. Movement is measured by geodetic methods or other means for measuring movement.

A unique concern with dams supported on timber pile foundations is the potential for deterioration of the piles, leading to settlement and structural damage. There is no established practice to measure deterioration of timber or concrete piles supporting concrete dams.

12.1.7 Small Earth Dams

Thousands of small [less than 40 ft (12.2 m) high] earth dams exist around the world. Small dams often do not have established surveillance and monitoring programs. Many may not have any instrumentation at all. Owners of small dams may not be aware of their responsibilities for dam safety. Complete or partial failure of a small dam could cause loss of property or life, depending on reservoir capacity and downstream development. Small dams may not receive the attention that larger dams enjoy from state and federal regulators because they pose little or no hazard to the public. As is mentioned in Chapter 2, some small-dam owners may not realize their duty to retain the reservoir. A surveillance and monitoring plan for a small dam needs to be simple but effective. A visit by an experienced dam safety engineer to observe how a dam was designed, operated, and maintained may yield positive results by helping an owner develop the surveillance and monitoring plan. That plan could include a suggested frequency of owner observation of the dam, highlighting areas of vulnerability and performance indicators such as sink holes, new seepage, changes in seepage rate or clarity, and slope sloughing. Instrumentation such as headwater and tailwater staff gauges or, if seepage exists at the toe of the structure, a weir or set of piezometers, or other appropriate means of periodic measurement may be recommended to complement visual surveillance.

12.2 APPURTENANT STRUCTURES

Failure of an appurtenant structure may result in loss of reservoir control. Spillway gates, outlet works, intakes, and conveyances all have responsibilities to retain the reservoir. Instrumented monitoring is seldom employed to measure the functionality of these appurtenant structures. The means and methods for measuring movements and leakage of appurtenant structures are the same as those described in Chapter 10:

- Spillway gates are maintained to ensure that they pass a flood when needed, and plunge pools are measured to ensure that scour does not jeopardize dam stability.
- Outlet works are maintained to ensure that they are functional in the event that they are required to lower the reservoir.

- Intakes are maintained with special attention to ensure that flow can be controlled in an emergency.
- Conveyances are maintained to ensure their ability to safely retain flow.
- Reservoir rims exhibiting instability are observed visually and occasionally instrumented to evaluate whether a landslide into the reservoir could initiate a failure sequence.

12.2.1 Spillways

Visual surveillance and regular maintenance are the most important duties to ensure that spillways are capable of safely discharging the inflow design flood. Instrumented monitoring is limited to periodic tests for trunnion friction, and strain measurements are occasionally performed to compute stress in gate members for comparison with design estimates.

Spillways discharging through a dam may deliver enough energy to scour the foundation progressively upstream to the point of undercutting the dam. Spillways discharging around a dam through a chute or tunnel may undercut a foundation as well. Routine scour measurements are important if foundations are erodible.

Stable transects are established across scour impact points. Geodetic techniques are useful for locating measurement points. Depths are measured at intervals along each transect to form a three-dimensional representation of scour, usually with a lead line or precision side-scan sonar. Figs. 12-19A and 12-19B illustrate the use of sonar and manual measurements, respectively, to check for scour from spillway discharge. Successive measurements are compared to estimate the rate of scour progression.

12.2.2 Outlet Works

Outlet works are located near a dam's foundation to permit discharge near the toe of accumulated reservoir sediment and to evacuate the reservoir in an emergency. Of interest is the depth of accumulated sediment, requiring a simple measurement.

The state of practice is for upstream control to leave conduits empty except when they are operated. Outlet conduits through concrete or bedrock typically are lined and are less problematic than conduits through the embankment fill. The primary areas of concern for monitoring outlet conduits through an embankment include development of seepage along the exterior of the conduit, which could initiate internal erosion, settlement, structural deformation, and collapse (Fig. 12-20).

Conduits may be instrumented, depending on identifying a performance indicator related to reservoir control. If pore-pressure measurements are required to evaluate stability, then hydraulic or electric piezometers can

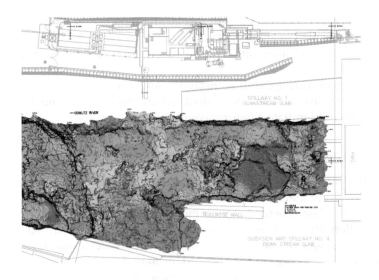

Fig. 12-19A. Multibeam side-scan sonar measurement
Source: Courtesy of City and Borough of Sitka, reproduced with permission.

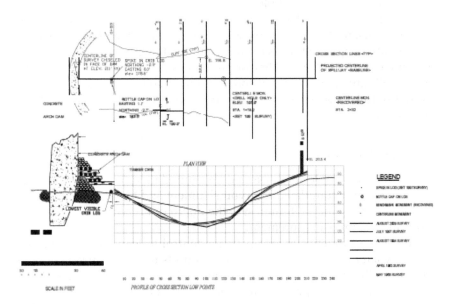

Fig. 12-19B. Manual measuring scour
Source: Courtesy of Lewis County PUD, reproduced with permission.

Fig. 12-20. Internal erosion along outlet conduit
Source: ASDSO.

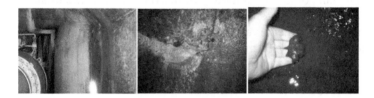

Fig. 12-21. Sediment from leaks into outlet works conduit

be placed alongside a conduit prior to backfill or during service by drilling through the wall of a conduit. Care must be taken that the tubing or cable does not form a preferential seepage path through the dam. When installed from the ground surface, piezometers should be situated at least one-conduit diameter away.

Outlet conduits are visually inspected to observe seepage at the downstream portal or at discrete locations (such as joints) along the length of the conduit. Periodic internal inspection includes checking alignment to judge whether movement is occurring, looking for sediment originating from the embankment, identifying structural cracking or corrosion that indicative of structural weakening.

Accumulation of sediment transported from an embankment into a conduit (Fig. 12-21) is a performance indicator that may require instrumented measurement because sediment transport into a conduit indicates a loss of ground outside that could initiate internal erosion in a failure sequence. Seepage at the outlet is measured using a weir or flume.

Fig. 12-22. Intake for outlet works

12.2.3 Intakes

Intakes may be free-standing for an embankment dam, integrated into a concrete dam, or located remotely from a dam (Fig. 12-22). Intakes supply flow directly to turbines or to a conveyance (e.g., penstock, flume, canal, or tunnel). If movement is a concern, then a typical measurement technique (e.g., geodetic survey, tiltmeter, or extensometer) is used.

12.2.4 Conveyances

Conveyances may be closed conduits such as penstocks or open channels such as flumes. Penstocks may be aboveground (supported on saddles) or belowground. Flumes and canals are aboveground. If leakage along a conveyance (Fig. 12-23) could be an initiator in a failure sequence resulting in a loss of reservoir control, then a typical measurement device (e.g., weir or flowmeter) is used. Similarly, if alignment of an aboveground conveyance (e.g., flume, penstock, or canal) requires checking, then instruments shown in Chapter 5 (such as strain gauges) may be required.

Scarps from active landslides and leaning trees are indicative of instability (Fig. 10-1).

Slope instability may initiate a landslide that blocks or damages a conveyance, causing a loss of capacity to deliver water. Geodetic techniques are used for surface measurements of movement, and inclinometers are used to measure movement in the subsurface. Heavy rainfall often causes soil to lose strength, but measuring rainfall may provide warning to suggest that enhanced visual surveillance is required (Fig. 12-24).

Rainfall is not the only initiator of slope instability above a conveyance.

Fig. 12-25 shows a section of a wooden flume that was destroyed by a rockfall that was triggered by an earthquake. Incipient landslide creep (Fig. 12-26) can also pose challenges to maintaining conveyance integrity.

Fig. 12-23. Leaky penstock

Geodetic surveys measuring alignment allow for adjustments to be made to avoid penstock movement and potential rupture.

A piezometer in a canal (Fig. 12-27) can protect against overtopping by measuring a pressure equivalent to water surface elevation and transmit that elevation to the inlet and outlet controls to avoid overtopping.

Visual surveillance is the key to identifying a performance indicator that may lead to a loss of reservoir control through a canal appurtenant structure (Fig. 12-28).

Periodic inspection of tunnels will locate areas where reinforcement is required (Fig. 12-29). Instrumented measurements usually are not required except in cases where ground squeezing may require convergence measurements by tape extensometers.

12.2.4.1 Reservoir Rim. The Vajont disaster in Italy in 1963 (Fig. 12-30) took more than 2,000 lives and devastated the valley downstream from the

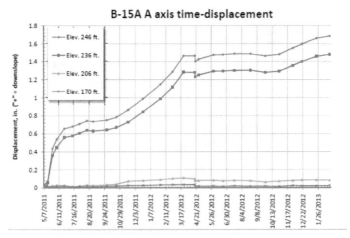

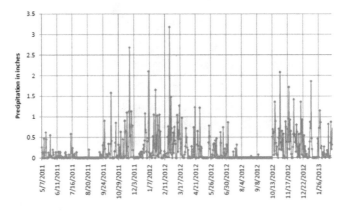

Fig. 12-24. Inclinometer response and rainfall
Source: Courtesy of PSE.

Fig. 12-25. Flume destroyed by rockfall initiated by an earthquake

Fig. 12-26. Penstock ring girder shoe to adjust for creep in slope above

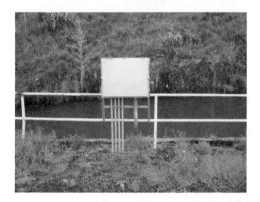

Fig. 12-27. Piezometer installation in power canal

Fig. 12-28. Canal buckle

Fig. 12-29. Partially lined tunnel in sandstone

Fig. 12-30. Vajont after landslide
Source: Wikimedia, https://en.wikipedia.org/wiki/Vajont_Dam (2018).

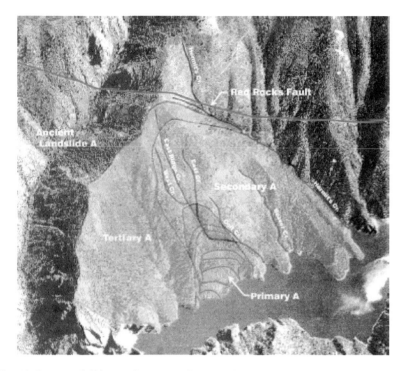

Fig. 12-31. Landslide moving toward reservoir
Source: U.S. Bureau of Reclamation.

dam. An enormous block of rock on the left abutment slid into the reservoir, displacing the entire reservoir over the crest of the dam into the valley below.

The incident raised awareness of the potential for a landslide into a reservoir initiating a failure sequence by generating a wave that could overtop a dam. Although the Vajont disaster is the only known case of overtopping by a landslide, there are numerous cases of reservoir slope instability, especially during first filling and then during a reservoir's service life. Some landslides are large enough to merit monitoring because the conditions favor developing enough energy in a slope failure to generate a wave large enough to overtop a dam (Fig. 12-31).

The previously mentioned landslides have been monitored over time with geodetic techniques measuring positions of monuments and slope stakes. A steep valley, deep reservoir, adverse jointing, low strength of soil or rock, and a straight valley in front of a dam present the most likely candidates for instrumented monitoring. New techniques employing airborne LiDAR and InSAR also offer opportunities for measuring landslide movement.

CHAPTER 13
SAMPLE DATA FORMS AND PLOTS

A picture is worth a thousand words.

English idiom

Chapter 8 discussed data recording and also presented some example formats for presenting seepage, pressure, and deformation data. This chapter presents sample reading recording forms, data tables, and sample data plots.

Many organizations use customized data forms and presentation formats that have been tailored to meet the needs of the organization. Many variations of data recording forms and data plots are suitable, but the use of standardized forms within an organization reduces recording errors and misinterpretation.

Best practice is to include the following elements on data recording forms:

- Instrument name,
- Date of the reading,
- Name of the person recording the data,
- Field reading, and
- Description of the field reading.

These elements provide the minimum information required to establish the value of the measured parameter at a specific point in space and time. The instrument name ties the field reading to a particular location and elevation when associated with the instrument attributes. The name of the person recording the data attests to the veracity of the reading and provides a contact point if there are any questions about the reading.

The data recording form may include additional optional information. The following list shows several types of information and explains how each type could be useful:

Time (hour and minute) of the reading—This may not be relevant if readings are taken far apart in time, for example, quarterly. If the reading interval is short, including the time can help produce more accurate time plots or help correlate the reading with a particular loading condition or event. If multiple readings are taken in a single day, the time is necessary.

Current weather conditions—Wet, dry, hot, cold, or stormy conditions can affect some field instrument readings. The response of some instrument sensors may drift at temperature extremes. Freezing conditions can affect water-level measurements. Variations in barometric pressure can affect pressure sensors if they are not properly vented to the atmosphere.

Plan view of the instrument location—A small plan view or map of the instrument location can help field personnel locate the instrument. Although each instrument is clearly marked, a plan view can help avoid confusion among similar instruments, especially if they are located in a group.

Coordinates of the instrument location—GPS coordinates can help field personnel locate the instrument. Although each instrument is clearly marked, geographic coordinates can provide additional confirmation that the correct instrument is being read.

Cross section of the instrument location—A cross section of the instrument installation can be useful in some situations. For example, if several hydraulic piezometers are installed in a single borehole but at different depths, the cross section may help detect situations in which they have been confused. The cross section may also alert field personnel that the reading does not seem reasonable and lead to a repeat measurement to confirm or correct the reading.

Instrument attributes—The instrument attributes may provide information about the surface characteristics of the instrument such as the height of a piezometer standpipe above the ground surface. If the field situation does not correspond to the attributes, the instrument may have been modified or damaged. If the attributes include the maximum depth of a hydraulic piezometer, it can help field personnel determine whether the instrument is dry or obstructed at an elevation above the bottom.

Most recent previous readings—The readings of most field instruments do not change dramatically unless some loading condition has changed. Comparing the current reading to a previous reading can help identify changed conditions or erroneous readings.

Identification of the reading device—Most instruments are read with equipment that needs to be checked or calibrated periodically. Identifying the reading device links the reading to the calibration records, increasing confidence in the accuracy of the reading or identifying readings that may have been taken with a faulty reading device.

Regardless of the details, well-designed forms and plots incorporate features that facilitate understanding of the instruments being read or evaluated. As such, best practices suggest that the forms and plots contain enough information about the instruments so that the user can understand the instruments and detect trends and anomalies. The following sections discuss a variety of forms and plots from actual projects.

13.1 READING RECORDING FORMS AND DATA REPORTS

13.1.1 Seepage

Table 13-1 shows a simple data report that tabulates weekly readings for several types of instruments. The form presents the reading dates, the instrument names, and the reading values. This form includes measurements of water level, flow rate, and pressure head from vibrating-wire piezometers (VWP in the table). Note that each data type includes the units of measurement.

13.1.2 Water Level

Table 13-2 presents a reading form for the manual collection of water-level data from a series of piezometers and wells. This was a standard form used for monthly data collection. Note that the preprinted form includes the list of instruments to read (column 1), the instruments' sensor elevations (column 3), and the elevations at the top of each instrument (column 4). The date in column 5 applies to all the readings. The field technician wrote the field reading, in this case the distance from the top of the pipe to the water level, in column 6 and then calculated the total head using the formula shown at the top of column 7. Any unusual aspects of the reading were noted as observations in column 8. Note, for example, that instrument L2-1 was contaminated and instrument P2-4 was dry.

In Table 13-2, the entries for the open hydraulic piezometers P2-4 and P2-5 should have indicated readings in column 6 for the depths measured even though no water was detected. This is considered a best practice and would have allowed a comparison of the elevation of the dry readings to the sensor elevations to distinguish between piezometers which were completely dry and ones that had obstructions in the standpipe at some elevation above the sensor.

13.1.3 Pressure

The data report in Table 13-3 presents pressure readings for VWPs installed in a standpipe. The readings were taken manually with a portable readout and then transferred to a spreadsheet to ease the conversion calculations

Table 13-1. Canal Flow and Pressure Record.

		Table 1 - LMS Pipe Flows and Drain VWP Head							
	Canal	LMS 1		LMS 2		LMS 3		LMS 4	
DATE	(ft – el)	Pipe (gpm)	VWP (ft)	Pipe (gpm)	VWP (ft)	Pipe (gpm)	VWP (ft)	Pipe (gpm)	VWP (ft)
6/14	602.3	3.7	-1.04	13.3	-0.83	7.1	-0.28	10.7	-0.33
8/16	602.2	2.4	-0.92	4.1	-0.72	2.6	-0.13	5.9	-0.18
8/23	601.4	2.5	-1.0	6.2	-0.79	3.7	-0.20	7.4	-0.28
8/30	602.2	2.6	-0.98	5.0	-0.80	3.7	-0.19	7.3	-0.25
9/6	603.4	2.5	-1.06	4.9	-0.88	3.3	-0.31	7.2	-0.36
9/13	602.1	3.4	-1.11	5.6	-0.94	4.9	-0.33	8.7	-0.37
9/20	602.0	3.9	-1.05	8.2	-0.87	7.1	-0.23	11.6	-0.32
9/21	601.9	4.0		7.8		5.7		10.8	
9/22	602.2	5.6		8.4		6.1		11.3	
9/27	602.2	3.4	-1.09	7.5	-0.94	5.7	-0.32	10.8	-0.36
10/5	602.2	4.8	-1.08	7.5	-0.94	5.7	-0.34	10.6	-0.37
	Canal	LMS 5		LMS 6		LMS 7		LMS 8	
DATE	(ft – el)	Pipe (gpm)	VWP (ft)	Pipe (gpm)	VWP (ft)	Pipe (gpm)	VWP (ft)	Pipe (gpm)	VWP (ft)
6/14	602.3	7.9	-0.13	8.8	-0.31	6.3	-0.20	6.1	-0.54
8/16	602.2	3.6	-0.01	3.0	-0.27	3.1	-0.12	2.9	-0.41
8/23	601.4	4.8	-0.11	3.9	-0.37	4.4	-0.21	4.7	-0.47
8/30	602.2	5.6	-0.05	3.4	-0.34	3.0	-0.20	4.5	-0.47
9/6	603.4	4.5	-0.18	3.0	-0.47	3.5	-0.35	3.2	-0.60
9/13	602.1	7.3	-0.17	6.0	-0.44	5.4	-0.31	12.9	-0.58
9/20	602.0	8.3	-0.12	6.6	-0.31	7.2	-0.24	8.1	-0.54
9/21	601.9	7.8		6.4		6.3		6.1	
9/22	602.2	8.8		7.2		8.7		33.1	
9/27	602.2	9.3	-0.18	8.7	-0.38	11.5	-0.13	10.3	-0.58
10/5	602.2	9.4	-0.15	7.6	-0.44	9.4	-0.28	9.4	-0.61
	Canal	LMS 9		LMS 10		LMS 11		Total Flow	
DATE	(ft – el)	Pipe (gpm)	VWP (ft)	Pipe (gpm)	VWP (ft)	Pipe (gpm)	VWP (ft)	(gpm)	
6/14	602.3	6.4	0.29	2.1	-0.13	11.6	0.13	83.9	
8/16	602.2	4.2	0.47	3.1	0.05	13.3	0.29	48.1	
8/23	601.4	6.2	0.40	3.9	0	15.9	0.22	63.7	
8/30	602.2	4.7	0.40	2.7	-0.04	11.9	0.25	54.5	
9/6	603.4	3.6	0.25	1.9	-0.28	9.4	0.10	47.1	
9/13	602.1	3.3	0.28	6.1	-0.13	5.9	0.15	70.5	
9/20	602.0	7.7	0.34	4.8	-0.04	25.8	0.27	99.3	
9/21	601.9	6.2		3.8		19.8		84.8	
9/22	602.2	6.0		4.1		19.9		119	
9/27	602.2	7.9	0.30	5.1	-0.06	13.5	0.14	93.6	
10/5	602.2	9.2	0.28	9.1	-0.11	19.7	0.18	102.4	

Source: Courtesy of Cowlitz County PUD.

Table 13-2. Water-Level Record.

OIL LEVEL: _15'04"_	AMUAY SURVEILLANCE PERIOD: _July 79_						
① INSTRUMENT	② LOCATION	③ SENSOR ELEVATION	④ TOP ELEVATION	⑤ DATE	⑥ READING	⑦·④-⑥ TOTAL HEAD	⑧ OBSERVATIONS
W2-1	FORS-2	- 0,594	+ 24,70	10-7-79	— • —	— • —	
W2-2		- 1,03	+ 20,30		17.29	+3.01	
W2-3		- 3,14	+ 6,50		4.76	+1.74	
L2-1		+ 6,52	+ 20,27		— • —	— • —	CONTAMINADO
P2-1		- 6,47	+ 20,10		17.61	+2.41	
P2-2		- 3,82	+ 25,36		21.82	+3.54	
P2-3		- 1,17	+ 24,68		21.31	+3.37	
P2-4		+ 10,68	+ 25,65		— • —	— • —	Dry.
P2-5		+ 12,70	+ 25,84		— • —	— • —	Dry
P2-6		- 6,05	+ 3,34		— • —	— • —	Revosado.
P2-7		- 15,70	+ 3,311		— • —	— • —	" "
P2-8		- 6,07	+ 3,395		2.13	+1.26	
P2-9		- 16,39	+ 3,395		4.43	-1.03	
P2-10		- 7,20	+ 3,352		5.96	-2.60	
P2-11		- 16,50	+ 3,352		5.98	- 2.62	
P2-12		- 6,17	+ 3,62		2.74	+ 0.88	
P2-13		- 23,08	+ 3,62		6.98	+3.36	
P2-14		- 8,873	+ 6,927		5.67	+1.26	

Source: Courtesy of T. W. Lambe, reproduced with permission.

from the raw units (B units = Hertz^2/1,000) to head pressure in KPa to head pressure in meters of water head (m H$_2$O) to piezometric elevation. The conversion formula, which takes initial readings, temperature changes, and barometric changes into account, is shown in the table.

13.1.4 Displacement

The data report in Table 13-4 presents weekly readings for joint meters that are being used to measure the movement of the joints along a concrete-lined canal.

Note that in addition to the reading dates and instrument names, the report includes the raw field reading (frequency), the measured temperature, which is used in the strain calculation, the calculated displacement, and the displacement from the initial and most recent prior readings. At each location, the displacement is being measured in three dimensions.

13.1.5 Inclinometers

Tables 13-5 and 13-6 illustrate inclinometer raw data as collected by an inclinometer system. The table is actually a CSV file exported from the readout

Table 13-3. VW Piezometer Record.

Piezo ID: PZ-E
Serial number: VW22654 (2 MPA range)
Installation Date: 29-Nov-11
Coordinates:
Collar Elevation: 1583.76 MASL
Borehole Depth: 136 M
Elev. Borehole Bottom: 1447.76 MASL
MASL: Meter ABOVE SEA LEVEL

Base Reading (29-November-2011)
Reading (B units): 9203.00 (Prior to installation)
Temperature (°C): 14.90 (Prior to installation)
Sensor Elevation: 1479.738 (MASL)
Sensor Depth: 104.03 M
Pressure $P(kPa)=C.F.[(Li-Lc)-[Tk-(Ti-Tc)]+[0.10(Bi-Bc)]$
Li, Lc = Initial and Current Reading (B units)
Ti, Tc = Initial and Current Temperature (°C)
Bi, Bc = Initial and Current Barometric pressure (millibars)

Calibration Factors
C.F. = 0.46396 kPa/B unit
Tk = 0.89474 kPa/°C rise　　B units = Hz2 / 1000 ie: 2700Hz = 7290 B units

Barometric Pressure
Initial (Bij): 1013.0 millibars

Date	Time	Collar Elevation (MASL)	Sensor Elevation (MASL)	Reading (B Units Hz)	Temperature (°C)	Barometric Pressure (millibars)	Calculated Head Pressure (kPa)	Calculated Head Pressure (m H2O)	Calculated Piezometric Elevation (MASL)	Observations
29/11/2011	17:00	1583.76	1479.74	7356.8	33.8	1013.0	974.77	99.365	1579.103	Initial reading after installation
30/11/2011	17:10	1583.76	1479.74	7327.2	34.0	1016.0	988.99	100.814	1580.552	
01/12/2011	16:25	1583.76	1479.74	7311.6	33.3	1021.0	996.10	101.539	1581.277	
02/12/2011	17:10	1583.76	1479.74	7367.0	32.7	1012.0	968.96	98.772	1578.510	
03/12/2011	9:20	1583.76	1479.74	7373.0	32.6	1011.0	965.98	98.469	1578.207	
05/12/2011	13:34	1583.76	1479.74	7395.6	32.4	1009.0	955.12	97.362	1577.100	
06/12/2011	16:30	1583.76	1479.74	7409.1	32.3	1015.0	949.37	96.775	1576.513	
07/12/2011	17:50	1583.76	1479.74	7420.8	32.3	1008.0	943.24	96.151	1575.889	
08/12/2011	17:20	1583.76	1479.74	7428.1	32.2	1012.0	940.16	95.837	1575.575	
09/12/2011	16:35	1583.76	1479.74	7436.9	32.2	1002.0	935.08	95.319	1575.057	
10/12/2011	10:10	1583.76	1479.74	7442.5	32.2	1005.0	932.78	95.085	1574.823	
12/12/2011	16:15	1583.76	1479.74	7459.4	32.1	1004.0	924.75	94.266	1574.004	
15/12/2011	15:50	1583.76	1479.74	7473.5	32.1	1015.0	919.31	93.711	1573.449	
19/12/2011	16:20	1583.76	1479.74	7497.4	32.0	1001.0	906.73	92.429	1572.167	
22/12/2011	10:00	1583.76	1479.74	7509.0	32.0	999.0	901.15	91.860	1571.598	
05/01/2012	11:12	1583.76	1479.74	7557.3	31.9	1014.0	880.15	89.720	1569.458	
09/01/2012	11:15	1583.76	1479.74	7569.7	31.9	1011.0	874.10	89.103	1568.841	
12/01/2012	17:55	1583.76	1479.74	7579.1	31.4	1005.0	868.69	88.551	1568.289	
16/01/2012	16:59	1583.76	1479.74	7588.9	31.8	1013.0	865.30	88.206	1567.944	
25/01/2012	17:58	1583.76	1479.74	7611.4	31.8	1011.0	854.66	87.121	1566.859	
01/02/2012	16:18	1583.76	1479.74	7626.1	31.8	1015.0	848.24	86.467	1566.205	
10/02/2012	10:21	1583.76	1479.74	7637.0	31.8	1016.0	843.28	85.962	1565.700	
15/02/2012	16:12	1583.76	1479.74	7641.5	31.8	1012.0	840.79	85.708	1565.446	
20/02/2012	15:51	1583.76	1479.74	7648.1	31.8	1013.0	837.83	85.406	1565.144	
23/02/2012	11:18	1583.76	1479.74	7651.8	31.8	996.0	834.42	85.058	1564.796	

Source: Courtesy of Seattle City Light, reproduced with permission.

Table 13-4. Record of Joint Meters.

Table 2 Intake / Canal Liner Joint Meters
Baseline, 1st Reading, and Recent 2006 Data

West Side Joint Meter

Channel Axis	Ch. 1 West X					Ch. 2 West Y					Ch. 3 West Z				
DATE	Frequency (Hz)	Temp. °C	Distance (mm)	Change from Prev. (mm)	Change from Base (mm)	Frequency (Hz)	Temp °C	Distance (mm)	Change from Prev. (mm)	Change from Base (mm)	Frequency (Hz)	Temp °C	Distance (mm)	Change from Prev. (mm)	Change from Base (mm)
Base Line Installation (6/21/06)	2860.9	N/A	40.81			2629	N/A	29.45			2580.6	N/A	28.68		
6/28/06	2857.1	10.9	40.65		0.16	2615.4	10.9	28.95		0.51	2565	11	28.11		0.57
9/6/06	2851.10	12.3	40.40	0.01	0.41	2605.60	12.3	28.59	0.03	0.87	2575	12.30	28.47	-0.16	0.21
9/13/06	2850.10	12	40.36	0.04	0.45	2604.70	12	28.55	0.03	0.90	2575.7	12.00	28.50	-0.03	0.18
9/20/06	2850.80	11.7	40.39	-0.03	0.43	2603.50	11.7	28.51	0.04	0.95	2577.3	11.80	28.56	-0.06	0.12
9/27/06	2851.60	11.2	40.42	-0.03	0.39	2604.00	11.2	28.53	-0.02	0.93	2578	11.20	28.58	-0.03	0.10
10/5/06	2849.00	13.4	40.31	0.11	0.50	2601.50	13.5	28.44	0.09	1.02	2578.2	13.50	28.59	-0.01	0.09

East Side Joint Meter

Channel Axis	Ch 1 East X					Ch. 2 East Y					Ch. 3 East Z				
DATE	Frequency (Hz)	Temp °C	Distance (mm)	Change from Prev. (mm)	Change from Base (mm)	Frequency (Hz)	Temp °C	Distance (mm)	Change from Prev. (mm)	Change from Base (mm)	Frequency (Hz)	Temp °C	Distance (mm)	Change from Prev. (mm)	Change from Base (mm)
Base Line Installation	2599.6	N/A	30.55			2820	N/A	39.27			2666.6	N/A	32.27		
6/28/06	2604.7	10.8	30.75		-0.20	2818.5	10.8	39.21		0.06	2615.3	10.8	30.28		1.99
9/6/06	2603.20	12.20	30.69	-0.02	-0.14	2813.10	12.2	38.98	0.01	0.29	2623.60	12.3	30.60	-0.14	1.67
9/13/06	2602.90	11.80	30.68	0.01	-0.13	2813.10	11.7	38.98	0.00	0.29	2624.10	11.8	30.62	-0.02	1.65
9/20/06	2603.78	11.60	30.72	-0.03	-0.16	2813.00	11.7	38.98	0.00	0.29	2625.20	11.8	30.66	-0.04	1.61
9/27/06	2604.70	11.20	30.75	-0.04	-0.20	2813.30	11.1	38.99	-0.01	0.28	2626.20	11.1	30.70	-0.04	1.57
10/5/06	2602.00	13.40	30.65	0.10	-0.09	2811.70	13.4	38.92	0.07	0.34	2626.00	13.4	30.69	0.01	1.58

Source: Courtesy of Eugene Water and Electric, reproduced with permission.

Table 13-5. Inclinometer Data Record.

Inclinometers	Location	Depth [Elevation] of Movement (feet)	Movements Recorded by Inclinometers			
			Prior to May 2000	May 2000 to May 2001	May 2001 to April 2002	April 2002 to April 2003
Forebay						
B-17	Mid-slope	13 [763]	1.0"(a)	--	--	--
F-1	Upper slope - west side	35 [744]	0.8"	0.0"	0.3"	0.2"
F-2	West side on canal road	(b) [--]	0.0"	0.0"	0.0"	0.0"
F-3	Upper slope - east side	25 [767]	1.0"	0.0"	-- (a)	--
F-4	East side on canal road	(b) [--]	0.0"	0.0"	0.0"	0.0"
F-5	West side above cut slope	53 [747]	0.3"(c)	0.0"	0.3"	0.2"
F-6	East side above cut slope	69 [756]	0.2"(c)	0.0"	0.4"	0.3"
East Percy						
B-9	Top of slide	19 [850]	0.2"	0.0"	0.0"	0.0"
Swafford						
B-14	Upper slide area (west side)	58 [769]	0.3"	0.0"	0.0"	0.0"
B-15	Top of steep slope (west side)	64 [762]	0.4"	0.0"	0.0"	0.0"
SF-1	Upper slide area (center)	80 [750]	0.3"	0.0"	0.0"	0.0"
SF-2	Upper slide area (east side)	28 [794]	0.5"	0.0"	0.0"	0.0"
"	" " " "	72 [750]	0.2"	0.0"	0.0"	0.0"
SF-3	Lower slide area (east side)	(b) [--]	0.0"	0.0"	0.0"	0.0"
McKenzie River						
B-19	Top of embankment (east)	21 [719]	0.2"	0.0"	0.0"	0.0"
B-21A	Top of embankment (center)	5 [735]	0.3"	0.0"	0.1"	0.0"
B-23	Top of embankment (west)	27 [713]	0.5"	0.1"	0.0"	0.0"
Canal Embankment						
C-1	Top of embankment (Sta. 246+00)	(b) [--]	--	--	0.0" (d)	0.0"
C-2	Top of embankment (Sta. 231+50)	(b) [--]	--	--	0.0" (d)	0.0"
C-3	Top of embankment (Sta. 220+00)	(b) [--]	--	--	0.0" (d)	0.0"

(a) Unable to read; slide movement has distorted casing too much for monitoring probe to pass through shear zone; (b) No movement to date; (c) Installed October 1999; and (d) Installed January 2002.

Source: Courtesy of Eugene Water and Electric, reproduced with permission.

Table 13-6. Inclinometer Data Record.

File Version 2.1
File Type Digital Inclinometer
Site Left Abutment
Borehole Borehole_1
Probe Serial# DP03530000
Reel Serial# DR08380000
Reading Date(m/d/y) 22/06/2008 12:16:32
Depth -41
Interval 0.5
Depth Units meters
Reading Units meters
Operator JJR
Comment: Initial Reading Inclo I-1
Offset Correction 0
ELEV. 448.176

Depth		Face A+	Face A-	Face B+	Face B-
-0.5	447.676	-0.000064	0.001299	-0.005706	0.006658
-1	447.176	-0.000515	0.001569	-0.006643	0.007637
-1.5	446.676	0.000611	0.000807	-0.006418	0.032546
-2	446.176	0.001157	0.000182	-0.006276	0.017232
-2.5	445.676	0.001795	-0.000153	-0.005835	0.042591
-3	445.176	0.000965	0.000484	-0.004605	0.029980
-3.5	444.676	0.000818	0.000416	-0.002506	0.007832
-4	444.176	-0.001470	0.002610	-0.001659	0.002635
-4.5	443.676	-0.001460	0.002549	-0.001119	0.002095
-5	443.176	-0.000250	0.001938	-0.000117	0.036616
-5.5	442.676	0.001478	-0.000355	0.001752	-0.000654
-6	442.176	0.004008	-0.002768	0.003622	0.001911
-6.5	441.676	0.005700	-0.004566	0.004526	-0.003575
-7	441.176	0.003817	-0.002654	0.003742	-0.002775
-7.5	440.676	0.003161	-0.002037	0.002923	-0.001932
-8	440.176	0.003267	-0.002053	0.003428	-0.002457
-8.5	439.676	0.002102	-0.000978	0.003078	-0.002118
-9	439.176	0.002342	-0.001146	0.003032	-0.002043
-9.5	438.676	0.003758	-0.002581	0.003806	-0.002661
-10	438.176	0.003635	-0.002448	0.004285	-0.002135

Source: Courtesy of Puget Sound Energy.

unit of the inclinometer system and shows readings in two perpendicular directions (A and B) that correspond to the orientation of the grooves in the inclinometer casing in which the inclinometer probe is run. In most systems, a first pass is done from bottom to top and readings are taken every 0.5 m (2 ft), followed by a second pass where the probe is turned 180° to eliminate systematic error bias that may be present in the probe. By convention, the readings of the first pass are labeled A+ and B+, and the readings of the second pass are labeled A– and B–. The actual reading that will be used in subsequent calculations will be 0.5 × [(A+) + (A–)] and 0.5 × [(B+) + (B–)], thereby eliminating the error bias. At the same time, calculating and plotting 0.5 × [(A+) – (A–)] and 0.5 × [(B+) – (B–)] display the error bias, usually called "checksums," and the checksums should be about constant for all readings in order to consider the set of readings as acceptable.

Additional calculations then need to be done in order to determine the profile of horizontal displacements in directions A and B, which is the objective of an inclinometer survey. These calculations are now always done by software, using either a spreadsheet or dedicated inclinometer software.

13.1.6 Temperature

Temperature per se is seldom measured; however, it is routinely measured if temperature can affect a measurement. As shown in Table 13-7, three separate temperatures are displayed (air, water, and gallery) to aid in the interpretation of the pressures in gravity dam monolith drains. The drain pressures respond to changes in the difference between air and water temperature and temperature in the gallery. Pressures rise during cold weather and decline in hot weather.

13.2 DATA PLOTS

This section presents sample data plots for various instrument types and highlights some of the key features and best practices of the plots.

Fig. 13-1 shows an instrument layout through a gravity dam, identifying the locations of the crest monuments, piezometers, and drains. The legend and the symbols identify the type of each instrument. This type of plan view is useful both to the field technician who needs to locate and read the instruments and to the engineer who needs to evaluate the significance of the readings.

Table 13-7. Temperature Relative to Drain Pressures.

	2:00 PM 1/5/2015	9:00 AM 1/6/2015	8:45 AM 1/7/2015	9:30 AM 1/8/2015	6:45 AM 1/9/2015	8:00 AM 1/12/2015
HEADWATER	561.2	560.5	560.1	560.4	561.1	560.3
TAILWATER	492.3	493.2	490.5	490.7	492.8	492.6
HEAD	68.9	67.3	69.6	69.7	68.3	67.7
AIR TEMP	33	38	36	35	36	37
WATER TEMP	40	40	40	40	40	40
GALLERY TEMP	52	52	51	50	50	50
DATE	2:00 PM 1/5/2015	9:00 AM 1/6/2015	8:45 AM 1/7/2015	9:30 AM 1/8/2015	6:45 AM 1/9/2015	8:00 AM 1/12/2015

DRAIN HOLE DATA:
STATION

Mono 4

Station		2:00 PM 1/5/2015	9:00 AM 1/6/2015	8:45 AM 1/7/2015	9:30 AM 1/8/2015	6:45 AM 1/9/2015	8:00 AM 1/12/2015
4-1.75	PRESSURE (PSI)	8.5	8.0	9.0	7.0	8.0	7.5
	STATIC WATER ELEV. (FT)	487.6	486.5	488.8	484.2	486.5	485.3
4-2.5	PRESSURE (PSI)	17.5	17.0	16.5	16.5	15.5	16.0
	STATIC WATER ELEV. (FT)	508.4	507.3	506.1	506.1	503.8	505.0
4-5.5	PRESSURE (PSI)	31.0	31.0	30.0	31.0	30.0	32.0
	STATIC WATER ELEV. (FT)	539.6	539.6	537.3	539.6	537.3	541.9
4-6.25	PRESSURE (PSI)	20.2	22.0	19.0	25.0	28.0	25.0
	STATIC WATER ELEV. (FT)	514.7	518.8	511.9	525.8	532.7	525.8

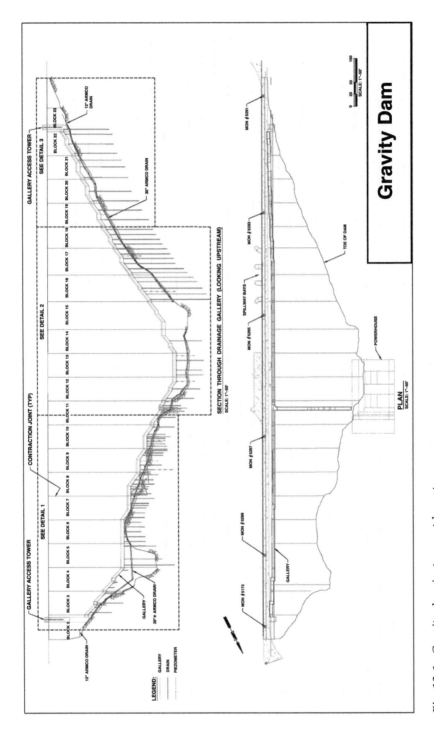

Fig. 13-1. Gravity dam instrument layout
Source: Courtesy of PSE.

13.2.1 Seepage

Fig. 13-2 presents a historical trend data plot of the flow rates measured in four toe drains associated with a hydroelectric dam. The data plot includes a separate graph of the reservoir and tailwater elevations to help evaluate whether changes in the toe drain flows correlate with changes in the water elevations. The data are continuous, indicating that they were obtained with a data-acquisition system, not manual readings. Several of the lines show dramatic variations in the flow rates over short time periods, which should be explained in the data evaluation report, confirms the high degree of scatter in the data ($R^2 = 0.59$), and quantifies the average long-term decline in the flow rate (decreasing by 0.12 gpm/day).

Fig. 13-3 summarizes the flow data from all the toe drains at a hydroelectric dam referenced in Fig. 13-1 over the same time period and the equation associated with the trend line.

13.2.2 Foundation Pressure

Fig. 13-4 shows a typical plot of piezometer response with time and headwater.

13.2.3 Pressure

Piezometer levels can indicate drain pressures and may be expressed as shown on Fig. 13-5.

13.2.4 Displacement

Fig. 13-6 illustrates a seasonal data plot of the horizontal movement of the center point in a concrete arch dam. Note the cyclical nature of the movements and their correlation to changes in temperature. Note also that the legend indicates the direction of the movements.

Fig. 13-7 presents another way of showing displacement. The vertical axes indicate the horizontal and vertical movements across a crack gauge across two monoliths and the reservoir elevation.

13.2.5 Temperature

Fig. 13-8 presents reservoir temperature versus time and concrete temperature at various depths versus time. From these two plots, the temperature differential can be extracted, allowing input to a thermal model for estimating stress in concrete.

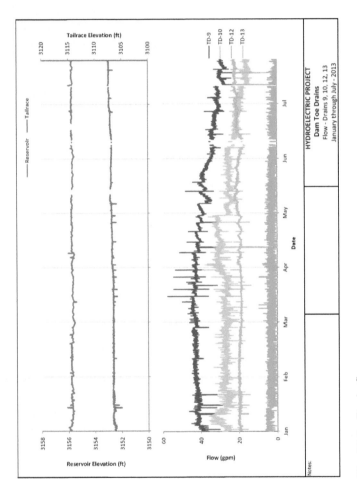

Fig. 13-2. Drain flows
Source: Courtesy of Seattle City Light, reproduced with permission.

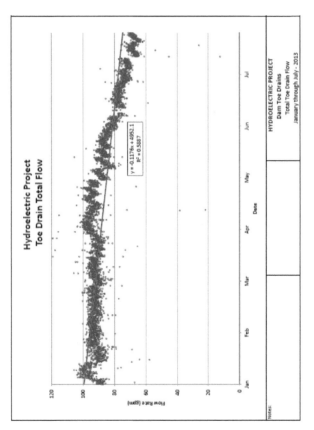

Fig. 13-3. Decreasing trend of drain flows
Source: Courtesy of Portland General Electric, reproduced with permission.

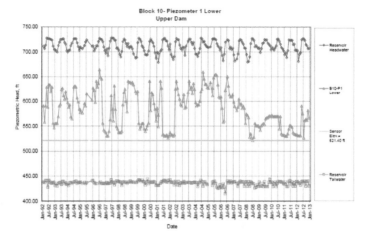

Fig. 13-4. Piezometer response with time and headwater
Source: Courtesy of PSE.

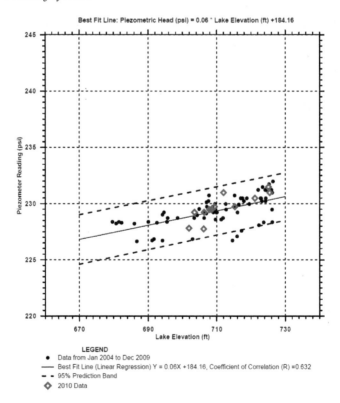

Fig. 13-5. Pressure versus headwater
Source: Courtesy of PSE.

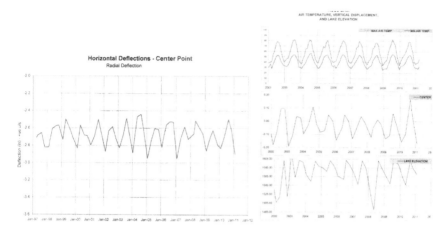

Fig. 13-6. Displacement versus time and temperature
Source: Courtesy of Seattle City Light, reproduced with permission.

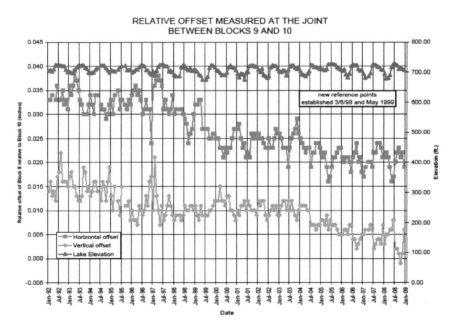

Fig. 13-7. Movement across a crack gauge
Source: Courtesy of PSE.

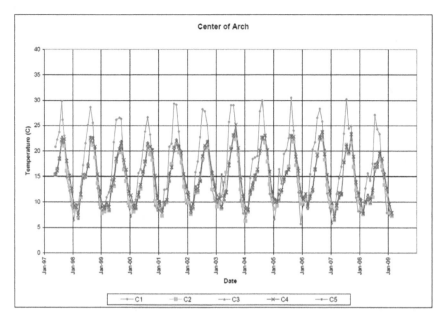

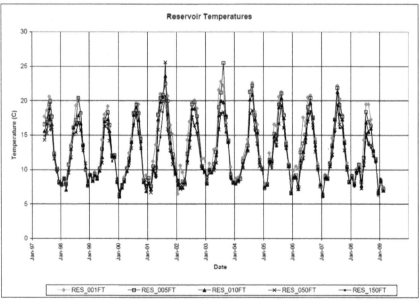

Fig. 13-8. Concrete and reservoir temperatures versus time
Source: Courtesy of Portland General Electric, reproduced with permission.

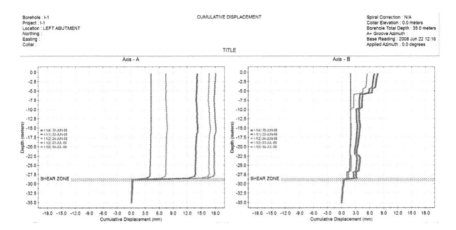

Fig. 13-9. Inclinometer profile showing horizontal displacement versus depth
Source: Courtesy of PSE.

13.2.6 Inclinometer Profile

Fig. 13-9 illustrates a typical inclinometer profile. This example is from a dam abutment in which excavation of the portal of a diversion tunnel triggered movement along an existing shear zone delimiting a large-size block. The movement was eventually stabilized by adding a massive concrete abutment to the tunnel portal. The profile shows the cumulative horizontal displacement in two perpendicular directions (A and B). The inclinometer casing is usually anchored at a depth where the ground is known to be stable, and displacements are calculated from the stable casing bottom.

CHAPTER 14
HISTORY OF INSTRUMENTATION
AND MONITORING

A people without the knowledge of their past history, origin and culture is like a tree without roots.

Marcus Garvey

The oldest surviving evidence of dam building is in Egypt. Sadd El Khafara was built across the Wadi El Garawi approximately 5,000 years ago. Constructed of rubble masonry in the form of an embankment dam, it was 37 ft (11.2 m) high, 265 ft (80.7 m) wide at the base, and 348 ft (106 m) long at the crest. The upstream face was lined with tightly fitting blocks of limestone. Apparently, it had an inadequate spillway, and it probably failed by overtopping.

Surviving evidence of achievements in the next two millennia B.C.E. is scarce. North of Baghdad, a dam known as Nimrod's Dam was built, and it diverted the entire flow of the Tigris into another channel of the river. Other projects followed not only along the Tigris but also along the Euphrates, Khost, and other rivers near Babylon. Alexander the Great is known to have commissioned a large canal joining the Tigris and Euphrates, probably in the fourth century B.C.E.

Romans are credited with the first use of curved dams in the south of France (Glanum) and in Tunisia (Kasserine) near where the Allies defeated Rommel in an epic World War II battle. These were curved gravity dams, and their shape is attributed to lengthening the crest to accommodate uncontrolled overflow. Two splendid examples of other Roman dams were constructed sometime in the first 500 years C.E. near Merida in Spain. Cornalvo (Fig. 14-1) and Proserpina Dams survive, and their outlet works still deliver water for beneficial use.

The Mongols invaded and defeated the Persians, and early in the second millennium, they built dams in Iran. Two of them, Kurit and Kebar (circa

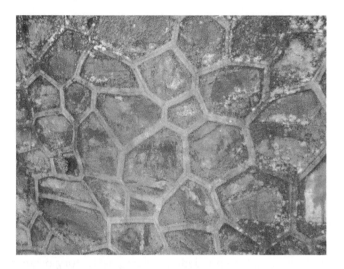

Fig. 14-1. Roman masonry at Cornalvo Dam after 1,000 years of service
Source: W. D. Schram (2018), reproduced with permission.

Fig. 14-2. Kebar Dam
Source: N. J. Schnitter (1994).

1280 C.E.), are worthy of mention because they were among the first true arch dams known to have been constructed; the Kebar Dam is shown in Fig. 14-2. The arch had nearly vertical upstream and downstream faces, with a constant radius of 125 ft (38.1 m) and it was 85 ft (25.9 m) high, 15 ft (4.57 m) wide at the crest, and 180 ft (54.8) long. They were keyed into trenches cut into rock on both abutments and built with a cemented (lime and ash mortar) core wall of rubble masonry faced both upstream and downstream with cemented dimension stones. Both had several levels of outlets presumed to be temporary diversion conduits during construction. Kurit has been raised several times, and a road across the crest still provides for vehicular traffic (de Rubertis 2011, Yang et al. 1999, Schnitter 1994, Smith 1971).

Early American dam building concentrated on providing a water supply and water power for mills, and timber was a common building material. Masonry construction was favored, and many of the structures built in the 1800s, such as Croton Dam, still are in use in New England. Embankment dams then began to make an appearance. The oldest surviving dam in the United States is Mill Pond Dam near Newington, Connecticut. It was constructed in 1676. The first recorded use of concrete was in 1890 at Lower Crystal Springs Dam, when interlocking concrete blocks were employed. The first concrete arch dam, Upper Otay, was built in 1888. Construction of Cheesman Dam on the South Fork of the Platte began in 1890, a beautiful masonry dam that still supplies water to the Denver metropolitan area. Hydraulic fill dams appeared to offer advantages in ease of construction and resistance to seepage in the fine-grained glacial soils of the East. Homogeneous embankments gradually gave way to zoned earthfill and rockfill dams (de Rubertis 2011).

Physical and visual inspections were part of these early projects from construction through initial filling and then through years of project operation. Formal instrumentation and monitoring systems came later. For example, surveyed elevation readings were known to be taken on the crest of Grosbois masonry dam in France in 1853. The dam was built between 1830 and 1838, and the readings were to measure the displacement of the dam's crest. In later years, surveyed elevation measurements became common practice for monitoring masonry dams.

In the late nineteenth century, open piezometers were used in India to study seepage flows under irrigation dams built on alluvium. In 1907, this type of instrument was employed by English engineers to determine the free surface of the nappe over a homogeneous earth dam.

In the United States, observation wells, water-level indicators, and hydrostatic-level indicators were some of the earliest measuring devices to record the water level or the pore water pressure on U. S. Bureau of Reclamation embankment dams. In 1911, observation wells, sometimes called saturation pipes, were installed at the Belle Fourche Dam in South Dakota. The wells were installed for the purpose of delineating the phreatic line and to help determine flow patterns through earth dams, abutments, and foundations. In 1935, water-level indicators were developed and installed at Hyrum Dam in Utah (Fig. 14-3) and Agency Valley Dam in Oregon. The water-level indicator apparatus was not very accurate because all the water in the system could not be displaced simultaneously. Moreover, the column of water in the system did not represent true pressure conditions because air

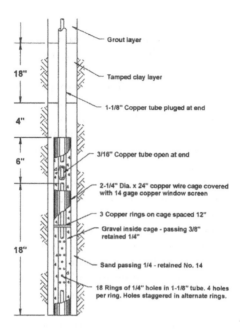

Grout layer

Tamped clay layer

1-1/8" Copper tube pluged at end

3/16" Copper tube open at end

2-1/4" Dia. x 24" copper wire cage covered
with 14 gage copper window screen

3 Copper rings on cage spaced 12"

Gravel inside cage - passing 3/8"
retained 1/4"

Sand passing 1/4 - retained No. 14

18 Rings of 1/4" holes in 1-1/8" tube. 4 holes
per ring. Holes staggered in alternate rings.

Fig. 14-3. Water-level indicator used at Hyrum Dam (Utah) in 1935
Source: ASCE Task Committee on Instrumentation and Monitoring Dam Perfor-
mance (2000).

was compressed in the soil voids and escaped into the larger tube. Extreme care was necessary during installation to prevent damage to the pressure point. Construction pore pressures could not be recorded because the water-level indicators could not be installed until after dam construction was completed. The water-level indicators were eventually abandoned. Concurrent with the development of the water-level indicators, hydrostatic pressure indicators were designed and installed in several dams. The 1938 and 1939 installation of hydrostatic pressure indicators at the Caballo Dam in New Mexico was typical for this period (Bartholomew 1987).

In 1922, the Engineering Foundation, an affiliation of the major engineering societies of the time, in cooperation with the Southern California Edison Company, constructed an extensively instrumented test arch dam to better understand the performance of these structures (Redinger 1986). The dam was constructed in a narrow canyon on a tributary of the San Joaquin River in the Sierra Mountains of California. Instrumentation included 140 "electric telemeters" buried in the concrete of the dam. These instruments consisted of a stack of carbon disks whose resistance changed with stress applied axially to the stack. Stress changes in the dam at different reservoir levels and temperature changes were measured by the telemeters and greatly added to the understanding of the behavior of arch dams.

In the United States, Roy Carlson, shown in Fig. 14-4, designed an apparatus that would measure pressure in concrete and earth dams by means of a sensor consisting of a stretched wire.

Carlson measured the variations in electrical resistance of the wire relative to the movement of two anchoring points using a Wheatstone bridge. This unbonded resistance-wire strain gauge was first produced commercially in 1932 and was later supplemented by piezometers, concrete stress cells, earth pressure cells, and joint meters.

Also in the United States in the 1930s, the U.S. Bureau of Reclamation began the use of controlled measurements to correlate the performance of completed dam structures with design assumptions and construction practices. This effort led to the design of water-level indicators and hydraulic twin-tube piezometers to measure pore pressure, as shown in Fig. 14-5 (Walker 1948, Daehn 1962).

An important milestone in the field was marked in 1931 in France when André Coyne (Fig. 14-6), founder of the French consulting engineering firm Coyne et Bellier and designer of more than 55 arch dams, obtained a patent for the vibrating-wire sensor (Fig. 14-7), then called an acoustic indicator (Coyne 1938).

With the major focus on the construction of new dams from the 1930s through the 1970s came the extensive utilization of dam instrumentation and monitoring programs. Many of the dam instrumentation and monitoring

Fig. 14-4. Dr. Roy Carlson (1900–1990)
Source: ASCE Task Committee on Instrumentation and Monitoring Dam Performance (2000).

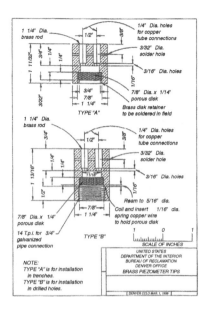

Fig. 14-5. Early twin-tube hydraulic piezometer tips developed by USBR in 1940
Source: ASCE Task Committee on Instrumentation and Monitoring Dam Performance (2000).

Fig. 14-6. André Coyne (1891–1960)
Source: ASCE Task Committee on Instrumentation and Monitoring Dam Performance (2000).

programs that were initiated during that era are still in existence today, although there have been many developments and improvements through the years.

Dam instrumentation and monitoring programs have undergone great changes in recent years because of progress in microcomputer technology. Developments in this field permit the acquisition and processing of data

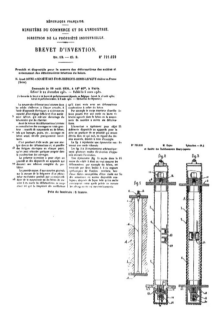

Fig. 14-7. André Coyne's patent for the vibrating-wire sensor registered in French as a témoin sonore (acoustic indicator)
Source: ASCE Task Committee on Instrumentation and Monitoring Dam Performance (2000).

that have changed the very nature of the results yielded by instrumentation. Instrumentation on dams in very remote areas of the world can be accessed via satellite, transmitting the data (GNSS) to offices on other continents. The data can instantly be reduced by a personal computer and plotted in a format that immediately allows engineers to identify deviations from the expected performance of the structure.

It is interesting to note that much of the design and manufacture of dam instruments has been conducted by practicing engineers and manufacturing companies specializing in the field. In some cases, industrial sensors have been adapted for use, yet a great many of the sensor designs used on dams today are unique to the field.

Equipment used in dam monitoring systems has to satisfy two challenging requirements. First, it must have a life span comparable to the design life of the dam. Second, it must be extremely stable with virtually no zero drift.

Scientific organizations, particularly the International Committee on Large Dams (ICOLD), have always given the utmost importance to collecting dam-related data from actual experience. The large number of reports on the subject, as well as the special ICOLD publications on dam instrumentation

(ICOLD 1988, 1989, 1992, 1994, 1999, 2000, 2005, 2009) clearly show the interest generated by changing instrumentation components and measuring techniques.

Agencies with global expertise and jurisdiction have also emphasized instrumentation and measurements with the publication of reference manuals and guidelines.

Although there have been many advances in data acquisition, processing, instrumentation components, and measuring techniques, neither the factors affecting dam performance nor the performance indicators have changed.

APPENDIX A
REFERENCES

American Congress on Surveying and Mapping (ACSM). (1989). "Definitions of surveying and associated terms." ASCE/ACSM, Gaithersburg, MD.

ASCE Task Committee on Instrumentation and Monitoring Dam Performance. (2000). *Guidelines for Instrumentation and measurements for monitoring dam performance*, ASCE, Reston, VA.

ASTM International. (2016). "Standard test method for low strain impact integrity testing of deep foundations." *ASTM D5882-16*, Gaithersburg, MD.

Balouin, T., Lahoz, A., Bolvin, C., and Flauw, Y. (2012). "Risk assessment required in the framework of new French regulation on dams. Methodology developed by INERIS." *Proc., 24th Congress and Int. Symp. on Dams for a Changing World*, International Commission on Large Dams (ICOLD), Paris.

Barker, M. (1998). *Australian risk approach for assessment of dams*, Australian National Committee on Large Dams, Hobart, Tasmania.

Bartholomew, C. L., and Haveland, M. L. (1987). *Concrete dam instrumentation manual: A water resources technical manual*, U.S. Department of the Interior, Bureau of Reclamation, Washington, DC.

Bartholomew, C. L., Murray, B. C., and Coins, D. L., eds. (1987). *Embankment dam instrumentation manual*, U.S. Department of Interior, Bureau of Reclamation, Washington, DC, 17.

Bartsch, M. (2004). *FMECA of the Ajaure Dam—A methodology study*, Thomas Telford Publishing, London, 187–196.

Bennett, V., Abdoun, T., Danisch, L., and Shantz, T. (2007). "Unstable slope monitoring with a wireless shape acceleration array system." *Proc., 7th Int. Symp. on Field Measurements in Geometrics*, September 24–27, Boston, FMGM.

Bordes, J. L., and Debreuille, P. J. (1985). "Some facts about long-term reliability of vibrating-wire instruments." *Transport. Res. Rec. No. 1004*: 20–26.

Central Water Commission, Dam Safety Organisation. (1986). "Report on dam safety procedures." Ministry of Water Resources, Government of India.

Choquet, P., Juneau, F., Debreuille, P. J., and Bessette, J. (1999). "Reliability, long-term stability and gate performance of vibrating wire sensors with reference to case histories." *Proc. 5th Int. Symp. Field Measurements in Geomechanics-FMGM99*, C. F. Leung, S. A. Tan, and K. K. Phoon eds., Singapore, December 1–3.

Choquet, P., Quirion, M., and Juneau, F. (2000). "Advances in fabry-perot fiber optic sensors and instruments for geotechnical monitoring." *Geotechnic. News*, March.

Chrzanowski, A., Avella, S., Chen, Y. Q., and Secord, J. M. (1992). "Existing resources, standards, and procedures for precise monitoring and analysis of structural deformations, Vol. 1." *TEC0025*, U.S. Army Corps of Engineers, Washington, DC.

Contreras, I. A., Grosser, A. T., and VerStrate, R. H. (2012). "Update of the fully grouted method for piezometer installation." J. Dunnicliff, ed. *Geotechnic. News* 30(2), 20–25.

Coyne, A. (1938). "Some acoustic monitoring results on concrete, reinforced concrete and metallic works." *Annales de l'Institut Technique des Batiments et des Travaux publics*, March (in French).

Daehn, W. W. (1962). "Development and installation of piezometers for the measurements of pore-fluid pressures in earth dams." *Proc., Symp. Soils and Foundation Engineering*, American Society for Testing and Materials, West Conshohocken, PA.

de Rubertis, K. (2011). "ASDSO. Dam engineering through the ages." *J. Dam Safety* 9(2), 31–39.

Dienum, P. J. (1987). "The use of tiltmeters for measuring arch dam displacements." *Water Power Dam Const.*, June.

Duffy, M. A., and Whitaker, C. (1998). "Design of a robotic monitoring system for the Eastside Reservoir in California." *Proc., American Congress on Surveying and mapping*, Vol. 1, Baltimore, 34–44.

Dunnicliff, J. (1988). *Geotechnical instrumentation for monitoring field performance*, Wiley, New York.

Dunnicliff, J. (1993). *Geotechnical instrumentation for monitoring field performance*, Wiley, New York.

FEMA. (2004). "Federal guidelines for dam safety, selecting and accommodating inflow design flood for dams." Interagency Committee on Dam Safety, Washington, DC.

FEMA. (2013). *Federal guidelines for inundation mapping of flood risks associated with dam incidents and failures, P-946*, Washington, DC.

FEMA. (2015). *Federal guidelines for dam safety risk management, P-1032*, Washington, DC.

FERC (U.S. Federal Energy Regulatory Commission). (2005). "Engineering guidelines for the evaluation of hydropower projects." In *Dam safety performance monitoring program*, Rev. 1, U.S. Federal Energy Regulatory Commission, Washington, DC.

FERC. (2007). "Dam safety performance monitoring program." In *Engineering guidelines for the evaluation of hydropower projects*, Rev. 1, U.S. Federal Energy Regulatory Commission, Washington, DC.

FERC. (2008a). "Dam safety surveillance and monitoring plan outline." In *Engineering guidelines for the evaluation of hydropower projects*, Rev. 2, Jan. 15, U.S. Federal Energy Regulatory Commission, Washington, DC.

FERC. (2008b). "Dam safety performance monitoring program." In *Engineering guidelines for the evaluation of hydropower projects*, Rev. 2, Jan. 15, U.S. Federal Energy Regulatory Commission, Washington, DC.

Holt, J., Poiroux, G., Lindyberg, R., and Cesare, M. (2013). "Detection without danger." *Civil Eng.* 83(1), 68–77.

Hydrometrics, Inc. (2011). *Guidelines for conducting a simplified failure mode analysis for Montana dams*, Montana Department of Natural Resources and Conservation, Helena, MT.

ICOLD (International Committee on Large Dams). (1972). "Dam monitoring—General consideration." *Bull. 60* (1988), revised and edited version of Bulletins 21 (1969) and 23, Paris.

ICOLD. (1988). "Dam monitoring general considerations." *Bull. 60*, Paris.

ICOLD. (1989). "Monitoring of dams and their foundations—State of the art." *Bull. 68*, Paris.

ICOLD. (1992). "Dam monitoring improvements." *Bull. 87*, Paris.

ICOLD. (1994). "Ageing of dams and appurtenant works." *Bull. 93*, Paris.

ICOLD. (1999). "Seismic observation of dams—Guidelines and case studies." *Bull. 113*, Paris.

ICOLD. (2000). "Automated dam monitoring systems—Guidelines and case histories." *Bull. 118*, Paris.

ICOLD. (2005). "Dam foundations, geologic considerations, investigation methods, treatment, monitoring." *Bull. 129*, Paris.

ICOLD. (2009). "General approach to dam surveillance." *Bull. 138*, Paris.

Inaudi, D., and Branko, G. (2006). "Reliability and field testing of distributed strain and temperature sensors." *Proc., SPIE Smart Structures and Materials Conf.*, International Society for Optics and Photonics (SPIE), Bellingham, WA.

ISRM (International Society for Rock Mechanics and Rock Engineering). (1981). "Suggested methods for pressure monitoring using hydraulic cells." In *The ISRM suggested methods for rock characterization, testing and monitoring*, Pergamon Press, Oxford, UK, 201–211.

Jacobs UK, Ltd. (2007). *Engineering guide to early detection of internal erosion, Jacobs Ref B2220300*, Department for Environment, Food and Rural Affairs, London.

Jansen, R. B. (1988). *Advanced dam engineering for design, construction and rehabilitation*, Van Nostrand Reinhold, New York.

Kane, W. F. (1998). "Embankment monitoring time domain reflectometry." *Proc., 5th Int. Conf. on Tailings and Mine Waste*, Colorado State University, Fort Collins, CO, 223–230.

Mazzanti P., Perissin D., and Rocca, A. (2015). "Structural health monitoring of dams by advanced satellite Sar interferometry: Investigation of past processes and future monitoring perspectives." *Proc., 7th Int. Conf. on Structural Health Monitoring of Intelligent Infrastructure*, Nhazca S.r.I., Rome.

McRae, J.-B., and Simmonds, T. (1991). "Long-term stability of vibrating-wire instruments: one manufacturer's perspective." *Proc., 3rd Int. Symp. on Field Measurements in Geomechanics*, Vol. 1, Rotterdam, Balkema, 283–293.

Mikkelsen, P. E., and Green, E. G. (2003). "Piezometers in fully grouted boreholes." *Proc., 6th Int. Symp. on Field Measurements in Geomechanics*, Oslo, Norway, September, 545–553.

Peltzer, G., Hudnut, K. W., and Feigl, K. L. (1994). "Analysis of coseismic surface displacement gradients using radar interferometry: New insights into the Landers earthquake." *J. Geophys. Res.* 99(19), 971–981.

Professional Association of Civil Engineers, Spanish National Committee on Large Dams. (2006). *Risk analysis applied to management of dam safety*, Madrid.

Redinger, D. H. (1986). *The story of Big Creek*, Trans-Anglo Books, Glendale, CA.

Regan, P. (2009). "An examination of dam failures vs. the age of dams." FERC, Washington, DC.

Schnitter, N. J. (1994). *A history of dams*, Balkema, Leiden, The Netherlands.

Schram, W. D. (2018). "Roman aqueducts." http://www.romanaqueducts.info/.

Sellers, J. B., and Taylor, R. (2008). "MEMS Basics." *Geotechnic. News*, March, 32–33.

Shakal, A. F., and Huang, M. J. (1996). "Strong motion instrumentation and recent data collected at dams." *Western Regional Technical Seminar*, Association of State Dam Safety Officials, Lexington, KY, 111–128.

Sherard, J. L., Woodward, R. J., and Gizienski, S. F. (1983). *Earth and earth–rock dams: Engineering problems of design and construction*, Wiley, New York.

Smith, N. (1971). *A history of dams*, Citadel Press, Secaucus, NJ.

Solinst. (1997). *Waterloo system*, Georgetown, Ontario, Canada.

Teixeira da Cruz, P., Materon, B., and Freitas, M. D., Jr. (2010). *Concrete face rockfill dams*, CRC Press, Boca Raton, FL.

Terzaghi, K., and Peck, R. B. (1968). *Soil mechanics in engineering practice*, 2nd Ed., Wiley, New York.

Thompson, P. M., Kozak, E. T., and Wuschke, E. E. (1990). "Underground geomechanical and hydrogeological instrumentation at the url." *Proc., Int. Symp. on Unique Underground Structures*, Denver, June 12–15.

Thuro, K., Wunderlich, T., and Heunecke, O. (2007). "Development and testing of an integrative 3D early warning system for alpine instable slopes (alpEWAS)." *Geotechnologien Sci. Rep.*, F. Wenzel and J. Zschau, eds., Springer, Berlin, 289–306.

U.S. Army Corps of Engineers (USACE). (1980). "Instrumentation for concrete structures." EM-1110-2-4300, September 15, Washington, DC.

USACE. (1995). "Instrumentation of embankment dams and levees." EM-1110-2-1908, June 30, Washington, DC.

USACE. (2002). "Structural deformation surveys." EM-1110-2-1009, June 1, Washington, DC.

USACE. (2011). "Engineering and design-safety of dams-policy and procedures." ER-1110-2-1156, October 28, Washington, DC.

U.S. Bureau of Reclamation. (1987). *Design of small dams*, U.S. Department of the Interior, Washington, DC.

U.S. Committee on Large Dams (USCOLD). (1989). *Strong motion instruments at dams*, USCOLD, Denver.

U.S. Department of Defense (DOD). (1974). "Military standard, procedures for performing a failure mode, effects, and criticality analysis." *MIL-STD-1629A*, Washington, DC.

U.S. Department of Transportation. (2008). "Reactive solutions: An FHWA technical update on alkali–silica reactivity." https://www.fhwa.dot.gov/pavement/concrete/reactive/issue03.cfm.

Walker, F. C. (1948). "Experience in the measurement of consolidation and pore pressures in rolled earth dams." *Proc., Int. Congress on Large Dams*, Stockholm, ICOLD, Paris.

Wikimedia Commons. (2018). "Standing section [of the St. Francis Dam]." https://commons.wikimedia.org/wiki/File%3AStanding_section.jpg.

Williams, S. C. (1985). *The elements of graphing data*, Wadsworth Advanced Books and Software, Monterey, CA.

Withiam, J. L., Fishman, K. L., and Gaus, M. P. (2002). "Recommended practice for evaluation of metal-tensioned systems in geotechnical applications." *NCHRP Report 477*. National Research Council, Washington, DC.

World Bank. (2002). *Regulatory frameworks for dam safety*, Washington, DC.

Yang, H., Haynes, M., Winzenread, S., and Okada, K. (1999). *The history of dams*, University of California–Davis, Davis, CA.

APPENDIX B
FAILURE MODE ANALYSES

It does not do to leave a live dragon out of your calculations, if
you live near him.

J. R. R. Tolkien

Failure mode analyses grew from some of technology's notable errors—
failures such as NASA's *Challenger* loss. Most countries with agencies regu-
lating dam safety encourage or require dam failure mode analyses to
systematically examine vulnerabilities and sequences of events that could
result in a loss of reservoir control. Results from the analyses identify monitoring requirements and actions needed to reduce or eliminate the probability that a progression of events will lead to a loss of reservoir control.

FMEA (failure mode effect analysis), FMECA (failure mode effect and
criticality analysis), PFMA (potential failure mode analysis), and FMA (failure mode analysis) are terms used describe a method of analysis, which
varies by jurisdiction; however, there are more similarities in the various
methods than there are differences. Most follow this general sequence:

- Research the records of design and history of performance.
- Compare the records with acceptance criteria for dam safety.
- Visually inspect the dam and appurtenant structures.
- Propose potential failure modes such as those described in Chapters
  3, 10, 11, and 12.
- For each potential failure mode, connect a required sequence of
  events, from initiation to resulting loss of reservoir control.
- Identify performance indicators that require surveillance and instru-
  mented measurements to judge whether a dam is performing as
  expected.

- Evaluate findings from surveillance and instrumented measurements for indications that would suggest or rule out a failure sequence.
- Examine opportunities for intervention.
- Decide whether or not action is required.
- Propose enhanced performance monitoring and remedial action as needed.
- Document the results of the analysis for regular update or as conditions change.

When evaluating potential failure modes for a dam, it is important to evaluate a dam's performance, not as an isolated structure but rather as part of an overall system that includes not only the dam structure itself but also

- Foundation and abutments,
- Pipelines,
- Penstocks,
- Spillways,
- Outlet works,
- Gates and valves,
- Electrical systems,
- Control and data-acquisition systems, and
- Personnel who maintain and operate the dam.

Any part of the system has the potential to introduce a condition that could lead to failure or could allow a failure to progress. The entire system demands scrutiny.

A failure mode can consist of multiple individual events, often represented by nodes in an event tree, which can eventually lead to failure. Each event in the failure mode event tree has an individual likelihood or probability of occurring. A failure mode is established by describing a series of events connecting the initial condition to the breach. Fully characterizing a failure mode assists in developing surveillance and monitoring options by exhibiting the different points in the process when intervention can interrupt the sequence and action can be taken to prevent conditions from progressing to a loss of reservoir control.

POTENTIAL FAILURE MODE IDENTIFICATION

Typically, a team of qualified individuals conducts the analysis. Prior to the analysis, the team is tasked with completing a thorough review of all relevant background information on the dam including geology, design, analysis, construction, flood and seismic loadings, operations, dam safety evaluations, and existing performance-monitoring documentation, looking for potential unknown defects that could introduce a mechanism of failure.

In addition to the review of background information, the team also participates in a thorough site inspection, looking for clues that could indicate vulnerabilities in the dam's structures or its foundation. Ideally, the core team includes dam operators.

Typically, a team member or a designated outside individual leads or facilitates the session. This individual should have substantial prior experience with dam engineering to provide direction and organization during the failure mode session. A session begins with team members brainstorming and offering candidate potential failure modes based on the information gained during the preliminary review. For each loading condition, a potential failure sequence is described from initiating condition to loss of reservoir control.

If the team judges that a potential failure mode cannot progress completely to reservoir release, it is considered not to be credible and documented as such and excused from further consideration. A failure mode analysis considers all potential paths, not just structural behavior of the dam and its foundation but also gate failures, penstock breaches, and operational issues.

After identifying credible potential failure modes, the team records factors that make each more or less likely to occur. For example, an overtopping failure during a flood might depend on the existence of redundant spillways to reduce the probability of failure. Conversely, if an embankment dam has minimal freeboard and only one small spillway gate, both conditions are factors that may make the failure more likely.

If additional information or analyses to properly evaluate a candidate failure mode, the team recommends that additional information be gathered to complete the review.

EVALUATING PRIORITIES FOR ACTION

After identifying potential failure modes, the next step is to prioritize them. Prioritizing actions associated with potential failure modes helps a dam owner allocate resources in the most efficient manner. Each potential failure mode depends on how likely that failure process is to occur and what the consequences would be if it were to occur. Potential failure modes with the highest likelihood of occurring and have high consequences require a larger share of resources dedicated either to remediating or monitoring.

Setting priorities for each potential failure mode is carried out by assigning it to a category based on its likelihood and consequences. This process can be as simple as the team considering the factors making each potential failure mode more or less likely and estimating the consequences associated with it. The team then uses this information to come to a consensus on prioritization of potential failure modes. The definitions and relative levels of the categories can vary depending on the agency and the analysis. In general, evaluation should include a "high" category to highlight failure modes with high likelihood and high consequence, a "lower" category for failure modes that are less likely and low consequence, and a "lowest" category for failure modes that have a very low likelihood of occurring and few consequences. Additionally, evaluation provides for an additional category for potential failure modes lacking sufficient information to categorize.

Different combinations of consequence and likelihood can result in higher or lower categorization. For instance, a failure mode with very high consequences (a community of 10,000 people directly downstream, for example) with a relatively low likelihood of failure would typically be categorized as "high" category because of the high consequence associated with failure. Sound professional judgment is necessary to evaluate the relative levels of consequence and likelihood of each failure mode so they can be appropriately categorized.

Whether a numeric value or relative category, the results provide the basis for understanding, communicating, prioritizing, and managing the risks associated with the identified potential failure modes for a dam. Development and implementation of a program of action to reduce the risk associated with failure modes can occur after identification and prioritization of potential failure modes. Risk reduction measures can consist of additional study, implementation of new mitigation measures, changes to operation or maintenance practices, changes to the emergency action plan and/or the development of a surveillance and monitoring program designed to provide early detection of the identified potential failure modes.

SURVEILLANCE AND MONITORING

Assessing the performance of a dam involves evaluating information collected by visual surveillance and recorded instrumentation data against the expected performance of the dam based on design predictions and the potential failure modes.

Surveillance and monitoring provide the means for the early detection of the identified potential failure modes. The basis for the design of surveillance and monitoring program for identified potential failure modes targets where and how a performance indicator of a potential failure mode can be detected. Early detection provides the best opportunity to intervene to prevent failure, or to provide adequate warning so that the population at risk can be evacuated when failure is imminent or occurs.

The development of surveillance and monitoring program takes into consideration

- At what point can unexpected behavior or failure mode initiation be detected?
- Can instrumentation detect change in the performance indicators before a failure sequence progresses?
- Can a developing failure be detected visually?
- Are trained personnel available to review collected data and information in order to evaluate performance indicators of initiation or progression of a failure mode?
- How quickly might a failure mode develop once detected?
- How frequently should monitoring be performed?
- Will there be time to take action if a failure mode is detected?
- What actions should be taken to reduce the risk identified by surveillance and monitoring?

In summary, understanding the complete sequence of identified potential failure modes provides the basis for selecting the right combination of surveillance and instrumentation to develop a dam performance–monitoring program.

APPENDIX C
RANGE, RESOLUTION, ACCURACY, PRECISION, AND REPEATABILITY

Watch every detail that affects the accuracy of your work.

Arthur C. Nielsen

Dance must have a precision without fault.

Arielle Dombasle

The 2000 *Guidelines for Instrumentation and Measurements for Monitoring Dam Performance* are contained in this appendix. They are repeated here to supplement Chapters 4 and 5 on planning and selecting instruments for measuring performance indicators.

To procure an appropriate system, the desired products need to be suitable for the intended use. Suitability can be determined by evaluating published performance specifications, which should take into consideration how to differentiate meaningful readings from meaningless noise. Understanding performance specifications is therefore critical for obtaining optimal results from a monitoring program.

Definitions of the many specifications that need to be considered are presented following. Among the most important are range, resolution, accuracy, precision, and repeatability. Because these are sometimes confused, their definitions and significance are discussed herein and illustrated in Fig. C-1. Performance specifications should be traceable to known standards, such as those maintained by the National Institute of Standards and Technology (NIST) in the United States.

Range specifies the highest and lowest values that the instrument is designed to measure. For example, the range limit for a piezometer might

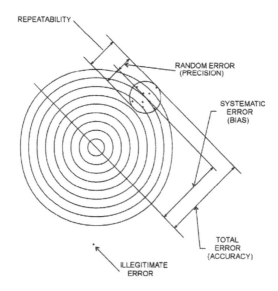

Fig. C-1. Precision, bias, error, and accuracy

be 0–50 psi. The expected range of the parameter to be measured must fall within the range limits of the instrument.

Span is related to range and is defined as the arithmetic difference between the upper and lower range limits. For example, a range of ±50 mm (1.96 in.) is equivalent to a span of 100 mm (3.93 in.), a range of –25° to +70°C (–13°F to 158°F) has a span of 95°C (203°F), and a range of 0–100 psi has a span of 100 psi.

Resolution is the smallest change in the measured parameter that can be distinguished by an instrument. The resolution is typically many times finer than the instrument's precision, repeatability, and accuracy. Resolution is the smallest change that can consistently and reliably be measured by an instrument.

Accuracy is the degree to which readings match an accepted standard (absolute) value and includes the combined effects of all sources of measurement error. Accuracy is written as a plus–minus (±) value, such as ±1 mm (0.03 in.), ±1% of reading, or ±1% of full span.

The accuracy number defines the maximum difference between the indicated value and the standard value when an instrument is used under specified operating conditions. These conditions include operation over the specified temperature range and input range of the instrument. The specified accuracy includes the combined effects of hysteresis, dead band, and repeatability and conformity errors. It is typically expressed in terms of the measured variable [e.g., ±2 mm (0.07 in.)] or as a percentage of the full-scale span (e.g., ± 5% of span).

Precision is the closeness of similar measurements to the arithmetic mean. It is expressed as a plus–minus ($\pm$) value. In dam monitoring, precision is often more important than accuracy because change, rather than absolute value, is of greatest interest.

Hysteresis is the dependence of the value of the output, for a given input, upon the history of prior excursions and the direction of the current traverse. It is usually determined by subtracting the value of the dead band from the maximum measured separation between upscale-going and downscale-going indications of the measured variable during a full-range traverse after transients have decayed.

Repeatability is the closeness of agreement of a number of consecutive measurements to one another under the same operating conditions. It is expressed in engineering units, such as 1 mm, or as a percentage of full-scale range. It is the closeness of agreement among a number of consecutive measurements of the output for the same value of the input under the same operating conditions approaching from the same direction for full-range traverses. It is usually measured as a non-repeatability and expressed as repeatability in percent of span. It does not include hysteresis.

The linearity is the closeness to which a curve (of instrument output versus input) approximates a straight line. It is usually measured as non-linearity (the maximum deviation between the curve and the straight line) and expressed as linearity, typically as a percentage of the full span of output of the instrument.

Conformity of a curve of actual output values is the closeness to which it approximates a specified curve (straight line, logarithmic, sinusoidal, etc.) of predicted output values. Linearity is a special case of conformity.

The dead band is the range through which an input can be varied without initiating observable response, usually expressed in percent of span.

For the output of a first-order system that is given a step input, the time constant is the time required to complete 63.2% of the total rise or decay.

Bias is the average indicated value of the measured variable that would result from an infinite number of measurements minus the actual value of the variable.

The calibration curve is a graphical representation of the measured output plotted versus known input values of the measured variable.

The scale factor is a number by which the measured output (millivolts, milliamps, frequency, polynomial, etc.) should be multiplied to compute the value of the measured variable.

Sensitivity is the ratio of the change in output magnitude to the change in the input that causes it after steady state has been reached. For a linear device, it is the inverse of the scale factor.

The temperature coefficient of the scale factor is the percentage change of scale factor per unit change of temperature of an instrument. This is also known as the temperature coefficient of the span.

The temperature coefficient of zero shift is the change from zero of measured output per unit change of instrument temperature in the absence of a change of input.

Long-term stability is an indication of the change in an instrument's output that would take place over a specified period of years in the absence of a change in the measured input (pressure, force, displacement, etc.). Because of the long time periods involved in gathering data and the diversity of factors that can contribute to output drift over these periods, precise answers can be elusive. However, publications discussing the long-term performance of most dam monitoring instruments are available and can be provided by the instrument manufacturers.

The electronic signal produced by the instrument determines how it will be read. The most common and robust analog signal types are DC voltage, DC current, and frequency. The most common digital signal types are RS232 serial output and RS485 serial output.

Specifications should include the output range of the signal, e.g., $+/-5$ V DC, from 4 to 20 mA, and other details that affect how the signal must be recorded.

Electronic instruments all require electrical power to operate. The power requirement specification should specify the acceptable voltage range of the input power, whether the voltage is AC or DC, the acceptable frequency range (if AC), and the current consumed by the instrument in amps or milliamps (mA). For DC-powered instruments, this specification should further state the amount of AC fluctuation or "ripple," typically expressed as a peak-to-peak voltage, that the instrument will tolerate and still function properly. The power specification should also state whether the instrument is protected from damage caused by power surges and power supply polarity reversal.

Furthermore, the power specification must state whether the instrument requires a regulated power supply, in other words, one that maintains a constant supply voltage, to obtain stable readings. For an instrument that requires regulated power, the output changes in proportion to the input voltage level. This change has nothing to do with the parameter being measured and can render a set of measurements useless. Many modern instruments regulate power with their own onboard electronics and therefore tolerate a wide range of input voltages. A "power requirements" specification for such an instrument might read, for example, "8–24 volts DC @ 8 mA, 250 mV peak-to peak ripple max., reverse polarity, short circuit and surge protected."

Whether an instrument requires regulated power or will operate stably with unregulated power is especially important in many dam applications where power must be supplied from variable power sources such as batteries or solar panels.

Although instrument specifications define instrument performance, the total performance of a monitoring system is also a function of the readout or data-acquisition equipment. For example, an instrument repeatability of 1 mm might be equivalent to an output repeatability of 1 mV. If the readout or data logger is unable to resolve changes smaller than 2 mV, the precision of the recorded data will be no better than 2 mm (0.07 in.). Electronic noises picked up in long cables or radio links between the instrument and the recording device can further degrade performance. It is important to evaluate the combined specifications of instrument and data-acquisition system and to suppress noise pickup to optimize results.

APPENDIX D
GLOSSARY OF TERMS

I am still learning.
Michelangelo

Term	Definition
Absorption	Any dissipative loss mechanism resulting in reduction of acoustic amplitude; characterized by an energy conversion process.
Abutment	Valley side against which a dam is constructed. Right and left abutments are the respective sides looking downstream.
Accelerograph	Accelerometer having provisions for recording the acceleration of a point on the earth during an earthquake.
Accuracy	Degree of conformity of a measured or calculated value to its definition or with respect to a standard reference (see Uncertainty).
Acoustics	Field of science that focuses on the application and detection of sound waves.
Active storage	Volume of a reservoir available for withdrawal.
Afterbay	Downstream pool or tailwater area of a dam or turbine outlet.
Air-vent pipe	Pipe admitting air to an outlet conduit to modulate pressure during release of water.
Alignment surveys	Measuring survey monuments.

Alkali–aggregate reaction (AAR)	Chemical reaction of an aggregate with the alkalis or carbonates in cement, resulting in swelling and cracking of the concrete.
Alkali–silica reaction (ASR)	Chemical reaction between the alkalis in portland cement and some silicas, resulting in significant expansion in concrete, causing damage such as mass cracking.
Ambient air temperature	Temperature of the surrounding air at a dam.
Anchors	Post-tensioned wire strands or solid bars installed and stressed to improve stability.
Anemometer	Device that measures air speed.
Appurtenant structures	Water-retaining structures such as an outlet, spillway, powerhouse, or tunnel.
Aqueduct	Channel for conveying water.
Aquifer	A permeable body of rock saturated with groundwater.
Aquitard	A bed of low permeability forming a groundwater barrier.
Arch buttress dam	Individual arched segments supported by buttresses.
Arch dam	Concrete or masonry dam that is curved to transmit load to the abutments.
Attenuation	Loss of energy away from an earthquake epicenter.
Autogenous growth	Self-generating growth produced without external influence.
Automated Data-Acquisition System (ADAS)	System for monitoring the performance of a dam that includes components for data collection that are permanently installed and programmed to operate without human intervention.
Auxiliary spillway	Spillway in addition to a service spillway.
Axis of dam	A straight or curved line along the centerline of the crest.
Barometer	Absolute pressure gauge to measure atmospheric pressure.
Bentonite	Clay largely composed of montmorillonite and beidellite formed from volcanic ash decomposition.
Berm	Horizontal step or bench in the sloping profile of an embankment dam.

Borros point anchor	Extensometer with a single anchor point in which the elevation of the top of a rod is measured with respect to a collar.
Bourdon tube	Mechanical measuring instrument employing a curved or twisted metal tube flattened in cross section as its sensor.
Buoyancy	Resultant vertical force exerted on a body by a static fluid in which it is submerged or floating.
Buttress dam	Dam consisting of a watertight upstream face supported at intervals on the downstream side by a series of buttresses.
Calibration	Determining by measurement or comparison with a standard the current value of each scale reading on a meter or other device.
Carlson meter	Any of a family of strain- or temperature-monitoring instruments that incorporate hand-wound nickel-chromium wire sensing elements in a half-bridge resistance measurement circuit.
Cavitation	Emulsification produced by disruption of a liquid into a liquid–gas two-phase system when the hydraulic pressure of the liquid is reduced to the vapor pressure.
CFM	Cubic feet per minute.
CFS	Cubic feet per second.
Cofferdam	Structure enclosing all or part of a construction area so that construction can proceed in a dry area.
Cohesionless	Granular soil in which strength is directly related to confining stresses.
Cohesive soil	Very fine-grained plastic soil in which strength depends upon moisture content.
Concrete lift	Vertical distance between successive horizontal construction joints.
Conduit	Closed channel for conveying water.
Consolidation	Process of increasing soil density either naturally or mechanically.
Consolidation grouting	Injection of grout along discontinuities to consolidate rock.
Construction joint	Bonded interface between two successive placements of concrete.

Constantan	Copper–nickel alloy known for electrical stability with temperature change that is used in common thermocouple compositions.
Core	Barrier of impervious material to retard seepage through the body of an embankment dam.
Crack meter	Device that measures change in the width of a crack.
Creep	Time-dependent strain of solids.
Crest length	Length of the top of a dam.
Crest of dam	Top of dam.
Crest wall	Parapet on dam crest.
Crib dam	Timber boxes filled with earth or rock.
Crustal zone	Outermost solid layer of the earth.
Culvert	Conduit under a road, railway, or embankment.
Cutoff wall	Wall of impervious material built into the foundation to reduce seepage under a dam.
CY	Cubic yard.
Data logging	Conversion of electrical impulses from instruments into digital data to be recorded in various types of memory.
Dead storage	The storage that lies below the invert of the lowest outlet and that cannot be withdrawn from the reservoir.
Deformation	Alteration of shape or dimensions.
Dental work	Removal of loose rock and soil from a dam, abutments, and foundation contact areas and the placement of backfill concrete.
Dewatered	Area previously containing water that has since been drained.
Differential movement	Difference in movement between two objects or points.
Dispersion	Two-phase fluid system in which one phase is finely dispersed (on colloidal dimensions) within a continuous liquid phase.
Dissolution	Dissolving.
Diurnal	Daily.
Diversion channel, canal, or tunnel	Waterway to divert water from the work.
Double-curvature arch dam	Arch dam that is curved vertically as well as horizontally (ellipsoidal shape).

Drainage area	Area that gathers water.
Drainage layer (blanket)	Layer of permeable material to facilitate drainage.
Drainage well (relief well)	Well to collect seepage through or under a dam.
Drawdown	Releasing water to lower the reservoir water surface.
Earth dam	Embankment dam in which more than 50% of the total volume is formed of compacted finer-grained materials than rockfill material.
Easting	Distance from one point to another in the easterly direction.
Elastic properties	Properties that allow a material to sustain deformation without permanent loss of size or shape.
Electrolytic tiltmeter	Tiltmeter that uses liquid-filled electrolytic spirit level as the sensing element.
Embankment dam	Dam constructed of natural materials (soil and rock).
Emergency gate	Standby or reserve gate used when the normal means of water control are not available.
Equipotential lines	Lines representing surfaces on which the water potential (pressure + elevation head) is the same at every point.
Error	Difference of a measured value from its known true or correct value (or sometimes from its predicted value).
Extensometer	Measures the change in distance between two anchored points.
Failure mode	Sequence of events exploiting weakness that progress until reservoir control is lost.
Failure surface	Surface upon which the strength of a material resisting shear is exceeded.
Fetch	Distance over a body of water traversed by waves without obstruction.
Flat jack	Hollow steel cushion made of two nearly flat disks welded all around the edge that can be inflated with oil under controlled pressure.
Flood surcharge pool	Volume of reservoir above the normal water surface and the maximum water surface during a flood.

Flume	Open channel constructed of concrete, steel, or wood used to convey water.
Forebay	Reservoir immediately upstream from an outlet.
Frequency	Number of peak-to-peak cycles per unit time characterizing a sinusoidal sound wave; typically given in units of millions of cycles per second or MHz.
Gage or gauge	An apparatus that measures something with a graduated scale.
Geodimeter	Geodetic distance meter.
Geoid	Shape of the entire earth at sea level.
Geo-synchronous satellite	Satellite that orbits the earth from west to east so as to remain fixed over a given place on the earth.
Geosynthetic	Membrane of woven or unwoven fabric.
Graded filter	Sand and gravel proportioned to permit passage of water without movement of finer material.
Gravity dam	Dam that depends on its weight for its stability.
Grout blanket	Shallow band of drill holes grouted with concrete to consolidate the foundation (consolidation grouting).
Grout curtain	Drilled holes filled with grout under pressure to form a cutoff under a dam.
Headwater	Elevation of the free water surface on the upstream side of the dam.
Headwaters	Source of waters in a river.
Homogeneous dam	Embankment dam constructed of similar earth material throughout.
Hydraulic fill dam	Embankment dam constructed of materials conveyed and placed by flowing water.
Hydraulic fracture	Fracture from pore water pressure exceeding soil strength.
Hydrogeology	Science dealing with groundwater.
Hydrograph	Graphical representation of stage, flow, velocity, or other characteristics of water at a given point as a function of time.
Hydrostatic pressure	Pressure at a point in a fluid from the weight of fluid above it.
Hysteresis plot	Plot describing changes in measurements relative to prior measurements.
Impermeable liner	Liner that will not permit water or other fluid to pass through.

Impervious soil	Soil with low permeability.
Inclinometer	Instrument for measuring the angle of deflection between a reference axis and a casing axis.
Inclinometer casing	Specially constructed casing fitted with slots or channels to prevent a probe-type inclinometer from rotating when lowered or raised.
Inflatable dam	Dam constructed of rubber or other watertight membrane that increases in vertical height when inflated.
Initial filling	First filling of a reservoir or other water-retaining structure.
In situ pore pressure	Pore pressure prior to any external influence.
Inverted pendulum	Pendulum anchored at its lowest point in rock or concrete to allow the upper floated end to move freely for the measurement of vertical change.
Joint meter	Device used to measure the movement of a joint in concrete or any other material.
kWh	Kilowatt-hour (1,000 watt hours).
Lateral translation	Horizontal movement.
Leakage	Rapid movement of water or other liquid through a porous medium such as through a dam, its foundation, or abutments.
Liquefaction	Change in saturated soil to a liquid or near-liquid state.
Longitudinal cracking	Cracking parallel with a dam axis.
Masonry dam	Dam constructed mainly of stone, brick, or concrete blocks jointed with mortar.
Maximum section	Cross section with the greatest difference between the base of a dam and its crest.
Measurement range	Highest and lowest values that an instrument is designed to measure.
Membrane dams	Dam with a watertight barrier on the upstream face.
Modulus of elasticity	Ratio of stress to strain in elastic media.
MPa	Mega-Pascal (1 million Pascals).
Multiplexer	Device for distributing many signals into a smaller number of measurement channels.
Multipoint extensometer	Extensometer that can measure movement between several anchored locations along its length.
Mw	Megawatt (1 million watts).
Nappe	Shape of water flowing over a barrier at its crest.

Northing	Distance from one point to another in the northerly direction.
Observation well	Well drilled to observe water level.
Offset	Difference between actual and reference values.
Open piezometer	Observation wells with subsurface seals that isolate the strata to be measured.
Overtopping	Water flowing over a dam.
Parshall flume	Calibrated device for measuring the flow of water in open conduits.
Pendulum (plumbline)	Rigid body mounted on a fixed horizontal axis, about which it is free to rotate under the influence of gravity.
Penstock	Pipe between the intake and the turbine of a hydroelectric unit.
Phase	Measure of a fraction of the period of a repetitive phenomenon measured with respect to some distinguishable feature of the phenomenon itself.
Phreatic line	Line below which soils are saturated.
Piezometer	Instrument for measuring fluid pressure (air or water) within soil, rock, or concrete.
Piping	Progressive development of movement of soil by seepage that progresses from downstream to upstream.
Pitot tube	Velocity-measuring device for a fluid stream.
Pneumatic piezometer	Piezometer with a porous filter connected to two tubes that measure pressure across a flexible diaphragm between them.
Poisson's ratio	Ratio of the horizontal to vertical strain under load.
Pore pressure	Interstitial pressure of fluid (air or water) within a mass of soil, rock, or concrete.
ppm	Parts per million.
Precision	Degree of mutual agreement among a series of individual measurements. Precision is often but not necessarily always expressed by the standard deviation of the measurements.
psi	Pounds of force per square inch.
Radiotelemetry	System used to provide a communication or control channel between two specific points.
Rain gauge	Instrument designed to collect and measure precipitation.

Relative density	Ratio of density of soil with reference to its maximum possible density for a given compaction effort.
Repeatability	Ability of a device to measure the same value repeatedly.
Reservoir drawdown	Reducing the water level in a reservoir.
Reservoir rim	Intersection of the reservoir water surface with the containing basin ground surface.
Resolution	Smallest increment of a measurable property.
Rockfill dam	Embankment dam in which more than 50% of the total volume comprises compacted or dumped pervious natural or crushed rock.
Saturation pipes	Groundwater monitoring well to determine the level of the groundwater.
Scour	Erosion caused by the flow of air, ice, or water.
Seepage	The slow movement of water or other liquid through a porous medium such as through a dam, its foundation, or abutments.
Seepage paths	The path that seepage follows.
Seepage quality	Level of turbidity and dissolved solids in the seepage.
Seepage quantity	Measurement of seepage volume.
Seismic response	Vibration of a dam from earthquake shaking.
Seismicity	Phenomena of the earth's tectonic movements.
Seismo-tectonics	Process of crustal plate movement that generates earthquakes.
Sensor	Device that responds to a physical stimulus and transmits a resulting signal.
Settlement	Decrease in elevation of a point on a dam's surface.
Settlement gauge	Instrument that measures elevation changes between two or more points.
Shear strength	Ability of a material to resist forces tending to cause movement along an interior planer surface.
Shear zone	Zone where the shear strength was exceeded and some permanent deformation (sliding) occurred.
Signal conditioning	Electronic circuitry used for converting transducer outputs into signals suitable for transmission over cable or radio and for recording by data loggers and other devices.

Slope failure	Downward and lateral movement of soil beneath a natural or constructed slope.
Snow course	Designated open area where measurements of snow cover are made to determine its water equivalent.
Sound velocity (ultrasonic velocity)	Speed at which sound waves travel, which depends on the physical properties of the material in which it is propagating; typically denoted by symbol c; sound velocity is usually measured by time-of-flight methods.
Stability	Resistance against movement routinely expressed as a factor of safety.
Staff gauge or stage recorder	Graduated scale placed in a position so that the level of water may be read directly.
Standards	Commonly used and accepted as an authority.
Standpipe	Vertical tube filled with water, also known as an open-well piezometer.
Strain gauge	Device that measures the change in distance between two points.
Strain rosette	Pattern of intersecting lines on a surface along which linear strains are measured to better define the three-dimensional distribution of strain about a point.
Stress meter	Instrument that measures stress directly.
Strong-motion accelerometer	Accelerometer designed to record ground shaking from strong earthquakes while remaining insensitive to smaller events.
Suspension	Two-phase system consisting of solid particles suspended in a liquid medium.
Tailwater	Elevation of the free water surface (if any) on the downstream side of a dam.
Tendon	*See* Anchors.
Theodolite	Optical instrument used in surveying. It consists of a sighting telescope mounted so that it is free to rotate around horizontal and vertical axes so that the angles can be measured.
Thermistor	Resistive circuit component having a high negative temperature coefficient of resistance, so that its resistance decreases as the temperature increases.

Thermocouple	Device consisting basically of two dissimilar conductors joined together at their ends measuring voltage as an indicator of temperature.
Thermometer	A device for measuring temperature in Fahrenheit or Celsius.
Tiltmeter	Instrument used to measure rotational movement with respect to gravity.
Total pressure cell	Gauge, usually consisting of a piezometer connected to an enclosed fluid cell that then measures the total load applied to the cell.
Total station	Surveying instrument that measures distance plus vertical and horizontal angles.
Transducers	Device or element that converts an input signal into an output signal of different form.
Trigonometric levels	Method of determining the difference of elevation between two points by using the principles of triangulation and trigonometric calculations.
Trilateration	Measurement of a series of distances between points on a surface of interest to establish their relative positions.
Trunnion friction	Friction created from the movement of a pin or pivot on bearings of gate trunnions.
Turbidity meter	Device that measures the loss of a light beam as it passes through a solution with particles large enough to scatter the light.
Uncertainty	Limits of the confidence interval between a measured and calculated quantity.
Undrained	State of strain that restricts pore water pressure dissipation from a body of soil.
Uplift	Upward pressure against the base of a dam.
Upstream blanket	Impervious layer placed on the reservoir floor upstream of a dam.
Vibrating-wire piezometer	Electrical piezometer that measures in situ pressure by changes in signal frequency.
Weir	Device in a channel that serves to regulate water level or measure flow.
Zoned embankment dam	Embankment dam composed of zones of selected materials having different degrees of porosity, permeability, and density.
Zoned rockfill dam	Embankment dam constructed of rock with internal core, filter, and drain zones.

INDEX

Page numbers followed by *f* and *t*
indicate figures and tables, respectively.

427